Nanotubes and Nanofibers

Advanced Materials Series

Series Editor:

Yury Gogotsi, *Drexel University, Philadelphia, Pennsylvania, USA*

series includes:

Carbon Nanomaterials
Yury Gogotsi

Nanotubes and Nanofibers
Yury Gogotsi

Nanotubes and Nanofibers

Edited By
Yury Gogotsi

**Drexel University,
Philadelphia, Pennsylvania, USA**

CRC Press
Taylor & Francis Group
Boca Raton London New York

CRC Press is an imprint of the
Taylor & Francis Group, an **informa** business
A TAYLOR & FRANCIS BOOK

This material was previously published in the *Nanomaterials Handbook* © 2006 by Taylor and Francis Group, LLC.

CRC Press
Taylor & Francis Group
6000 Broken Sound Parkway NW, Suite 300
Boca Raton, FL 33487-2742

First issued in paperback 2019

© 2006 by Taylor & Francis Group, LLC
CRC Press is an imprint of Taylor & Francis Group, an Informa business

No claim to original U.S. Government works

ISBN-13: 978-0-8493-9387-7 (hbk)
ISBN-13: 978-0-367-39068-6 (pbk)

Library of Congress Cataloging-in-Publication Data

Nanotubes and nanofibers / Yury Gogotsi, editor.
    p. cm.
   Includes bibliographical references and index.
   ISBN 0-8493-9387-6 (978-0-8493-9387-7)
    1. Nanotubes. 2. Nanostructured materials. I. Gogotsi, IU. G., 1961-

TA418.9.N35N3577 2006
620'.5--dc22
2006012338

Visit the Taylor & Francis Web site at
http://www.taylorandfrancis.com

and the CRC Press Web site at
http://www.crcpress.com

*This book is dedicated to my father, Professor George Gogotsi,
who inspired me to become a scientist.*

Preface

Nanomaterials, which are materials with structural units on a nanometer scale in at least one direction, is the fastest growing area in materials science and engineering. Material properties become different on the nanoscale: for example, the theoretical strength of materials can be reached or quantum effects may appear. One-dimensional and quasi-one-dimensional materials such as nanotubes and nanowires demonstrate many extreme properties that can be tuned by controlling their structure and diameter. Nanotubes, nanowires, and nanofibers are not only excellent tools for studying one-dimensional phenomena, but they are also certainly among the most important and promising nanomaterials and nanostructures. The role of nanomaterials in industries is growing. Nanofibers are already used for insulation and reinforcement of composites, and many materials and structures incorporating nanotubes and nanowires are under development. Extensions to other tubular and rodlike or wirelike structures provide the scope for new discoveries and novel applications.

This book describes a large variety of nanotubes, nanofibers, and related structures such as whiskers and nanowires. This area is very broad; it is impossible to cover all fibrous nanomaterials in a single volume. Carbon nanotubes (Chapters 1 and 2) receive special attention in this book because carbon is as important for nanotechnology as silicon is for electronics. Sumio Iijima's discovery of carbon nanotubes in 1991 stimulated the development of the whole nanotechnology field. Since then, a dramatic progress in synthesis and purification of nanotubes has been achieved and many applications have emerged. After carbon, many other materials have been produced in the tubular shape with nanometer diameters.

Designed specifically to provide an overview of nanotubes and nanofibers for today's scientists, graduate students, and engineering professionals, the book will offer a treatment of subjects using terms familiar to a materials scientist or engineer. The book consists of eight chapters selected from the recently published *Nanomaterials Handbook* and is written by the leading researchers in the field. Providing coverage of the latest material developments in the United States, Asia, Europe, and Australia, the book describes both commercially available and emerging materials.

Finally, I would like to acknowledge all the people who have been helpful in making this book possible. My family was very patient and understanding, and my students and postdocs did a great job allowing me to concentrate on the book project. The staff of Taylor & Francis helped immensely.

Yury Gogotsi
Philadelphia, PA

Preface

Editor

Dr. Yury Gogotsi is professor of materials science and engineering at Drexel University in Philadelphia, Pennsylvania. He also holds appointments in the Departments of Chemistry and Mechanical Engineering at Drexel University and serves as director of the A.J. Drexel Nanotechnology Institute and associate dean of the College of Engineering. He received his M.S. (1984) and Ph.D. (1986) degrees from Kiev Polytechnic and a D.Sc. degree from the Institute of Materials Science, Ukrainian Academy of Science, in 1995. His research group works on carbon nanotubes, nanoporous carbide-derived carbons, and nanofluidics. He has also contributed to the areas of structural ceramics, corrosion of ceramic materials, and pressure-induced phase transformations, creating a new research field called high pressure surface science and engineering. He has coauthored 2 books, edited 9 books, obtained more than 20 patents, and authored about 200 journal papers and 12 book chapters. He has advised a large number of M.S., Ph.D., and post-doctoral students at Drexel University and University of Illinois at Chicago.

Gogotsi has received several awards for his research, including I.N. Frantsevich Prize from the Ukrainian Academy of Science, S. Somiya Award from the International Union of Materials Research Societies, G.C. Kuczynski Prize from the International Institute for the Science of Sintering, and Roland B. Snow Award from the American Ceramic Society (twice). He has been elected a fellow of the American Ceramic Society, academician of the World Academy of Ceramics, and full member of the International Institute for the Science of Sintering.

Contributors

Ying Chen
Department of Electronic Materials
  Engineering
Research School of Physical Sciences and
  Engineering
The Australian National University
Canberra, Australia

Svetlana Dimovski
Department of Materials Science and
  Engineering
Drexel University
Philadelphia, Pennsylvania

Fangming Du
Department of Chemical and Biomolecular
  Engineering
University of Pennsylvania
Philadelphia, Pennsylvania

John E. Fischer
Department of Materials Science and
  Engineering
University of Pennsylvania
Philadelphia, Pennsylvania

Yury Gogotsi
Department of Materials Science and
  Engineering
Drexel University
Philadelphia, Pennsylvania

Frank K. Ko
Department of Materials Science and
  Engineering
Drexel University
Philadelphia, Pennsylvania

Eduard G. Rakov
D.I. Mendeleev University of Chemical
  Technology
Moscow, Russia

Jonathan E. Spanier
Department of Materials Science and
  Engineering
Drexel University
Philadelphia, Pennsylvania

R. Tenne
Department of Materials and Interfaces
Weizmann Institute
Rehovot, Israel

Karen I. Winey
Department of Materials Science and
  Engineering
University of Pennsylvania
Philadelphia, Pennsylvania

Hongzhou Zhang
Department of Electronic Materials
  Engineering
Research School of Physical Sciences and
  Engineering
The Australian National University
Canberra, Australia

Table of Contents

Chapter 1 Carbon Nanotubes: Structure and Properties..1

John E. Fischer

Chapter 2 Chemistry of Carbon Nanotubes 37

Eduard G. Rakov

Chapter 3 Graphite Whiskers, Cones, and Polyhedral Crystals...............................109

Svetlana Dimovski and Yury Gogotsi

Chapter 4 Inorganic Nanotubes and Fullerene-Like Materials of Metal Dichalcogenide
and Related Layered Compounds...135

R. Tenne

Chapter 5 Boron Nitride Nanotubes: Synthesis and Structure..................................157

Hongzhou Zhang and Ying Chen

Chapter 6 Nanotubes in Multifunctional Polymer Nanocomposites............................179

Fangming Du and Karen I. Winey

Chapter 7 One-Dimensional Semiconductor and Oxide Nanostructures.........................199

Jonathan E. Spanier

Chapter 8 Nanofiber Technology..233

Frank K. Ko

Index 245

1 Carbon Nanotubes: Structure and Properties

John E. Fischer
Department of Materials Science and Engineering,
University of Pennsylvania, Philadelphia, Pennsylvania

CONTENTS

1.1 Introduction 1
1.2 Structure 3
    1.2.1 Single-Wall Tubes, Bundles, and Crystalline Ropes 3
    1.2.2 Multiwall Tubes 5
    1.2.3 Macroscopic Nanotube Materials 5
    1.2.4 Fibers 7
    1.2.5 Filled Tubes 7
    1.2.6 Nanotube Suspensions 10
1.3 Physical Properties 11
    1.3.1 Mechanical Properties 12
    1.3.2 Thermal Properties 13
    1.3.3 Electronic Properties 18
    1.3.4 Magnetic and Superconducting Properties 30
1.4 Summary and Prospects 31
Acknowledgments 31
References 31

1.1 INTRODUCTION

Carbon nanotubes were discovered in 1991 as a minor by-product of fullerene synthesis [1]. Remarkable progress has been made in the ensuing 14 years, including the discovery of two basic types of nanotubes (single-wall and multiwall); great strides have been taken in their synthesis and purification, elucidation of the fundamental physical properties, and important steps are being taken toward realistic practical applications.

Carbon nanotubes are long cylinders of 3-coordinated carbon, slightly pyramidalized by curvature [2] from the pure sp^2 hybridization of graphene, toward the diamond-like sp^3. Infinitely long in principle, a perfect tube is capped at both ends by hemi-fullerenes, leaving no dangling bonds. A single-walled carbon nanotube (SWNT) is one such cylinder, while multiwall tubes (MWNT) consist of many nested cylinders whose successive radii differ by roughly the interlayer spacing of graphite (see Figure 6.1 in the chapter by Du and Winey [3]). The minimum diameter of a stable freestanding SWNT is limited by curvature-induced strain to ~0.4 nm [4]. MWNT may have outer shells >30 nm

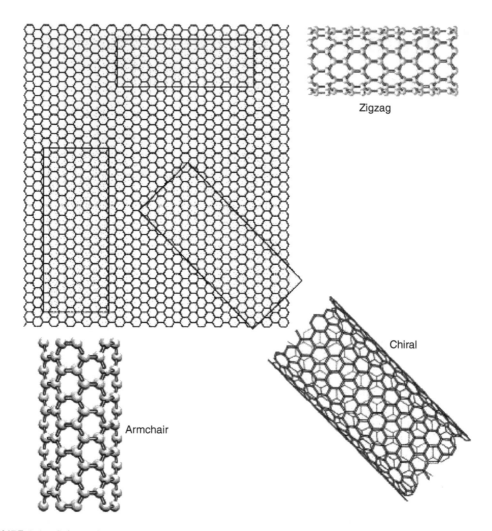

FIGURE 1.1 Schematic representation of the relation between nanotubes and graphene. The three rectangles can be rolled up into seamless nanotubes; the short side, referred to as the roll-up vector $\mathbf{R}$, becomes the circumference. $\mathbf{R} = n\mathbf{a_1} + m\mathbf{a_2}$, is a graphene 2D lattice vector, where $\mathbf{a_1}$ and $\mathbf{a_2}$ are unit vectors. Integers n and m uniquely define the tube: diameter, chirality, metal vs. semiconducting nature, and band gap, if semiconducting. In a bulk polydisperse sample consisting of a distribution of diameters, the larger the average diameter, the greater the number of n, m pairs that will satisfy the seamless roll-up condition.

in diameter, with varying numbers of shells, affording a range of empty core diameters. Lengths up to 3 mm have been reported [5]. Nanotubes are distinguished from less-perfect quasi-one-dimensional carbon materials by their well-developed parallel wall structure. Other elements too can be made into nanotubes, so one often encounters the term "SWCNT" to distinguish them from noncarbon tubes, i.e., BN, BCN, metal dichalcogenides [6]. The unique feature of carbon nanotubes is that they exist in both metallic and semiconducting varieties, a blessing or a curse depending on the context.

SWNTs can be conceptualized as seamless cylinders rolled up from graphene rectangles (cf. Figure 1.1), or as cylindrical isomers of large fullerenes. C_{70} is the smallest nanotube; compared with C_{60} it contains an extra belt of hexagons normal to the fivefold axis of the hemi-C_{60} caps. Adding more belts leads to longer tubes of the metallic (5,5) armchair category as defined later. According to theory, cylindrical fullerene isomers are less stable than the more nearly spherical ones because the 12 pentagons necessary to ensure closure are localized on the two caps. This results in strain concentrations at the ends of closed tubes, which in turn makes it easier to perform additional chemistry on the ends than on the sidewalls [7].

There exist many quasi-one-dimensional all-carbon structures, which are neither SWNT nor MWNT. Composite materials reinforced with carbon or graphite fibers are often used in sporting goods, high-performance aircraft, and other applications where high stiffness and lightweight are required. Companies such as Hyperion sell commodity quantities of vapor-grown carbon fibers as conductivity additives for paints and plastics. Carbon nanohorns have received a good deal of attention lately as catalyst supports, fuel cell, and battery electrodes [8]. Such materials lack the atomic perfection of nanotubes, but are nonetheless crucially important in science and industry.

The discovery and rapid evolution of carbon nanotubes has played a major role in triggering the explosive growth of R & D in nanotechnology. Many of the early lessons learnt carried over to rapid developments in inorganic semiconductor nanowire science and engineering, in particular, field effect transistor (FET)-like switching devices, and chemical and biological sensors. The nanotube field *per se* has fanned out to encompass molecular electronics, multifunctional composites, flat-panel display technology, high-strength lightweight structural materials, nanoscale metrology (mass, heat, functional scanning probe tips, etc.), and others. In this chapter, I attempt to provide a broad-brush introduction to the materials responsible for all this excitement. My selection of sources is personal. Apologies in advance to friends, colleagues, and others whose work I fail to mention.

This chapter is a survey of the physical properties of carbon nanotubes, with emphasis on macroscopic assemblies of engineering interest. The important subjects of synthesis, purification, and composite materials are covered elsewhere in this volume [3,7]. Contact is made with single-tube properties where appropriate; an obvious issue is the extent to which properties of macroscopic material approach those of ideal individual tubes. Materials scientists will immediately note with dismay the paucity of information relating to defects and their influence on properties. In this important arena theory and simulations are far ahead of experiments. Atomic-scale defects may be at the resolution limit of high-resolution transmission electron microscopy (HRTEM). Their presence can be inferred from the breakup of individual tube device characteristics into multiple quantum dots defined by defect-related internal barriers [9].

1.2 STRUCTURE

We start the discussion by considering a single isolated tube. Is it a molecule? If so, we might be able to dissolve it in order to perform high-resolution ^{13}C NMR, the method which proved that the carbons in C_{60} were all equivalent, thus confirming the soccer-ball structure. So far, no true solvent for pristine nanotubes has been found. The surfactants, or chemical functionalization, necessary to obtain stable suspensions must perturb the intramolecular structure however slightly. Worse yet, we still lack even minute samples consisting of a single type of nanotube, so even if we had an NMR spectrum, it would be impossible to interpret. HRTEM is a borderline technique for resolving individual carbon atoms 0.14 nm apart. Electron diffraction from a single tube is possible in principle. The most convincing images are from scanned probe microscopy at low temperatures. The example shown in Figure 1.2(a) has a screw axis and "handedness" like DNA, and is referred to as a chiral SWNT.

1.2.1 SINGLE-WALL TUBES, BUNDLES, AND CRYSTALLINE ROPES

An SWNT can be envisioned as a narrow rectangular strip of nanoscale graphene "chicken wire" with carbon atoms 0.14 nm apart at each apex, rolled up into a seamless cylinder 1–10 nm in diameter and as long as several micrometers. "Graphene" refers to a monolayer of sp^2-bonded carbon atoms. Several possibilities for legal strips, those that will roll up seamlessly, are shown in Figure 1.1. Because the length and width of legal strips are "quantized," so too are the lengths and diameters of the tubes. The short side of the rectangle becomes the tube diameter and therefore is "quantized" by the requirement that the rolled-up tube must have a continuous lattice structure. Similarly, the rectangle must be properly oriented with respect to the flat hexagonal lattice, which allows only a finite number of roll-up choices. The longer the short side, the larger the tube diameter *and* the larger the number of choices. Two of them correspond to high-symmetry SWNT; in "zigzag" tubes (top), some of the C–C bonds lie parallel to the tube axis, while in "armchair" tubes (bottom left), some bonds are perpendicular to the axis. Intermediate orientations of the rectangle produce chiral tubes when wrapped, as in Figure 1.2(a).

The different wrappings have profound consequences on the electronic properties. The allowed electron wave functions are no longer those of the unwrapped infinite two-dimensional graphene; the rolling operation imposes periodic boundary conditions for propagation around the circumference, the consequences of which depend on the symmetry. This is the reason why SWNT can either be metallic or insulating.

SWNT in close proximity can self-assemble into more or less close-packed parallel arrays, referred to as ropes or bundles according to whether the arrays are well ordered or not, respectively. These can be directly visualized in the electron microscope when a rope accidentally curves upward from the grid such that the focal plane cuts normally through the rope, as shown in Figure 1.2(b). The driving force is the van der Waals (vdW) attraction, amplified by the considerable lengths involved. Figure 1.2(b) shows a well-ordered rope consisting of ~100 tubes with similar diameters. In principle there is no limit to the number of tubes per rope, while the perfection of the 2D organization is directly related to the diameter polydispersity and thus to the synthesis method. The 2D triangular lattice implied by Figure 1.2(b) has been studied quantitatively using x-ray, neutron, and electron diffraction. Data analysis for materials containing highly ordered ropes has advanced to the point that diameter dispersivity, finite size, and the filling of the interior lumen can all be accounted for.

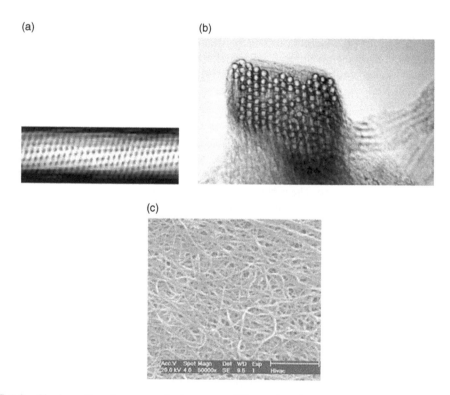

FIGURE 1.2 Single-wall carbon nanotube images at different length scales. (a) Scanning tunneling microscope image of a chiral SWNT (image by Clauss, W.); (b) HRTEM image of a nanotube rope (from Thess, A. et al., *Science* **273**, 483 (1996), with permission); (c) tangled spaghetti of purified SWNT ropes and bundles (Smalley, R. E., website).

Long-range order in three dimensions is generally frustrated. The 2D close-packed lattice imposes threefold rotational symmetry, while the point symmetry normal to the molecular axes will be modulo 3 only in special cases. This observation has important consequences for the nature of electronic tube–tube interactions in macroscopic materials.

Crystalline ropes are observed in some raw nanotube soots; others show TEM evidence for bundles, but with little or no 2D diffraction intensity. Crystallinity can often be improved by anaerobic thermal annealing, which is generally interpreted as minor rearrangements of misaligned tubes in a bundle, as opposed to grain growth by long-range migration of individuals accreting onto a small primordial rope. Purified bulk material resembles a spaghetti of nanotube ropes depicted in the SEM image in Figure 1.2(c).

Van der Waals bonding among the tubes in a rope implies that the equation of state should be closely related to those of graphite and solid C_{60} (similar bonding in 2, 1, and 3 directions, respectively [10]). The compressibility in a hydrostatic diamond anvil-cell environment, as determined by x-ray diffraction, is not consistent with the trend established by the other two carbon isomorphs, suggesting that the tubes become polygonalized into hexagonal cross sections at rather modest pressures [11]. The volume compressibility has been measured up to 1.5 GPa as 0.024 GPa^{-1}. The deformation of the nanotube lattice is reversible up to 4 GPa, beyond which the nanotube lattice is destroyed. Similarly, the thermal expansion of the 2D triangular lattice [12] of the rope does not follow the pattern of interlayer graphite and FCC C_{60}. One possibility is that the circular cross section of a tube is only an idealization, with the time-averaged cross-sectional shape changing with temperature due to thermally driven radial fluctuations.

1.2.2 MULTIWALL TUBES

Multiwall tubes have two advantages over their single-wall cousins. The multishell structure is stiffer than the single-wall one, especially in compression. Large-scale syntheses by enhanced chemical vapor deposition (CVD) processes are many, while for single-wall tubes, only the Rice HiPco process appears to be scaleable. The special case of double-wall tubes is under intense investigation; the coaxial structure of two nested tubes is well defined by the synthesis conditions, and large enhancement in stiffness over the single wall has already been achieved [13]. Multiwall structures are intrinsically frustrated since it is impossible to satisfy epitaxy simultaneously, or compatibility of wrapping indices (n, m) on adjacent shells, with the very strong tendency to maintain the intershell spacing close to that of the graphite interlayer spacing of 3.35 Å. Experimentally, the intershell correlations in MWNT are characteristic of turbostratic graphite in which the ABAB stacking order is severely disrupted [14].

Are multiwall tubes really concentric "Russian doll" structures, or are they perhaps scrolled? A minority of TEM studies suggests scrolling by virtue of observing n and $n+1$ graphene layers on opposite sides of the MWNT. One such example [14] is shown in Figure 1.3. On the other hand, attempts to intercalate bulk MWNT samples result in destruction of the cylindrical morphology, which would not be the case if the MWNT were a continuous graphene sheet rolled into a seamless scroll [15].

Several unique features of multiwall tubes should be noted here. Hydrothermal synthesis results in water-filled tubes, a natural model system for studying 1D nanofluidics and exploiting these as nanoreactors [16]. Synthetic methods can be adapted to allow direct fiber production from the primordial soot or the hot reactor plume [17,18], or by twisting fibers from vertically aligned MWNT grown as a carpet-like array on a substrate [19,20]. Current pulses, through individual MWNT, controllably remove one shell at a time, permitting the custom construction of devices to order, with metallic or semiconducting outer shells [21].

1.2.3 MACROSCOPIC NANOTUBE MATERIALS

Individual nanotubes can be used to construct electronic devices, gas and biosensors, sensitive nanobalances, NEMS resonators, scanning probe tips, etc. [22]. As the cost of production decreases

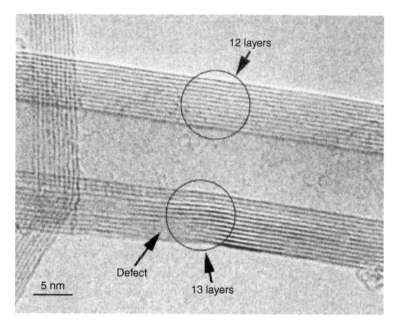

FIGURE 1.3 HRTEM image of an MWCNT. The number of graphene layer images on the two edges of the tube differs by 1, implying a scroll structure rather than the assumed-to-be-universal Russian doll morphology. (From Lavin, J. G. et al., *Carbon* **40**, 1123–1130 (2002), with permission.)

with time, applications involving macroscopic nanotube-derived materials become feasible. Such materials should obviously be strong and lightweight. Prospects for replacing traditional materials rest on opportunities to design in multifunctional combinations of mechanical, electrical, thermal, dielectric, magnetic, and optical properties.

Macroscopic nanotube materials are available in many forms: compressed random mats of raw or purified soot, filter-deposited foils ("bucky paper") [23], spin-coated or solvent-cast films [24], and various forms of all-nanotube [25,26], or composite [3,27] fibers. Property optimization for these complex materials depends on many factors. For example, the degree of preferred nanotube orientation within a macroscopic body determines the extent to which one takes advantage of the excellent but anisotropic intrinsic nanotube properties. Another major factor is the macroscopic density. This is important from several aspects. Empty space means missing material in the body, such that most properties will never be optimal. On the other hand, porosity on different length scales can be useful, e.g., for perm-selective membranes, fast diffusion for ionic conduction in electrochemical devices [28], and tailored nanoporosity for filtering, storage, or sequestration of specific molecular-scale analytes.

Materials of SWNT can be partially aligned by mechanical shear [15], anisotropic flow [3,26], gel extrusion [29], filter deposition from suspension in strong magnetic fields [30], or the application of electric fields during or after growth [31]. Fibers exhibit axially symmetric alignment with mosaic spreads of a few degrees in composites [3], and as low as 20° for all-SWNT post-stretched gel fibers [25]. Ordinary filter-deposited or solvent-cast films exhibit substantial adventitious alignment of tube axes in the film plane due simply to the rather gentle landing of rigid sticks on a flat surface [32].

The most complete characterization of the degree of alignment is obtained by combining the best features of Raman scattering [33] and traditional x-ray fiber diagrams [34]. With two independent experimental parameters, texture can be modeled as a "two-phase" system consisting of an aligned fraction, characterized by the angular distribution of tube axes about the preferred direction, plus a completely unaligned fraction to account for SWNT aggregates, which are insufficiently dispersed to respond to the aligning field. The model distribution function, shown schematically in Figure 1.4, is represented by a Gaussian "bugle horn" for the axially aligned fraction sitting atop a right-circular cylinder, which accounts for the unaligned fraction. X-ray two-dimensional fiber diagrams unambiguously give the Gaussian width from the azimuthal dependence of either a Bragg intensity or form factor diffuse scattering if the material is not sufficiently crystalline. A similar method was previously applied to texture studies of oriented films and fibers of conjugated polymers [35]. This approach can be applied to thin films cut into strips and carefully stacked in a capillary as well as to fibers. Both geometries are shown schematically in Figure 1.5. Angle-dependent polarized Raman scattering, using the x-ray-derived distribution width as input unambiguously gives the unaligned fraction. This is because the resonantly enhanced Raman contribution from the SWNT contribution dominates over all the other constituents of the sample [36], even at the 1% loading level [37]. An example of this combined fitting approach is shown in Figure 1.6 for nanotube fibers extruded from strong acid solution with no post stretching [34]. Here we deduce a mosaic distribution width of 55° full-width at half-maximum (FWHM) and an aligned fraction of 90%. In the case of magnetically aligned buckypapers, these parameters can be correlated with measurements of electrical resistivity ρ and thermal conductivity κ parallel and perpendicular to the alignment direction [30]. Differences between 7 and 26 T aligning magnetic fields give some clues about the alignment mechanism and strategies for improving the process.

1.2.4 FIBERS

The best alignment achieved so far is in polymer–SWNT composites [3,27]. Here the "dope" consists of either a polymer melt with SWNT dispersed at concentrations up to 10 wt%, or mixtures of polymer and SWNT in the same organic "solvent." This is extruded through a small orifice after twin-screw mixing and then stretched by take-up spindles. The extensional flow through the orifice, combined with shear flow during stretching, combine to yield FWHMs of the order of a few degrees in the best cases [3]. SWNT–(poly)vinyl alcohol (PVA) composite fibers exhibit modest alignment as extruded, but this can be greatly improved by controlled elongation prior to removal of the PVA by heat treatment [25,38]. Nanotube fibers consisting mainly of MWNT can be obtained directly during carbon arc synthesis [18], or by agglomeration as the tubes are pulled off a substrate [19]. Nothing quantitative is known yet about the degree of alignment in these and related materials [20].

1.2.5 FILLED TUBES

Carbon soot produced by arcing in helium or laser ablation contains both fullerenes (mainly C_{60}) and nanotubes, the proportions varying with the amount of transition metal catalysts in the graphite

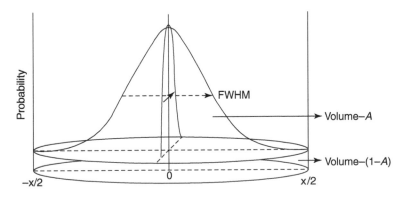

FIGURE 1.4 Two-phase model for the axially symmetric distribution of nanotube axes in oriented fibers. The inverted horn is the aligned fraction, characterized by an FWHM. The pillbox on top of which the horn is positioned is the unaligned fraction, characterized by its volume relative to the total.

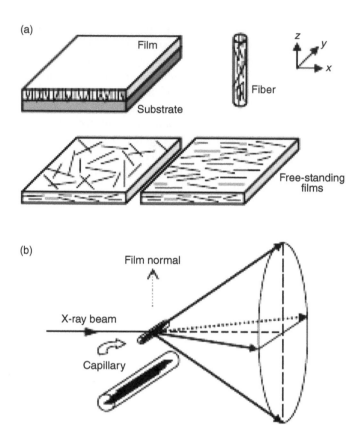

FIGURE 1.5 Schematic of the experimental setup for measuring out-of-plane mosaic. The film plane is parallel to the incident x-ray beam. Out-of-plane preferred orientation results in azimuth (χ)-dependent anisotropic scattering within the 2D detector plane; (a) and (b) refer to films and fibers, respectively.

electrode or target, as discussed previously. Acid purification of nanotube-rich soot reveals the presence of "peapods" — SWNT with chains of C_{60} inside [39], in substantial amounts [40], denoted by "C_{60}@SWNT." It is now understood that acid etching opens holes in the SWNT, either by decapitation or creation of sidewall defects. Fullerenes and other species can diffuse in and be trapped by annealing after filling [41]. Rational synthesis now achieves >80% filling routinely [42]. The process has been extended to higher fullerenes [43], endofullerenes [44] and other substances [7]. We speculated early on that host–guest interactions might favor commensurate packings, in which case the ideal container for C_{60} would be the (10,10) tube. The interior lumen is a good fit for C_{60} (vdW diameter ~1 nm), and, with armchair symmetry, exactly four "belts" of hexagons would be commensurate with one C_{60} [45]. This turned out not to be important after all, implying very weak interactions between the pea and the pod. There are indeed strong steric constraints on the tube diameter, which will accept a perfect 1D chain [46], while there seems to be no effect of symmetry on filling probability. Global filling fractions depend strongly on diameter dispersivity; HiPco tubes have broad diameter distributions and generally give at best 50–60% filling.

Films of crystalline peapods exhibit unique "fiber diagrams" using the 2D detector (cf. Figure 1.5). Figure 1.7 shows an example of a detector image from such a film. Diffraction peaks from the 1D lattice of close-packed C_{60} peas are concentrated in the horizontal direction, perpendicular to those from the 2D pod lattice. This separation is a consequence of the out-of-plane preferred orientation; the sample is neither a "perfect powder" nor a single crystal, but rather a "mosaic crystal" with identical orientation distributions exhibited by [0 0 L] and [H K 0] families of reflections [32]. The absence of mixed [H K L] indicates that the peas in different pods "float" along the tube axis

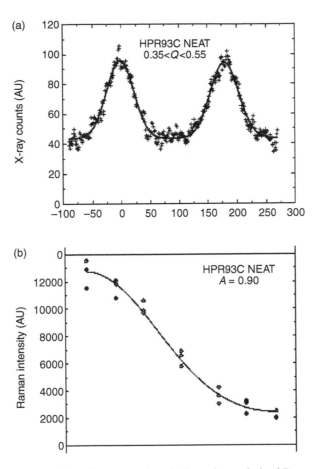

FIGURE 1.6 Combining x-ray fiber diagrams and angle-dependent polarized Raman scattering to solve the texture problem in SWNT materials. (a) Background-subtracted X-ray counts, summed over different Q intervals, every $1°$ in χ. Symbols represent data; smooth curves are fits to two Gaussians plus a constant. (b) Angle-dependent polarized Raman data (open circles) and fits (solid curves). A two-parameter model was used, with one (FWHM) fixed at the value determined from (a). (From Zhou, W. et al., *J. Appl. Phys.* **95**, 649 (2004), with permission.)

with respect to each other; there are no 3D correlations among peas in different pods [32,41]. Also shown are radial cuts through 2D images of an unfilled control sample $[0\ 0\ L]$ I, the filled sample $[0\ 0\ L]$ II, and $[H\ K\ 0]$ III. Note that the filling of C_{60} into nanotubes significantly changes the diffraction profile. The $(0\ 0\ 1)$ and $(0\ 0\ 2)$ peaks from the 1D C_{60} chains are easily detected. Filling fractions can be obtained from detailed curve fits; these seem to be limited only by the diameter dispersion of the starting material, i.e., a real-world sample will always contain some tubes that are too small to be filled [42,46].

One of the most significant developments in the peapod arena is their exploitation as precursors to the formation of highly perfect double-wall nanotubes [13]. Thermal decomposition inside the original SWNT container leads to rearrangement of broken bonds on near-neighbor tubes into a smaller seamless cylinder whose diameter respects the usual vdW spacing requirement. This process takes place at a temperature considerably below those of the amorphization of FCC C_{60} [47] and enormously below the gas-phase decomposition temperature [48]. Clearly, the coordination and packing environment in the condensed phase leads to facile destruction of the icosahedral molecule by rather violent near-neighbor collisions, more effective in the 12-coordinated 3D solid than in the 2-coordinated chain structure of the "peas."

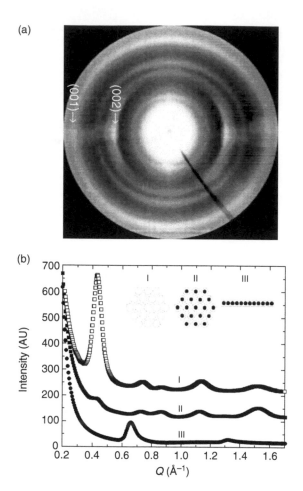

FIGURE 1.7 (a) Detector image from a C_{60}@SWNT peapod film. Diffraction peaks from the 1D C_{60} lattice are concentrated in the direction perpendicular to those from the 2D pod lattice, a consequence of the out-of-plane preferred orientation. (b) X-ray diffraction patterns from starting SWNT film (control sample), and C_{60}@SWNT film (peapod sample). Note that the filling of C_{60} into nanotubes significantly changes the diffraction profile. The (0 0 1) and (0 0 2) peaks from the 1D C_{60} chains are easily detected. (From Zhou, W. et al., *Appl. Phys. Lett.* **84**, 2172 (2004), with permission.)

The extent to which nanotube properties are affected by the endohedral doping is controversial and will be discussed later.

1.2.6 NANOTUBE SUSPENSIONS

It is appropriate to conclude our discussion of structure in nanotube materials with a few words about tubes in solution or suspension. These are important for fundamental and technological reasons. Absorption spectroscopy of isolated tubes [49] reveals important aspects of their presumed simple electronic structure and its dependence on tube diameter and symmetry as well as the importance of excitonic and other higher order effects [50]. Nanotube suspensions also provide a unique system in which to study rigid rod-phase behavior [51] and network formation at higher concentrations. Furthermore, an improved understanding of suspension structure should lead to better control of solution processes such as phase separation, chemical derivatization, control and improvement of polymer/nanotube composite properties, and optimization of the "dope" from which fibers are spun.

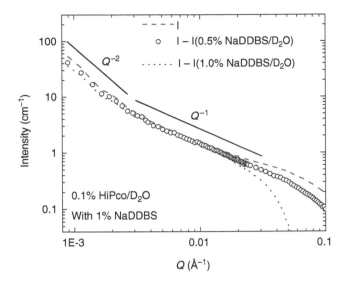

FIGURE 1.8 Analysis of SANS data from nanotubes in surfactant/heavy water suspension. Subtraction of surfactant contribution from total intensity for HiPco tubes. Dashed and dotted lines correspond to the extreme assumptions of no excess surfactant and no surfactant on the tubes, respectively. A Q^{-1} slope is clearly visible from 0.003 to 0.02 A^{-1}, no matter what fraction of surfactant intensity is subtracted. There is also a crossover to -2 exponent at 0.004 A^{-1}, suggesting the formation of a dilute network with a mesh size ~160 nm. (From Zhou, W. et al., *Chem. Phys. Lett.* **384**, 185–189 (2004), with permission.)

Individual SWNTs with diameters 1–2 nm and lengths 100 nm to several micrometers can be envisioned as rigid rods, and if sufficiently dilute, should display a correspondingly simple structure factor in scattering experiments (light, x-rays, neutrons, etc.). In particular, the scattered intensity, I, from a suspension of isolated rigid rods with diameter D and length L should follow a Q^{-1} law for scattering wave vectors $2\pi/L < (Q = 4\pi \sin \theta/\lambda) < 2\pi/D$ [52]. This is generally not observed; suggesting that even at very low concentrations, the SWNTs tend to form loose networks with node spacings of the order of the tube length. One exception is the dilute suspensions using the ionic surfactant sodium dodecylbenzene sulfonate (NaDDBS) and its relatives [52]. It is believed that the benzenoid moiety provides π stacking interactions with the graphene-like tube wall structure, of sufficient strength to adhere to single tubes and thus preventing them from aggregating. Figure 1.8 shows small-angle neutron scattering data for HiPco SWNT in 1% NaDDBS/D_2O suspensions. After correcting for scattering from NaDDBS micelles and incoherent proton background, one indeed observes a linear regime at intermediate values of Q. There is a crossover from Q^{-1} to Q^{-2} behavior at about 0.004 A^{-1}, corresponding to average tube length ~160 nm. This is precisely what is found from atomic force microscopy (AFM) image analysis of >1000 tubes, confirming that we have truly isolated tubes at this very low concentration, i.e., 10^{-3} wt%. Linear regimes are not observed at any practicable concentration using Triton-X, polymer melts, or superacids. Fiber spinning requires much higher concentrations, so we must accept some degree of aggregation as the dope is being extruded out of the orifice or spinneret. How this limits the properties of the ensuing fibers has not been determined as yet.

1.3 PHYSICAL PROPERTIES

We focus on the properties of macroscopic nanotube materials. These ultimately derive from the intrinsic attributes of single tubes, vastly complicated by defects, impurities, preferred orientation, network connectivity, large-scale morphology, etc. Theory does a reasonable job of predicting, rationalizing, and explaining single-tube properties, while structure–property relations in macroscopic assemblies are still in a rudimentary state. For composites in particular, properties of the simplest materials show large deviations from the rule of mixture behavior.

To achieve viability in practice, any class of new materials must have some property that surpasses the state of the art sufficiently to justify replacement. Alternatively, new materials may provide some new capability otherwise unavailable. It seems likely that the first large-scale applications for nanotubes will arise from multifunctionality, for example, stiff lightweight structural parts, which also conduct electricity and have low thermal expansion.

1.3.1 MECHANICAL PROPERTIES

The strength of the carbon–carbon bond gives rise to the extreme interest in the mechanical properties of nanotubes. Theoretically, they should be stiffer and stronger than any known substance. Simulations [53] and experiments [54] demonstrate a remarkable "bend, don't break" response of individual SWNT to large transverse deformations; an example from Yakobson's simulation is shown in Figure 1.9. The two segments on either side of the buckled region can be bent into an acute angle without breaking bonds; simulations and experiments show the full recovery of a straight perfect tube once the force is removed.

Young's modulus of a cantilevered individual MWNT was measured as 1.0 to 1.8 TPa from the amplitude of thermally driven vibrations observed in the TEM [55]. At the low end, this is only ~20% better than the best high-modulus graphite fibers. Exceptional resistance to shock loads has also been demonstrated [56]. Both the modulus and strength are highly dependent on the nanotube growth method and subsequent processing, due no doubt to variable and uncontrolled defects. Values of Young's modulus as low as 3 to 4 GPa have been observed in MWNT produced by pyrolysis of organic precursors [57]. TEM-based pulling and bending tests gave more reasonable moduli and strength of MWNT of 0.8 and 150 GPa, respectively [58].

Multiwall nanotubes and SWNT bundles may be stiffer in bending but are expected to be weaker in tension due to "pullout" of individual tubes. In one experiment, 15 SWNT ropes under tension broke at a strain of 5.3% or less. The stress–strain curves suggest that the load is carried primarily by SWNT on the periphery of the ropes, from which the authors deduce breaking strengths from 13 to 52 GPa [59]. This is far less than that reported for a single MWNT [58]. On the other hand, the mean value of tensile modulus was 1 TPa, consistent with near-ideal behavior. Clearly, the effects of nonideal structure and morphology have widely different influences on modulus and strength. There is always some ambiguity in choosing the appropriate cross-sectional area to use in evaluating stress–strain data. On a density-normalized basis the nanotubes look much better [22]; modulus and strength are, respectively, 19 and 56 times better than steel.

Figure 1.10 shows the formation of a remarkable nanotube yarn by pulling and (optionally) twisting material from a "forest" of vertically aligned MWNT grown by a CVD process [19]. The untwisted yarns are very weak; if they accidentally touch a surface while being pulled off the substrate, they immediately break. On the other hand, twisted single-strand yarns exhibited strengths in the range 150 to 300 MPa; this improved to 250 to 460 MPa in the two-ply yarns. Further gain to

FIGURE 1.9 MD simulation of a large-amplitude transverse deformation of a carbon nanotube, apparently beyond the elastic limit. In fact, the tube snaps back once the computer-generated force is removed, and there is no plastic deformation or permanent damage. (From Yakobson, B. I., *Appl. Phys. Lett.* **72**, 918–920 (1998), with permission.)

(a)

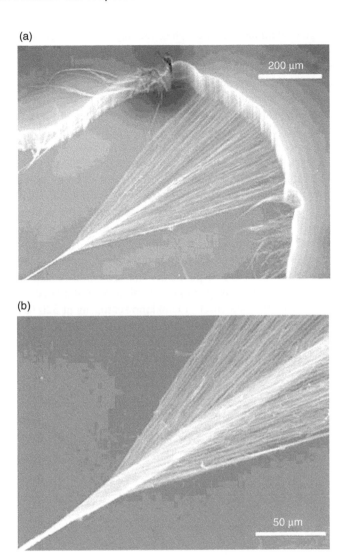

(b)

FIGURE 1.10 *Ex situ* SEM snapshots of a carbon nanotube yarn in the process of being drawn and twisted. The MWNTs, ~10 nm in diameter and 100 μm long, form small bundles of a few nanotubes each in the forest. Drawing fiber normal (a) and parallel (b) to the edge of the forest of nanotubes. (From Zhang, M. et al., *Science*, **306**, 1358 (2004), with permission.)

850 MPa was achieved by infiltration with PVA [19], which also improved the strain-to-failure to 13%. Toughness, the so-called "artificial muscle" [60], is a major issue in the optimization of nanotube actuators, where the figure of merit is the work done per cycle.

Another form of nanotube material useful for sensors and actuators are thin films or buckypapers. A typical result for solution-cast film with random SWNT orientations in the plane [32] is described in [61]. The tensile modulus, strength, and elongation-to-break values are 8 GPa (at 0.2% strain), 30 MPa, and 0.5%, respectively, much lower than what can be routinely achieved in fibers. This suggests that failure in the films occurs via interfibrillar slippage rather than fracture within a fibril.

1.3.2 THERMAL PROPERTIES

The thermal conductivity κ of carbon materials is dominated by atomic vibrations or phonons. Even in graphite, a good electrical conductor, the electronic density of states is so low that thermal transport via "free" electrons is negligible at all temperatures. Thus, roughly speaking, $\kappa(T)$ is given by the product of (1) the temperature-dependent lattice specific heat C_p, a measure of the density of occupied phonon modes at a given temperature; (2) the group velocity of the phonon modes (speed of sound V_s for acoustic branches, not strongly T-dependent); and (3) a mean free path accounting for elastic and inelastic, intrinsic and extrinsic phonon-scattering processes. It was conjectured early on that nanotubes would be excellent heat conductors [62]; the axial stiffness conduces to large V_s, the 1D structure greatly restricts the phase space for phonon–phonon (Umklapp) collisions, and the presumed atomic perfection largely eliminates elastic scattering from defects. Calculations of the lattice contribution to κ, pioneered by the Tomanek group [63], yield values in the range of 2800 to 6000 W/mK.

Experiments on individual tubes are extremely difficult, and so far have been limited to MWNT [64]. Figure 1.11 shows the temperature-dependent thermal conductance of a 14-nm-diameter, 2.5-µm-long MWNT; $\kappa(T)$ based on an effective cross-sectional area is shown in the inset. The salient features are a peak value ~3000 W/mK at 320 K, in good agreement with the perfect-tube calculations, and a roll-off in κ above 320 K signaling the onset of Umklapp scattering. For bulk MWNT foils, κ is only 20 W/mK [65], suggesting that thermally opaque junctions between tubes severely limit the large-scale diffusion of phonons. The temperature dependence below 150 K reflects the effective dimensionality of the phonon dispersion, a subtlety that is more pronounced in SWNT and will be discussed later. The onset of Umklapp scattering at 320 K is highly significant. Knowing C_p and V_s from other experiments and theoretical estimates, the mean free path is ~500 nm, comparable to the length of the sample. This means that phonon transport at low T is essentially ballistic; on average a phonon scatters only a few times as it traverses the sample. There is no corresponding rollover in the data from bulk nanotube samples.

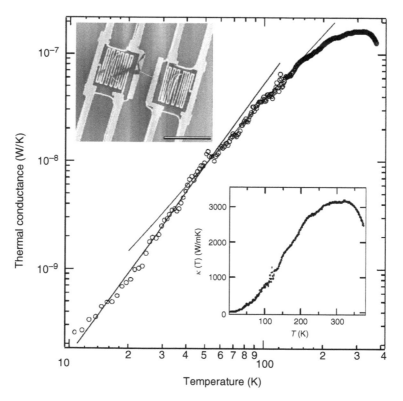

FIGURE 1.11 Thermal conductance of an individual MWNT of 14 nm diameter. The power law slope decreases from ~2.5 to ~2.0 above ~5 K. Saturation occurs at 340 K, the thermal conductance decreasing at higher T. Lower inset shows the thermal conductivity, for which some assumptions about effective area are required. (From Kim et al., *Phys. Rev. Lett.* **87**, 215502 (2001), with permission.)

One of the most important concepts in nanotube physics is quantization of the electronic states by the restricted 1D geometry. Analogous 1D phonon quantization is difficult to observe because of the far smaller energy/temperature scale. In general, the smaller the object, the higher the temperature at which quantum confinement effects can be detected, so SWNT with radii ~2 nm raise the temperature to practicable values. Temperature-dependent heat capacity of purified SWNT shows direct evidence for 1D quantized phonon subbands [66] (see Figure 1.12). The sample is ~40 mg of purified HiPco SWNT, for which the 2D triangular lattice is not well developed. This material was chosen in the hope that the 1D physics would not be obscured by tube–tube coupling. However, a good fit to the data (solid dots) requires an anisotropic, weakly 3D Debye model, shown in the inset, which accounts for weak coupling between SWNT in a rope (black curve). Below 8 K, only the acoustic branch contributes (blue curve), while above 8 K the first optic subband becomes significant (red curve). The key point is that this subband "turns on" at a temperature high enough to be detected, yet low enough not to be buried by graphene-like 2D contributions. The relevant fitting parameters are 4.3 meV for the $Q = 0$ subband energy (in reasonable agreement with predictions for this diameter) and 1.1 meV for the transverse Debye energy, signaling very weak intertube coupling. It would be of interest to repeat this experiment on samples with weaker or stronger tube–tube coupling, the former, for example, by intercalation with large dopant ions [67]. Above 25 K, C_p of bulk SWNT is indistinguishable from that of graphite. Extension of the measurement on the same sample down to the milliKelvin regime [68], confirmed the importance of 3D effects at the lowest temperatures; a definitive analysis was impeded by a strong nuclear hyperfine contribution from residual Ni catalyst.

Both SWNT and MWNT materials and composites are being actively studied for thermal management applications, either as "heat pipes" or as alternatives to metallic or alumina particle additives to low thermal resistance adhesives. The individual MWNT results are promising; similar data for SWNT are not yet available. In the case of polymer composites, the important limiting factors are quality of the dispersion and interphase thermal barriers [3]. What limits κ in all-nanotube

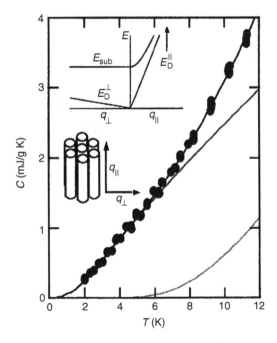

FIGURE 1.12   Specific heat vs. T of SWCNTs (dots) fitted to an anisotropic two-band Debye model that accounts for weak coupling between SWNT in a rope (solid curve). The rope structure and ensuing phonon dispersion are shown as insets. (From Hone, J. et al., *Science* **289**, 1730 (2000), with permission.)

fibers, sheets, etc.? Tube wall defects can be minimized by avoiding long sonication and extended exposure to acids in the fabrication process. In one example, partial alignment of nanotube axes in buckypaper yielded anisotropic thermal conductivities, with $\kappa_{//}/\kappa_{\perp}$ ~5 to 10 [69].

Experimental determinations of κ are subject to many pitfalls, one of which is the problem of defining the cross-sectional area through which the heat is flowing. This is similar to the dimensional ambiguity we encountered with Young's modulus. There are many well-tested methods for measuring thermal conductance G, but translating that into thermal conductivity κ is not straightforward. This is illustrated by two different results from the magnetic field-aligned buckypaper mentioned above. In the first method, sample dimensions were deduced from a measurement of electrical resistance of the κ sample and scaling to the resistivity of a different sample, with well-defined geometry, taken from the same batch. This assumed that the resistivity is uniform over centimeter length scales in the original piece of buckypaper, and gave κ parallel to the average tube axis ~250 W/mK. A second measurement in which the actual dimensions were carefully measured gave a more modest 60 W/mK.

Figure 1.13 shows the effect of stretch alignment on the thermal conductivity of SWNT fibers gel extruded from water/PVA suspensions [25]. Here the fibers are statically stretched in a PVA/water bath by suspending a weight on the end of a length of pristine fiber. Vacuum annealing after elongation eliminates the PVA, so the fiber properties are entirely due to aligned nanotubes. Reducing the mosaic by increasing the elongation, form 23° to 15° FWHM, gives an almost threefold increase in κ (300 K). In contrast to the individual MWNT data (Figure 1.11), there is no evidence for $\kappa(T)$ saturation or rolloff at high temperature. This almost certainly means that κ is limited by interparticle barriers and not by the Umklapp mean free path.

Figure 1.14 shows $\kappa(T)$ for similar fibers extruded from superacid suspension but not stretched [34]. Here the correlations are with nanotube concentration and syringe orifice diameter. Concentration dictates the phase of the rigid rod suspension [26] and controls the viscosity, while the orifice diameter affects the degree of flow-induced alignment. From the three examples shown, and from other observations, the authors find that (1) annealing to remove acid residues and the attendant p-type metallic doping (see below) has no effect on κ at any temperature; (2) reducing the

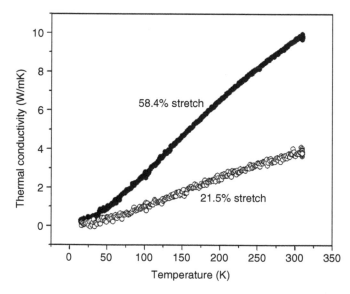

FIGURE 1.13 Thermal conductivity vs. temperature for annealed nanotube fibers, which have been stretch-aligned to 21% (lower curve) and 58% (upper curve). Additional alignment of the latter yields a factor of ~2 further enhancement of κ. The temperature dependence is dominated by that of the lattice specific heat, as is generally observed. (From Badaire, S. et al., *J. Appl. Phys.* **96**, 7509 (2004), with permission.)

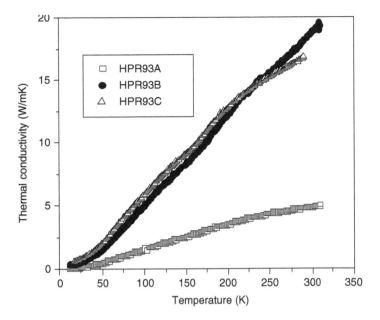

FIGURE 1.14 Thermal conductivity κ vs. temperature of three HiPco fibers. HPR93 exhibits the broadest mosaic distribution of nanotube axes, as determined from the x-ray fiber diagrams.

concentration from 8 to 6%, or reducing the syringe diameter from 500 to 250 μm improves κ (300 K) fourfold; and (3) further syringe diameter reduction to 125 μm yields no further improvement. The syringe diameter effects are clearly due to improved alignment; the smallest orifice gives the smallest mosaic, 45° FWHM compared to 65° with the 500 μm one. The best κ value is still orders of magnitude below graphite and ideal MWNT, most probably due to thermal barriers. It may be significant that the highest macroscopic κ was obtained for materials made from pulsed laser vaporization (PLV) SWNT without extended exposure to strong acids. PLV tubes made thus are known to be highly perfect, while acid immersion can lead to defects and ultimately to very short tubes. Either way, the phonon mean free path in such materials may be quite low.

Thermal conductivity of peapods is enhanced relative to that of empty tubes, and the enhancement exhibits interesting temperature dependence [70,71]. Difference data, κ(filled)-κ(empty), from 10 to 285 K are shown in Figure 1.15. Three distinct regions are evident. Region I is dominated by the contribution from the very soft C_{60}–C_{60} acoustic modes. A 1D Debye temperature of 90 K is estimated for these modes using a Lennard–Jones potential. At T_D all the LA modes are excited, the excess heat capacity saturates and $\Delta\kappa$ levels off to a constant value (region II). The filling fraction of this sample is well below 50%, and the theory predicts the existence of C_{60} capsules separated by lengths of empty tubes to account for the unfilled sites [46]. We imagine the onset of a new thermal transport mechanism (region III), beginning at temperatures sufficiently high to break the weak vdW bond between a terminal C_{60} and its capsule. This will happen more efficiently at the hot end, and, once liberated, the free C_{60} will thermally diffuse to, and join, the end of a cold capsule, as shown schematically in the inset. The internal degrees of freedom represent an additional thermal transport process accompanying the mass transport inside the tube. The rate of this process increases with increasing T and will saturate at a very high T, which transforms the assembly of capsules to a 1D gas. This scenario is speculative at this point since it has been observed only in one sample. A rigorous test of the mechanism would be to find out if the effect disappears in a completely filled sample. Other endohedral species may give larger enhancements, and these materials may then become candidates for thermal management applications such as additives to conducting adhesives.

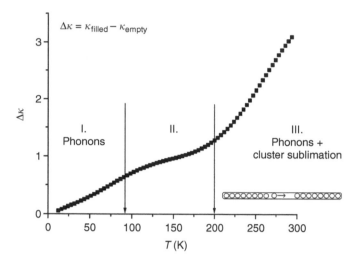

FIGURE 1.15 Excess thermal conductivity in a partially filled C_{60}@SWNT buckypaper. Enhancement in regime I attributed to the 1D LA phonons on the C_{60} chains. Saturation in regime II after $T > T_{Debye} \sim 90$ K of the 1D molecular modes. Further enhancement in regime III assigned to hot C_{60} molecules subliming off the end of a capsule, diffusing to the end of a colder capsule while carrying internal vibrations as a combined heat and mass transfer.

1.3.3 ELECTRONIC PROPERTIES

At the last count Google.scholar gave 4400 hits on this rubric, so this section will be highly selective.

Theory led experiment in this important aspect of nanotube properties, and the most important results emerge from a simple idea. One monolayer of graphene has two energy bands $E_n(k)$ which cross at the Fermi energy E_F, so the system is on the cusp of a metal–insulator (MI) transition. The electronic ground state can therefore be described either as a zero-gap semiconductor or as a metal with infinitesimal density of states at the Fermi energy $N(E_F)$. In 3D graphite, the interlayer interaction causes a transition to a semimetal which conducts well, because the $\sim 10^{18}$ charge carriers/cm^3 have very high mobility. On the other hand, rolling a single sheet into a tube is a symmetry-breaking transition, which leads to semiconducting or metallic ground states, depending on the choice of wrapping (Figure 1.1). The band structure of a single tube consists of unusual 1D subbands resulting from the requirement that the radial wave function be commensurate with the circumference of the tube. The larger the tube, the smaller the subband spacing; in the 2D graphene limit, there are only the two bands mentioned above. The longitudinal dispersion is linear for a range of energies close to zero.

As noted earlier, bulk properties are generally dominated by extrinsic effects, although some of the 1D subtleties remain, especially in oriented material. The properties of the principal techniques are resistivity and its temperature dependence, density of states spectroscopy using electron energy loss and photoemission, optical absorption, and reflectivity. In a few cases, thermoelectric power and conduction electron spin resonance have also been applied to nanotube physics.

The electrical properties of bulk nanotube materials have evolved considerably since the first experiments 10 years ago. Early samples consisted of pressed "mats" of nanotube soots from carbon arc or laser ablation processes, which contained only 10 to 30 wt% tubes. Major impurities were amorphous carbon, metal catalyst particles, and fullerenes, mainly C_{60}. Mat densities were as low as 1% of the crystallographic value computed from a triangular lattice of 1.4 nm diameter SWNT. On the other hand, long high-quality tubes could be found in these soots. This is because they had not been subjected to acid treatment or sonication in attempts to remove impurities, both of which are now known to create sidewall defects and ultimately to cut tubes into shorter lengths.

The prevailing attitude at the time was that only single-tube measurements would give useful information. Nevertheless, some valuable information, still valid today, was obtained from four-probe measurements on these mats. First, correcting for the missing material, ρ (300 K) values within 1–2 decades of in-plane graphite were obtained, suggesting that bulk SWNT materials could be developed into strong, lightweight synthetic metals [72]. Second, the very modest temperature dependence in the range 80 K $< T <$ 350 K changed from nonmetallic to weakly metallic at ~200 K, suggesting a phonon-scattering mechanism at high temperatures. Third, experiments extended to very low T showed a strong divergence in ρ (T) as $T > 0$, indicating strong localization of carriers [73]. Finally, chemical doping with bromine or potassium [67,74] showed that like graphite, bulk SWNT material acted as an amphoteric host for redox doping. All of these phenomena have since been revisited using purified samples with SWNT content approaching 100%.

Figure 1.16 shows the anisotropy of ρ (T) measured parallel and perpendicular to the alignment direction for the case of buckypaper deposition in a magnetic field [30]. The temperature dependence is weakly nonmetallic and quite similar for both orientations. Furthermore, the anisotropy is independent of temperature. These observations indicate that we are not observing the intrinsic anisotropy of aligned SWNT; transport in the perpendicular direction is dominated by misaligned bundles or fibrils, which short out the larger intertube/interbundle resistance. The nonmetallic T-dependence is consistent with previous results on unoriented buckypaper after similar acid purification and high-temperature vacuum annealing. These processes are presumed to reduce the mean free path for electron–phonon scattering via defect creation or tube cutting, and thus eliminate the high-temperature regime of positive dρ/dT. At 295 K, the measured $\rho_{//}$ for 7 T field alignment is 0.91 mΩ cm. Accounting for gross porosity raises this to 0.41 mΩ cm.

The weak anisotropy can be accounted for by a simple model of 1D paths in the plane of the sample, each sample containing on average n elements (ropes, tubes) of fixed length and resistance. The resistance of each path is linearly proportional to n and, since for a fixed number of elements, the number of paths is inversely proportional to n, the resistance of the ensemble of paths in parallel is

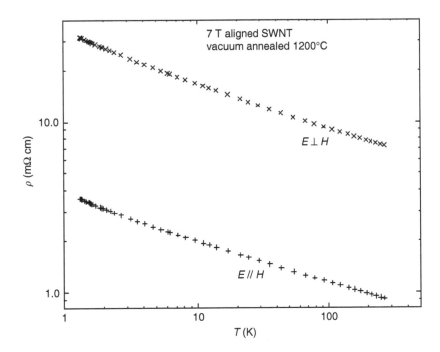

FIGURE 1.16 Resistivity vs. temperature for annealed 7 T-aligned buckypaper, measured with current perpendicular and parallel to the average alignment axis on two different samples with in-line four-probe contacts. (From Fischer, J. E. et al., *J. Appl. Phys.* **93**, 2157 (2003), with permission.)

proportional to n^2. The average number of elements n required for current to flow through the sample is equal to L/λ, where L is the length of the sample and λ is the mean projection of the element length in the current direction: λ depends on the distribution width and aligned fraction. Using x-ray and Raman data as input, very good agreement is obtained between predicted and measured $\rho_{//}/\rho_{\perp}$ for 7 and 26 T data.

How does the experimental $\rho_{//}$ (300 K) compare with that of an assembly of ballistic conductors? Assume that each metallic tube in a perfectly aligned sample is comprised of finite-length ballistic conductors in series, the length being the mean free path λ for electron–phonon backscattering. Each tube has two transport channels corresponding to the two bands degenerate at E_F. Taking $\lambda = 300$ nm [75], $\rho_{//} = $ (area/tube)/($2G_o \times 300$ nm) $= 1.5 \times 10^{-5}\ \Omega$ cm, where G_o is the conductance quantum $12.6\ \mathrm{K}\ \Omega)^{-1}$. Assuming further that tube growth is stochastic with respect to wrapping indices, only one third of the tubes will be metallic at 300 K and so $\rho_{//} \sim 50\ \mu\Omega$ cm, roughly twice the value for graphite. This is a surprising result because there are many factors which will increase this number in real samples: finite distribution width of tube axes, unaligned tubes, empty volume (porosity), junction resistance between tube segments and between ropes, incoherent intertube scattering within a rope, and elastic scattering from tube ends, defects and impurities. These may be partially offset by p doping of the semiconducting tubes by acid residues from purification and by atmospheric oxygen.

Pure nanotube fibers offer the prospect of even lower resistivities since they are readily aligned during extrusion and can be further aligned by stretching. We discuss next the results of transport measurements on the two sets of fibers described above: syringeextruded from oleum suspensions and PVA/water mixtures.

Figures 1.17(a) and (b) show the effects of extrusion conditions and post-extrusion annealing on the temperature-dependent resistivity of fibers extruded from anhydrous sulfuric acid, or oleum [34]. Varying degrees of alignment were obtained from HPR93 fibers extruded using three different combinations of SNWT concentration and orifice diameter. The texture results of these have been discussed above. Annealing was carried out in vacuum or flowing argon at 1100°C for 24 h. In the neat state, all three fibers exhibit low resistance with metallic temperature dependence above 200 K, as shown in Figure 1.17(a). The best alignment is obtained for HPR93B, which correlates nicely with the lowest ρ (300 K) = 0.24 mΩ cm, about a factor 10 less than graphite. For all three neat fibers, both the small values and the weak temperature dependence are due to the strong redox doping effect of bisulfate from the acid suspension. The nondivergent low-T behavior in the neat state can be ascribed to interparticle tunneling induced by thermal fluctuations [76].

The effect of annealing on ρ (T) is shown in Figure 1.17(b), which displays the results of a series of successively higher annealing temperatures at constant time intervals. All three annealed samples show large increases in resistivity at all T, in addition to notably steeper nonmetallic temperature dependence. In general, annealing removes dopant molecules and the fibers become more resistive with higher annealing temperatures. This effect is more pronounced at low T. Note that for HPR93B, $d\rho/dT$ still becomes more negative with decreasing T at our lower limit of 1.3 K, unlike the nondivergent behavior of the neat fibers. These results suggest that removing dopant molecules leads to localization of charge carriers within the ropes.

A similar study of fibers spun from PVA–water–surfactant [29] focused on the effect of post-extrusion stretching while the green "gel" fiber still contained ~50% PVA. The effect on ρ (300 K) is shown in Figure 1.18 [25]. An initial decrease by a factor ~4 up to 21% stretch is followed by saturation beyond ~35% stretch. Surprisingly, the x-ray-derived-distribution width is narrowing continuously over the whole range of stretching (up to 80%), suggesting that above an intermediate degree of alignment, ρ (297 K) is limited by some other factor which does not improve with further stretching. Note also that for these composite gel fibers, ρ (300 K) is 20 to 30 times larger than for the neat oleum-based fibers. Here, the insulating PVA impedes long-range transport while the presence of trace p-type dopants enhances the neat conductivity of the oleum fibers.

Selected fibers were annealed in H_2 at 1000°C after stretching, to remove the insulating PVA and thereby obtain more highly conducting material. This process leads to a strong reduction in

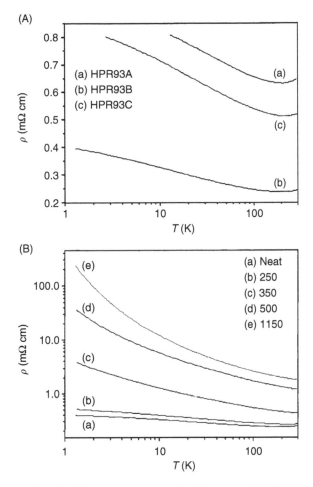

FIGURE 1.17 (a) Four-point resistivity vs. temperature for three neat SWNT fibers extruded from oleum suspension. ρ decreases at all T as alignment improves. Nonmetallic behavior at low T levels off as $T \to 0$ (non-divergent behavior) while metallic behavior is observed above 200 K. (b) Effect of vacuum annealing on $\rho(T)$ for HPR93B; note log–log scale. ρ increases as the bisulfate p-dopants are removed, especially at very low T (factor ~500 at 1.3 K). Metallic behavior is lost, and the slope $d\rho/dT$ continues to increase as $T \to 0$ (divergent behavior). (From Zhou, W. et al., *J. Appl. Phys.* **95**, 649 (2004), with permission.)

ρ (300 K) and a flattening of the $\rho(T)$ curve to a much weaker T dependence than in the green gel fiber. The data for green PVA/SWNT composite fiber and for the PVA-free annealed one are shown in Figure 1.19. Both have been stretch-aligned as indicated in the figure legend; note the log–log scale. At 300 K, ρ for the composite fiber exceeds that of the annealed one by 40 to 50 times. Here, the elimination of PVA removes high-resistance interbundle contacts so the macroscopic carrier transport is more efficient. At low temperature the difference is more dramatic, suggesting that the presence of PVA enhances the effects of carrier localization by disorder. The composite fiber resistance becomes too high to measure at the lowest temperature. Extrapolating to 2 K, we deduce a resistance ratio of at least 11 decades, strongly suggesting totally different conduction mechanisms in the two morphologies. It is obvious from the data that disorder dominates charge transport in both states. Furthermore, the disorder is most likely to encompass all three dimensions since the fibers consist of coupled and intertwined objects. In particular, we expect that the disorder is neither limited to 1D carrier localization on individual tubes or bundles, nor to defect scattering/trapping on single tubes. The huge effect of PVA, which permeates the interbundle volumes in the composite

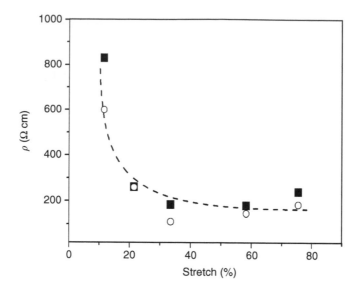

FIGURE 1.18 Four-probe electrical resistivity at room temperature vs. stretch ratio for PVA composite nanotube fibers. The experiments have been conducted on two series of samples (black squares, open circles). Accurate cross-sectional areas were obtained from SEM image analysis. The resistivity decreases with stretch ratio by a factor between three and four for the two sets of data. The absolute value is intermediate between all-SWNT fibers and polymer composites. The dashed line is a guide to the eye. (From Badaire, S. et al., *J. Appl. Phys.* **96**, 7509 (2004), with permission.)

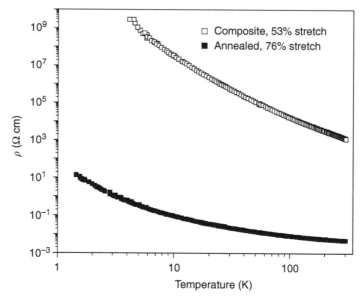

FIGURE 1.19 Resistivity vs. temperature for composite (53% stretch) and annealed (76% stretch) nanotube fibers (note log–log scale). Note that $\rho(T)$ for the composite fiber is more strongly divergent as $T \to 0$ as compared to the annealed one. (From Badaire, S. et al., *J. Appl. Phys.* **96**, 7509 (2004), with permission.)

fibers, cannot possibly be 1D. It is interesting to note that for both fiber series, one ends up with about the same ρ (300 K) after annealing, whereas this value is approached from opposite directions in the two different processes.

The qualitatively different $\rho(T)$ behaviors in the two sets of fibers can be attributed to an MI transition as a function of doping level [77]. On the metallic side for heavily doped tubes, ρ tends to a finite value as T approaches zero, and the effective band gap, $-d\ln\rho/dT$, vanishes as T approaches zero; examples of this behavior are shown in Figure 1.20. On the insulating side for PVA-rich gel fibers or annealed oleum fibers, ρ diverges as T approaches zero, and the exponential T dependence can be ascribed to strong localization and either 3D Mott or Coulomb gap variable-range hopping [77]. The MI transition is revealed by systematic measurements of resistivity and transverse magnetoresistance (MR) in the ranges 1.9 to 300 K and 0 to 9 T, as a function of p-type redox doping. The observed changes in transport properties are explained by the effect of doping on semiconducting SWNTs and tube–tube coupling.

How does the presence of endohedral C_{60} affect the electronic properties of peapods? Despite predictions based on work function differences between peas and pod, there is no evidence for charge transfer redox doping by the endofullerenes. Figure 1.21 shows $\rho\,(T)$ data for a peapod sample and an unfilled control [70,71]. Near 300 K the resistivities are the same within experimental error, and the peapod temperature dependence is steeper suggesting stronger disorder effects compared with the empty tubes.

More fundamental than electron transport properties is the spectrum of allowed electron energies or density of states $N(E)$. This can be measured on individual tubes using scanning tunneling spectroscopy, which yields much important information. For bulk materials, one measures the related quantities such as electron energy loss function, using electron energy loss spectroscopy (EELS), absorption spectroscopy using thin films, or reflection spectroscopy if flat and reasonably smooth surfaces are available. We close this section by giving a few examples.

Figure 1.22 shows the loss function of a 100-nm-thick film of ~60% SWNT deposited on a TEM grid [78]. Both energy and momentum of the transmitted electrons can be determined, so the observed transitions can be separated into dispersing and nondispersing, which correspond to "collective" (plasmon) and "localized" interband processes, respectively. The inset shows the loss function over a wide energy range, in which the plasmons representing collective excitations of the π and $\pi + \sigma$ electrons can be clearly seen at 5.2 and 21.5 eV, respectively, in agreement with the theory. Features in the loss function at 0.85, 1.45, 2.0, and 2.55 eV are independent of q and are therefore assigned to interband transitions. Their nature is revealed by a Kramers–Kronig transform

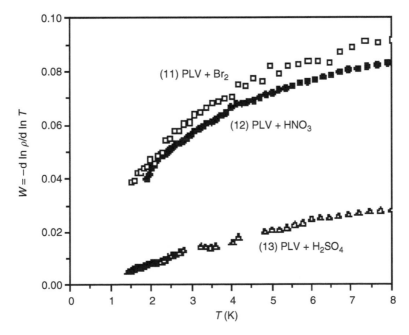

FIGURE 1.20 Reduced activation energy W vs. T for the most conductive p-doped SWNT samples. Extrapolating from the lowest data point 1.3 K to $T = 0$ shows that W, and any possible energy gap, vanishes, signaling a true metallic state. (From Vavro, J. et al., *Phys. Rev. B* **71**, 155410 (2005), with permission.)

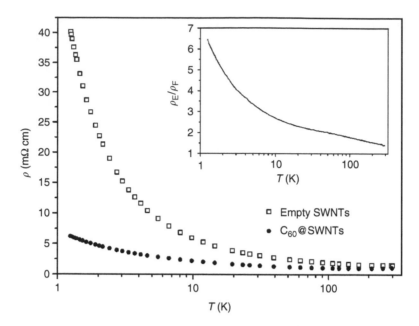

FIGURE 1.21 Four-point resistivity vs. T for C_{60}@SWNT (filled circles) and empty nanotubes (open squares). The ratio empty/filled is shown in the inset. At ambient temperature the resistivities are indistinguishable. (From Vavro, J. et al., *Appl. Phys. Lett.* **80**, 1450 (2002), with permission.)

to obtain real and imaginary parts of the dielectric function, ε_2 being directly proportional to the optical absorption coefficient. The results are shown in Figure 1.23, which reveals three interband transitions at 0.65, 1.2, and 1.8 eV. These are the energy separations of the 1D van Hove singularities, slightly broadened by diameter dispersivity in bulk samples. We now understand the sequence of transitions as E_{11}^S, E_{22}^S, and E_{11}^M where n is the band index and S and M denote semiconducting and metallic tubes, respectively.

EELS also gives important information about the effects of redox doping on electronic properties [79]. Figure 1.24 shows how the loss function evolves with potassium concentration. The upshift in the π plasmon near 6.5 eV is due to the gradual addition of an extra electron per n carbons to the previously empty π^* conduction band. The extra delocalized electron is charge-compensated by K° oxidizing to K^+ during intercalation. Figure 1.24(b) shows this relation explicitly, with K/C determined from core-level spectroscopy. The disappearance of fine structure below 2 eV signals the upshift in E_F, which quenches the van Hove interband transitions since the final states in the neutral material (Figure 1.23) are now occupied by doping.

Optical absorption and reflectance are complementary to EELS. The energy resolution is higher, polarization and selection rule information is accessible, and the spectra extend to lower energies which would otherwise be obscured in EELS by the tail of the ~100 kV incident electron beam. The lower energy cutoff permits accurate measurements of the so-called Drude plasmon, a collective excitation of the delocalized electrons, and how it evolves with doping.

Figure 1.25 shows the reflectance of an unoriented undoped film [80]. The solid curve is a model fit, including a free carrier Drude term and several interband Lorentz oscillators. These results confirm that undoped material consists of a mixture of conducting and semiconducting nanotubes, typical for bulk samples. The interband transition energies are consistent with electronic structure calculations for 1.4-nm-diameter tubes, and with EELS described above. The theory predicts that because these transitions involve 1D van Hove singularities, they should be completely polarized along the nanotube axis, i.e., they should vanish when the incident and transmitted beams are cross polarized. This was proved in a clever experiment on magnetically aligned SWNT suspensions with

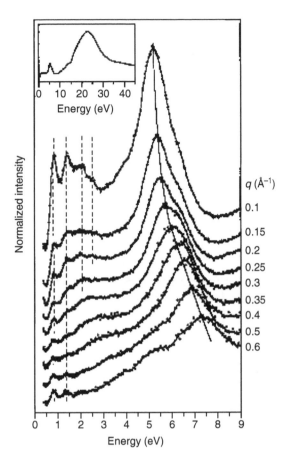

FIGURE 1.22 The loss function of purified SWNT from EELS in transmission for the different q values is shown. The contributions from the elastic peak have been subtracted. The inset contains the loss function over an extended energy range, showing the π plasmon and the $\pi + \sigma$ plasmon at around 5 and 22 eV, respectively. (From Pichler, T. et al., *Phys. Rev. Lett.* **80**, 4729 (1998), with permission.)

the alignment locked in place by gelation [81]. Polarized Raman scattering was used to quantify the degree of alignment. Combined with absorbance data from the partly aligned sample, the intrinsic polarized absorbance is recovered. The fitted Drude plasma frequency ω_p is 0.29 eV, characteristic of a small concentration of conduction electrons or holes.

Chemical doping is revealed in absorption and reflectance spectroscopy by several mechanisms. The higher concentration of free carriers shifts the Drude edge to higher energy. The Drude scattering time may be affected by doping-induced disorder or dimensional crossover. Low-energy interband transitions may be quenched by the E_F shift through valence or conduction band singularities. Examples of these phenomena are evident in the reflectance spectra shown in Figure 1.26 for potassium-doped purified SWNT buckypaper [82]. These experiments were performed in sealed quartz tubes to avoid air exposure; consequently, the results differ significantly from those quoted in an earlier publication in which such precautions were not taken [83]. Data recorded in the range 0.07 to 4 eV show clearly that with increasing K concentration, the 1D van Hove transitions disappear and the Drude edge shifts progressively to higher energies into the visible spectrum [84]. Colors from purple to golden-brown, reminiscent of alkali graphite compounds, are observed as the K concentration increases to saturation. The Drude edge blue-shifts strongly, signaling a large increase in conduction electron concentration after doping. Now the fitted value of σ_p is 1.85 eV. In the free-electron model for metals, the DC conductivity is proportional to ω_p^2. This implies a ~40×

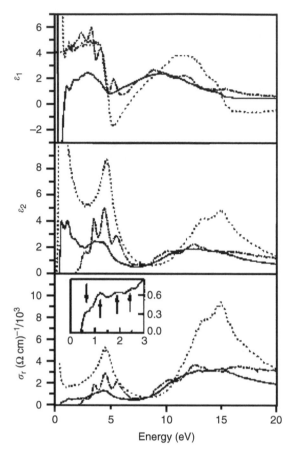

FIGURE 1.23 Real and imaginary parts of the dielectric function (upper panels) and the real part of the optical conductivity for SWNT (solid curves), C_{60} (dot–dash curves) and graphite (dotted curves). (From Pichler, T. et al., *Phys. Rev. Lett.* **80**, 4729 (1998), with permission.)

enhancement in σ by saturation doping, in excellent agreement with direct measurements. Reference [83] reports a much smaller enhancement, underscoring the importance of avoiding air exposure. Similar experiments with electron acceptor doping show a maximum upshift to only 1.2 eV. The weaker effect on ω_p with acceptors is again familiar from graphite intercalation compounds, in which the fraction of a free hole per intercalated molecule is considerably less than the fraction of a free electron per alkali ion.

A key parameter in characterizing doped bulk SWNT samples is the position of the Fermi energy, above or below the neutrality point for donors and acceptors, respectively. This is straightforward in electrochemical experiments, in which temperatures are limited to near ambient and the sample is not optically accessible at all wavelengths. An alternative is provided by the phonon drag, or electron–phonon scattering contribution to the thermopower. The basic idea is quite simple [85], and is shown schematically in Figure 1.27(a). The Fermi surface of a 1D metal consists of two points separated in k by a wave vector of amplitude $|Q|$. Momentum-conserving (inelastic) electron–phonon scattering thus requires participation of acoustic phonons with wave vector Q, which are absorbed or emitted as electrons scatter from the left- to right-moving branch of $E(k)$. Thus, the temperature dependence will reflect the temperature dependence of the heat capacity, and phonon drag is quenched below a characteristic temperature T_o given approximately by $0.2\pi\hbar\omega_Q/k_B$ (cf. Figure 1.27(a)). $|Q|$ is set by the chemical potential μ, which in turn is controlled by the doping level; h and k_B are the Planck and Boltzmann constants, respectively. Operationally (Figure 1.27(b)), one records thermopower vs. temperature, locates T_o from the peak in the derivative, and backs out E_F from the

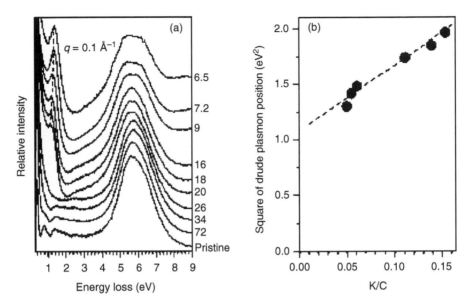

FIGURE 1.24 (a) The loss function of pristine and potassium-intercalated SWNT; the potassium concentration increases from bottom to top. The dashed line indicates the shift of the position of the charge carrier plasmons with increasing concentration. (b) The quantitative relation between the square of the energy position of the charge carrier plasmon and the relative potassium concentration. (From Liu, X. et al., *Phys. Rev. B* **67**, 125403 (2003), with permission.)

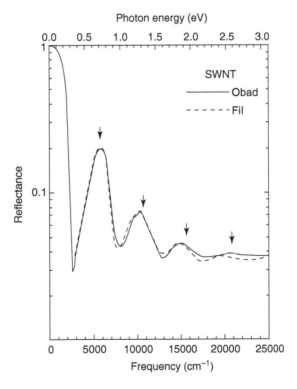

FIGURE 1.25 Reflectance of an unaligned SWNT mat measured from 25 to 25,000 cm^{-1} (lower scale), corresponding to the energy interval 0.003–3 eV (upper scale). *R* approaches 1 at *E* = 0, characteristic of a metal. Four interband transitions involving 1D van Hove singularities are identified by the downward arrows. (From Hwang, J. et al., *Phys. Rev. B* **62**, R13310 (2000), with permission.)

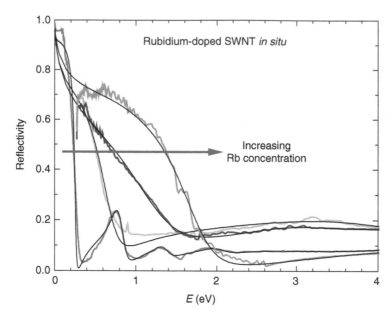

FIGURE 1.26 Reflectance vs. photon energy for four SWNT buckypaper samples with increasing carrier concentration from alkali doping. Leftmost curve is for undoped SWNT. Doping causes a strong blueshift of the free carrier Drude edge, and quenches the interband transitions as E_F increases through the first two-conduction band van Hove singularities. (From Zhou, W. et al., *Phys. Rev. B* **71**, 205423 (2005), with permission.)

relevant dispersion relations. In the example shown, we see the sharp contrast between undoped and sulfuric acid-doped data from which a Fermi level depression of 0.5 eV is obtained.

The phonon drag method for locating E_F suffers one major drawback — it relies on a model for the energy band dispersion $E(k)$. The original one-electron tight-binding models have been extremely useful [86], while the importance of higher order effects is becoming evident [87]. For studies of the metallic state of doped nanotube materials, it would be helpful to have a characterization technique which directly measures the free carrier concentration (Hall effect), or at least the density of states at the Fermi energy $N(E_F)$. The Pauli paramagnetism of conduction electrons provides one option. This is difficult to observe directly in magnetization measurements due to strong diamagnetic corrections, which are not well quantified. This limitation is overcome by conduction electron spin resonance (CESR), which in essence is the Zeeman effect of delocalized electrons. Evidence for Pauli spins in raw SWNT soot was reported 10 years ago [88]. This was controversial because (1) it was believed that the random orientation combined with anisotropic metallic properties would broaden the CESR beyond recognition, and (2) assuming stochastic growth such that only approximately one third of the tubes is metallic, and given the theoretical result that $N(E_F)$ per mole is only approximately one fourth that of 3D graphite, these few intrinsic metallic spins might be undetectable. The original interpretation was supported by subsequent measurements of the temperature dependence of integrated CESR linewidth [89]. While the more typical paramagnetic resonance associated with localized spins obeys a Curie law temperature dependence, susceptibility $\chi \sim 1/T$, the nanotube χ was independent of T in the range 70 to 300 K. The same sample exhibited the aforementioned shallow minimum in resistivity vs. T at about 200 K, so the T independence of χ_{Pauli} ruled out an MI transition in the undoped material.

CESR was extended to alkali-doped buckypaper using *in situ* electrochemical doping to follow the resonance vs. K concentration [90]. Figure 1.28(a) shows the evolution of the full profile, which consists of a narrow Pauli contribution, magnified in Figure 1.28(b), superposed on a broad anti-ferromagnetic resonance (AFMR) associated with residual transition metal catalyst. The Pauli

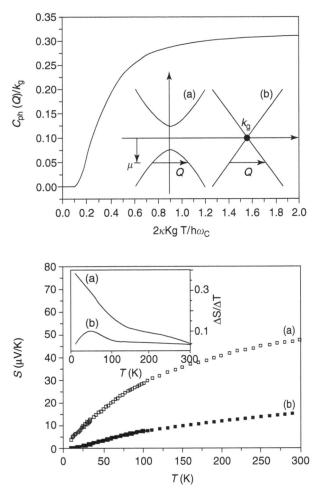

FIGURE 1.27 Phonon drag in SWNT. Top: calculated phonon heat capacity vs. T for a p-type doped nanotubes with chemical potential μ applicable to semiconducting or metallic tubes is shown in the inset. The cutoff in occupied phonon energies leads to a weak threshold in thermopower S. Bottom: $S(T)$ and its derivative for SWNT fibers extruded from sulfuric acid (b) and after annealing to remove the acid and return μ toward zero. The peak in the derivative (b) is a measure of the doping-induced downshift in E_F. (From Vavro, J. et al., *Phys. Rev. Lett.* **90**, 065503 (2003), with permission.)

CESR grows in intensity with increasing K concentration, while the position and width are constant. The AFMR intensity diminishes as the sample becomes more conducting and the microwave skin depth decreases. Figure 1.28(b) shows that χ_{Pauli}, proportional to the double integral of the CESR line, increases monotonically with K/C. Now that an absolute scale is established, the absence of CESR in this undoped material is surprising since there is ample sensitivity.

It also calls into question the origin of the T-independent signal observed previously in unpurified soot [89]. Figure 1.29 shows that χ_{Pauli} is independent of T for two saturation-doped samples, from 30 to 290 K. This proves beyond doubt that alkali-doped SWNTs are true metals. The absolute value 5×10^{-8} emu/g translates to $N(E_F) = 0.015$ states/eV per carbon per spin, approximately five times smaller than a theoretical estimate [91]. The anomalously small value in doped material, and the absence of a true ESR signal in clean undoped material, remain to be resolved.

The independence of the CESR linewidth on temperature is unusual. The inverse of the width is proportional to the spin relaxation rate, which is usually dominated by spin–orbit interaction with

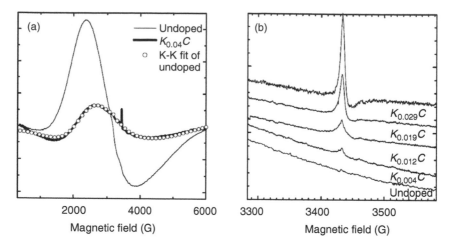

FIGURE 1.28 Conduction electron spin resonance in K-doped SWNT. (a) Ferromagnetic resonance of Ni catalyst particles in undoped and fully doped SWNT (light and heavy curves, respectively). The narrow CESR peak is not seen in the undoped sample, and the Ni resonance changes line shape as the sample becomes more conducting and the skin depth decreases. (b) The CESR line grows continuously with increasing K concentration, while the spin relaxation rate (linewidth) and g factor (position) remain constant. (From Claye, A. S. et al., *Phys. Rev. B* **62**, R4845 (2000), with permission.)

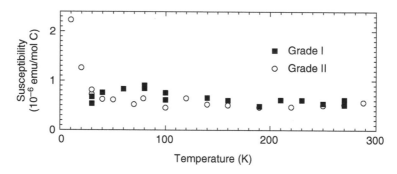

FIGURE 1.29 The Pauli spin susceptibility vs. T for saturation-doped SWNT is independent of T above 30 K, characteristic of a metallic state. (From Claye, A. S. et al., *Phys. Rev. B* **62**, R4845 (2000), with permission.)

the dopants. The only explanation is inhomogeneous doping; small saturation-doped regions are created at the outset, and their number density increases with global K/C without affecting their size. This model is supported by accompanying x-ray diffraction data [90].

1.3.4 MAGNETIC AND SUPERCONDUCTING PROPERTIES

We close with a brief mention of these collective phenomena in bulk nanotube materials. An excellent entrée to the magnetism controversy is the recent paper by Cesvedes et al. and references therein [92]. The authors prove convincingly that there is no bulk magnetism in a clean multiwall nanotube sample. On the other hand, when placed on a flat ferromagnetic substrate, fringing fields can be observed by magnetic force microscopy. Thus, the prospect for "contact-induced magnetism" and the application of carbon nanotubes to nanoscale spintronic devices remain open. Magnetic contrast is observed for carbon nanotubes placed on cobalt or magnetite substrates, but is absent on silicon, copper, or gold. Spin transfer of about 0.1 μB per contact carbon atom is obtained.

The dialog about nanotube superconductivity began with a very simple argument [93]: if one could tune the chemical potential to the peak of a van Hove singularity, either chemically or electrostatically, one should have a high $N(E_F)$ value, a prerequisite for BCS superconductivity. The obvious problem in applying this to bulk material is the diameter dispersivity. Even in single tubes, higher order interactions will broaden the square-root energy divergences. This has not prevented some from conjecturing T_c above 400 K for nanotube superconductors. Super-current flow through an SWNT with low-resistance contacts was reported in 1999 [94]; analogous to contact-induced magnetism, no claims were made for bulk nanotube superconductivity. Evidence for a bulk anisotropic Meissner effect below 20 K in aligned 0.4-nm-diameter SWNT was claimed and justified by the argument that small-diameter tubes will be the stiffest, and thus, the average phonon energy in the BCS equation will be favorable [95]. This dramatic result has not been reproduced by other researchers since it appeared 4 years ago. As with bulk magnetism, it would appear that the jury is still out concerning bulk superconductivity in carbon nanotubes and other nanostructures.

1.4 SUMMARY AND PROSPECTS

It is difficult to counter the argument that carbon is the most amazing element in the Periodic Table. In the space of a mere 30 years, we have experienced the discovery of (or renaissance in) three material families: graphite intercalation compounds (GICs), fullerene solids, and nanotubes. GICs provided a wealth of detailed chemical and physical information, while immediate applications were frustrated by cost, lack of air stability, and most importantly, the lack of new or greatly improved applications and properties. Though the foundations of physics and chemistry remain unshaken, the Li ion battery industry has certainly benefited greatly from this store of knowledge. Fullerene solids fared better, and may yet surprise us in the practical world; while no large volume application has yet emerged, the elucidation of electronic properties in this highly correlated system has made invaluable contributions to fundamental science. Nanotubes may offer the best prospects yet; money is already being made on some small-scale applications. The combination of enhanced properties encourages the drive to multifunctional materials, and also affords an excellent laboratory for studying 1D phenomena. Extensions to other tubular and nanowire-like materials provide ample scope for new discoveries.

ACKNOWLEDGMENTS

The nanotube phase of my research life has been enriched by delightful interactions with numerous students, postdocs, colleagues, and collaborators. This chapter draws heavily from the work of recently awarded theses of Roland Lee, Zdenek Benes, Agnes Claye, Norbert Nemes, Juraj Vavro, and Marc Llaguno, and soon to be awarded theses of Wei Zhou and Csaba Guthy. Funding from the National Science Foundation, the Office of Naval Research, and especially the Department of Energy are gratefully acknowledged. I am grateful as well to Linda Fischer for a most thorough proofreading.

REFERENCES

1. S. Iijima, Helical microtubules of graphitic carbon, *Nature* **354**, 56–58 (1991).
2. R. C. Haddon and S.-Y. Chow, Hybridization as a metric for the reaction coordinate of the chemical reaction: concert in chemical reactions, *Pure Appl. Chem.* **71**, 289–294 (1999).
3. F. Du and K. I. Winey, this volume.
4. N. Wang, G. D. Li and Z. K. Tang, Mono-sized and single-walled 4 Å carbon nanotubes, *Chem. Phys. Lett.* **339**, 47 (2001).
5. G. Eres, A. A. Puretzky, D. B. Geohegan and H. Cui, In situ control of the catalyst efficiency in chemical vapor deposition of vertically aligned carbon nanotubes on predeposited metal catalyst films, *Appl. Phys. Lett.* **84**, 1759–1761 (2004).
6. R. Tenne, this volume.

7. E. G. Rakov, this volume.
8. T. Yoshitake, Y. Shimakawa, S. Kuroshima, H. Kimura, T. Ichihashi, Y. Kubo, D. Kasuya, K. Takahashi, F. Kokai, M. Yudasaka and S. Iijima, Preparation of fine platinum catalyst supported on single-wall carbon nanohorns for fuel cell application, *Physica B* **323**, 124–126 (2002).
9. M. Freitag, A. T. Johnson, S. V. Kalinin and D. A. Bonnell, Role of single defects in electronic transport through carbon nanotube field-effect transistors, *Phys. Rev. Lett.* **89**, 216801 (2002).
10. J. E. Fischer, P. A. Heiney, A. R. McGhie, W. J. Romanow, A. M. Denenstein, J. P. McCauley Jr. and A. B. Smith III, Compressibility of solid C_{60}, *Science* **252**, 1288 (1991).
11. J. Tang, L.-C. Qin, T. Sasaki, M. Yudasaka, A. Matsushita and S. Iijima, Compressibility and polygonization of single-walled carbon nanotubes under hydrostatic pressure, *Phys. Rev. Lett.* **85**, 1887 (2000).
12. Y. Maniwa, R. Fujiwara, H. Kira, H. Tou, H. Kataura, S. Suzuki, Y. Achiba, E. Nishibori, M. Takata, M. Sakata, A. Fujiwara and H. Suematsu, Thermal expansion of single-walled carbon nanotube (SWNT) bundles: X-ray diffraction studies, *Phys. Rev. B* **64**, 241402 (2001).
13. S. Bandow, G. Chen, G. U. Sumanasekera, R. Gupta, M. Yudasaka, S. Iijima and P. C. Eklund, Diameter-selective resonant Raman scattering in double-wall carbon nanotubes, *Phys. Rev. B* **66**, 075416 (2002).
14. J. G. Lavin, S. Subramoney, R. S. Ruoff, S. Berber and D. Tomanek, Scrolls and nested tubes in multiwall carbon nanotubes, *Carbon* **40**, 1123–1130 (2002).
15. L. Jin, C. Bower and O. Zhou, Alignment of carbon nanotubes in a polymer matrix by mechanical stretching, *Appl. Phys. Lett.* **73**, 1197 (1998).
16. Y. Gogotsi, J. A. Libera, A. Güvenç-Yazicioglu, and C. M. Megaridis, *In situ* multiphase fluid experiments in hydrothermal carbon nanotubes, *Appl. Phys. Lett.* **79**, 1021–1023 (2001).
17. H. W. Zhu, C. L. Xu, D. H. Wu, B. Q. Wei, R. Vajtal and P. M. Ajayan, Direct synthesis of long single-walled carbon nanotube strands, *Science* **296**, 884 (2002).
18. Y.-L. Li, I. A. Kinloch and A. H. Windle, Direct spinning of carbon nanotube fibers from chemical vapor deposition synthesis, *Science* **304**, 276 (2004).
19. M. Zhang, K. R. Atkinson and R. H. Baughman, Multifunctional carbon nanotube yarns by downsizing an ancient technology, *Science* **306**, 1358 (2004).
20. K. Hata, D. N. Futaba, K. Mizuno, T. Namai, M. Yumura and S. Iijima, Water-assisted highly efficient synthesis of impurity-free single-walled carbon nanotubes, *Science* **306**, 1362 (2004).
21. P. G. Collins, M. S. Arnold and P. Avouris, Engineering carbon nanotubes and nanotube circuits using electrical breakdown, *Science* **292**, 706–709 (2001).
22. R. H. Baughman, A. A. Zakhidov and W. A. De Heer, Carbon nanotubes — the route toward applications, *Science* **297**, 787–792 (2002).
23. A. G. Rinzler, J. Liu, P. Nikolaev, C. B. Huffman, F. J. Rodriguez-Macias, P. J. Boul, A. H. Lu, D. Heymann, D. T. Colbert, R. S. Lee, J. E. Fischer, A. M. Rao, P. C. Eklund and R. E. Smalley, Large-scale purification of single-wall carbon nanotubes: process, product, and characterization, *Appl. Phys. A* **67**, 29 (1998).
24. T. V. Sreekumar, T. Liu, S. Kumar, L. M. Ericson, R. H. Hauge and R. E. Smalley, Single-wall carbon nanotube films, *Chem. Mater.* **15**, 175 (2003).
25. S. Badaire, C. Zakri, P. Poulin, V. Pichot, P. Launois, J. Vavro, M. Chen and J. E. Fischer, Correlation of properties with preferred orientation in extruded and stretch-aligned single-wall carbon nanotubes, *J. Appl. Phys.* **96**, 7509 (2004).
26. L. M. Ericson, H. Fan, H. Peng, V. A. Davis, J. Sulpizio, Y. Wang, R. Booker, W. Zhou, J. Vavro, C. Guthy, S. Ramesh, C. Kittrell, G. Lavin, H. Schmidt, W. W. Adams, M. Pasquali, W.-F. Hwang, R. H. Hauge, J. E. Fischer and R. E. Smalley, Macroscopic, neat, single-walled carbon nanotube fibers, *Science* **305**, 1447 (2004).
27. R. Haggenmueller, W. Zhou, J. E. Fischer and K. I. Winey, Production and characterization of polymer nanocomposites with highly aligned single-walled carbon nanotubes, *J. Nanosci. Nanotech.* **3**, 105 (2003).
28. A. S. Claye, J. E. Fischer, C. B. Huffman, A. G. Rinzler and R. E. Smalley, Solid-state electrochemistry of the Li single-wall carbon nanotube system, *J. Electrochem. Soc.* **147**, 2845 (2000).
29. B. Vigolo, A. Penicaud, C. Coulon, C. Sauder, R. Pallier, C. Journet, P. Bernier and P. Poulin, Macroscopic fibers and ribbons of oriented carbon nanotubes, *Science* **290**, 1331 (2000).

30. J. E. Fischer, W. Zhou, J. Vavro, M. C. Llaguno, C. Guthy, R. Haggenmueller, M. J. Casavant, D. E. Walters and R. E. Smalley, Magnetically aligned single-wall carbon nanotube films: preferred orientation and anisotropic transport properties, *J. Appl. Phys.* **93**, 2157 (2003).

31. S. Huang, X. Cai and J. Liu, Growth of millimeter-long and horizontally aligned single-walled carbon nanotubes on flat substrates, *J. Am. Chem. Soc.* **125**, 5636–5637 (2003).

32. W. Zhou, K. I. Winey, J. E. Fischer, S. Kumar and H. Kataura, Out-of-plane mosaic of single-wall carbon nanotube films, *Appl. Phys. Lett.* **84**, 2172 (2004).

33. H. H. Gommans, J. W. Alldredge, H. Tashiro, J. Park, J. Magnuson and A. G. Rinzler, Fibers of aligned single-walled carbon nanotubes: polarized Raman spectroscopy, *J. Appl. Phys.* **88**, 2509 (2000).

34. W. Zhou, J. Vavro, C. Guthy, K. I. Winey, J. E. Fischer, L. M. Ericson, S. Ramesh, R. Saini, V. A. Davis, C. Kittrell, M. Pasquali, R. H. Hauge and R. E. Smalley, Single-wall carbon nanotube fibers extruded from super-acid suspensions: preferred orientation, electrical, and thermal transport, *J. Appl. Phys.* **95**, 649 (2004).

35. J. E. Fischer, Q. Zhu, X. Tang, E. M. Scherr, A. G. MacDiarmid and V. B. Cajipe, Polyaniline fibers, films, and powders — X-ray studies of crystallinity and stress-induced preferred orientation, *Macromolecules* **27**, 5094 (1994).

36. M. A. Pimenta, A. Marucci, A. Empedocles, M. G. Bawendi, E. B. Hanlon, A. M. Rao, P. C. Eklund, R. E. Smalley, G. Dresselhaus and M. S. Dresselhaus, Raman modes of metallic carbon nanotubes, *Phys. Rev. B* **58**, R16 016 (1998).

37. R. Haggenmueller, H. H. Gommans, A. G. Rinzler, J. E. Fischer and K. I. Winey, Aligned single-wall carbon nanotubes in composites by melt processing methods, *Chem. Phys. Lett.* **330**, 219 (2000).

38. P. Launois, A. Marucci, B. Vigolo, A. Derre and P. Poulin, Structural characterization of nanotube fibers by x-ray scattering, *J. Nanosci. Nanotechnol.* **1**, 125–128 (2001).

39. B. W. Smith, M. Monthioux and D. E. Luzzi, Encapsulated C-60 in carbon nanotubes, *Nature* **396**, 323 (1998).

40. B. Burteaux, A. Claye, B. W. Smith, M. Monthioux, D. E. Luzzi and J. E. Fischer, Abundance of encapsulated C-60 in single-wall carbon nanotubes, *Chem. Phys. Lett.* **310**, 21 (1999).

41. B. W. Smith, R. M. Russo, S. B. Chikkannanavar and D. E. Luzzi, High-yield synthesis and one-dimensional structure of C_{60} encapsulated in single-wall carbon nanotubes, *J. Appl. Phys.* **91**, 9333 (2002).

42. X. Liu, T. Pichler, M. Knupfer, M. S. Golden, J. Fink, H. Kataura, Y. Achiba, K. Hirahara and S. Iijima, Filling factors, structural, and electronic properties of C_{60} molecules in single-wall carbon nanotubes, *Phys. Rev. B* **65**, 045419 (2002).

43. P. M. Rafailov, C. Thomsen and H. Kataura, Resonance and high-pressure Raman studies on carbon peapods, *Phys. Rev. B* **68**, 193411 (2003).

44. T. Okazaki, K. Suenaga, K. Hirahara, S. Bandow, S. Iijima and H. Shinohara, Electronic and geometric structures of metallofullerene peapods, *Physica B: Cond. Matt.* **323**, 97–99 (2002).

45. B. W. Smith and J. E. Fischer, unpublished.

46. M. Hodak and L. A. Girifalco, Ordered phases of fullerene molecules formed inside carbon nanotubes, *Phys. Rev. B* **67**, 075419 (2003).

47. M. R. Stetzer, P. A. Heiney, J. E. Fischer and A. R. McGhie, Thermal stability of solid C60, *Phys. Rev. B* **55**, 127 (1997).

48. E. Kolodny, B. Tsipinyuk and A. Budrevich, The thermal stability and fragmentation of C_{60} molecule up to 2000 K on the milliseconds time scale, *J. Chem. Phys.* **100**, 8542 (1994).

49. S. M. Bachilo, M. S. Strano, C. Kittrell, R. H. Hauge and B. R. Weisman, Structure-assigned optical spectra of single-walled carbon nanotubes, *Science* **298**, 2361–2366 (2002).

50. C. L. Kane and E. J. Mele, Ratio problem in single-carbon nanotube fluorescence spectroscopy, *Phys. Rev. Lett.* **90**, 207401 (2003).

51. V. A. Davis, L. M. Ericson, A. N. G. Parra-Vasquez, H. Fan, Y. Wang, V. Prieto, J. A. Longoria, S. Ramesh, R. K. Saini, C. Kittrell, W. E. Billups, W. W. Adams, R. H. Hauge, R. E. Smalley and M. Pasquali, Phase behavior and rheology of SWNTs in superacids, *Macromolecules* **37**, 154–160 (2004).

52. W. Zhou, M. F. Islam, H. Wang, D. L. Ho, A. G. Yodh, K. I. Winey and J. E. Fischer, Small angle neutron scattering from single-wall carbon nanotube suspensions: evidence for isolated rigid rods and rod networks, *Chem. Phys. Lett.* **384**, 185–189 (2004).

53. B. I. Yakobson, Mechanical relaxation and "intramolecular plasticity" in carbon nanotubes, *Appl. Phys. Lett.* **72**, 918–920 (1998).

54. M. R. Falvo, G. J. Clary, R. M. Taylor II, V. Chi, F. P. Brooks, J. R. Washburn and R. Superfine, Bending and buckling of carbon nanotubes under large strain, *Nature* **389**, 582–584 (1997).

55. M. Treacy, T. W. Ebbesen and J. M. Gibson, Exceptionally high Young's modulus observed for individual carbon nanotubes, *Nature* **381**, 678–680 (1996).

56. Y. Q. Zhu, T. Sekine, T. Kobayashi, E. Takazawa, M. Terrones and H. Terrones, Collapsing carbon nanotubes and diamond formation under shock waves, *Chem. Phys. Lett.* **287**, 689–693 (1998).

57. F. Li, H. M. Cheng, S. Bai, G. Su and M. S. Dresselhaus, Tensile strength of single-walled carbon nanotubes directly measured from their macroscopic ropes, *Appl. Phys. Lett.* **20**, 3161–3163 (2000).

58. Z. L. Wang, R. P. Gao, P. Poncharal, W. A. De Heer, Z. R. Dai and Z. W. Pan, Mechanical and electrostatic properties of carbon nanotubes and nanowires, *Mater. Sci. Eng. C* **16**, 3–10 (2001).

59. M.-F. Yu, B. S. Files, S. Arepalli and R. S. Ruoff, Tensile loading of ropes of single-wall carbon nanotubes and their mechanical properties, *Phys. Rev. Lett.* **84**, 5552 (2000).

60. O. Inganäs and I. Lundström, Carbon nanotube muscles, *Science* **284**, 1281–1282 (1999).

61. T. V. Sreekumar, Tao Liu, S. Kumar, L. M. Ericson, R. H. Hauge and R. E. Smalley, Single-wall carbon nanotube films, *Chem. Mater.* **15**, 175–178 (2003).

62. J. Hone, M. Whitney, C. Piskoti, and A. Zettl, Thermal conductivity of single-walled carbon nanotubes, *Phys. Rev. B* **59**, R2514 (1999).

63. S. Berber, Y. K. Kwon and D. Tomanek, Unusually high thermal conductivity of carbon nanotubes, *Phys. Rev. Lett.* **84**, 4613 (2000).

64. P. Kim, L. Shi, A. Majumdar and P. L. McEuen, Thermal transport measurements of individual multi-walled nanotubes, *Phys. Rev. Lett.* **87**, 215502 (2001).

65. W. Yi, L. Lu, D. L. Zhang, Z. W. Pan and S. S. Xie, Linear specific heat of carbon nanotubes, *Phys. Rev. B* **59**, R9015 (1999).

66. J. Hone, B. Batlogg, Z. Benes, A. T. Johnson and J. E. Fischer, Quantized phonon spectrum of single-wall carbon nanotubes, *Science* **289**, 1730 (2000).

67. J. E. Fischer, Chemical doping of single-wall carbon nanotubes, *Acc. Chem. Res.* **35**, 1079 (2002).

68. J. C. Lasjaunias, K. Biljacović, Z. Benes, J. E. Fischer and P. Monceau, Low-temperature specific heat of single-wall carbon nanotubes, *Phys. Rev. B* **65**, 113409 (2002).

69. J. Hone, M. C. Llaguno, N. M. Nemes, J. E. Fischer, D. E. Walters, M. J. Casavant, J. Schmidt and R. E. Smalley, Electrical and thermal transport properties of magnetically aligned single-wall carbon nanotube films, *Appl. Phys. Lett.* **77**, 666 (2000).

70. J. Vavro, M. C. Llaguno, B. C. Satishkumar, D. E. Luzzi and J. E. Fischer, Electrical and thermal properties of C-60-filled single-wall carbon nanotubes, *Appl. Phys. Lett.* **80**, 1450 (2002).

71. J. Vavro, M. C. Llaguno, B. C. Satishkumar, R. Haggenmueller, K. I. Winey, D. E. Luzzi, J. E. Fischer, G. U. Sumanasekera and P. C. Eklund, Electrical and thermal properties of C-60-filled single-wall carbon nanotubes, in *Molecular Nanostructures*, H. Kuzmany, J. Fink, M. Mehring and S. Roth, Eds., *AIP Conf. Proc.* **633**, 127 (2002).

72. J. E. Fischer, H. Dai, A. Thess, R. Lee, N. M. Hanjani, D. DeHaas and R. E. Smalley, Metallic resistivity in crystalline ropes of single-wall carbon nanotubes, *Phys. Rev. B* **55**, R4921 (1997).

73. M. S. Fuhrer, M. L. Cohen, A. Zettl and V. Crespi, Localization in single-walled carbon nanotubes, *Solid State Commun.* **109**, 105 (1998).

74. R. S. Lee, H. J. Kim, J. E. Fischer, A. Thess and R. E. Smalley, Conductivity enhancement in single-walled carbon nanotube bundles doped with K and Br, *Nature* **388**, 255 (1997).

75. M Radosavljevic, Improving carbon nanotube nanodevices: ambipolar field effect transistors and high current interconnects, PhD thesis, University of Pennsylvania (2001).

76. P. Sheng, Fluctuation-induced tunneling conduction in disordered materials, *Phys. Rev. B* **21**, 2180 (1980).

77. J. Vavro, J. M. Kikkawa and J. E. Fischer, Metal–insulator transition in doped single-wall carbon nanotubes, *Phys. Rev. B* **71**, 155410 (2005).

78. T. Pichler, M. Knupfer, M. S. Golden, J. Fink, A. Rinzler and R. E. Smalley, Localized and delocalized electronic states in single-wall carbon nanotubes, *Phys. Rev. Lett.* **80**, 4729 (1998).

79. X. Liu, T. Pichler, M. Knupfer and J. Fink, Electronic and optical properties of alkali-metal-intercalated single-wall carbon nanotubes, *Phys. Rev. B* **67**, 125403 (2003).

80. J. Hwang, H. H. Gommans, A. Ugawa, H. Tashiro, R. Haggenmueller, K. I. Winey, J. E. Fischer, D. B. Tanner and A. G. Rinzler, Polarized spectroscopy of aligned single-wall carbon nanotubes, *Phys. Rev. B* **62**, R13310 (2000).

81. M. F. Islam, D. E. Milkie, C. L. Kane, A. G. Yodh, and J. M. Kikkawa, Direct measurement of the polarized optical absorption cross section of single-wall carbon nanotubes, *Phys. Rev. Lett.* **93**, 037404 (2004).

82. N. M. Nemes, J. E. Fischer, K. Kamarás, D. B. Tanner and A. G. Rinzler, Synthesis, isolation and characterisation of new alkaline earth endohedral fullerenes, in *Molecular nanostructures*, H. Kuzmany, J. Fink, M. Mehring and S. Roth, Eds., *AIP Conf. Proc.* **633**, 259 (2002).

83. B. Ruzicka, L. Degiorgi, R. Gaal, L. Thien-Nga, R. Bacsa, J.-P. Salvetat and L. Forró, Optical and dc conductivity study of potassium-doped single-walled carbon nanotube films, *Phys. Rev. B* **61**, R2468 (2000).

84. W. Zhou, J. Vavro, N. M. Nemes, J. E. Fischer, F. Borondics, K. Kamarás and D. B. Tanner, *Phys. Rev. B* **71**, 205423 (2005).

85. J. Vavro, M. C. Llaguno, J. E. Fischer, S. Ramesh, R. K. Saini, L. M. Ericson, V. A. Davis and R. E. Smalley, Thermoelectric power of p-doped single-wall carbon nanotubes and the role of phonon drag, *Phys. Rev. Lett.* **90**, 065503 (2003).

86. C. L. Kane and E. J. Mele, Size, shape, and low-energy electronic structure of carbon nanotubes, *Phys. Rev. Lett.* **78**, 1932 (1997).

87. C. L. Kane and E. J. Mele, Ratio problem in single-carbon nanotube fluorescence spectroscopy, *Phys. Rev. Lett.* **90**, 207401 (2003).

88. A. Thess, R. Lee, P. Nikolaev, H. Dai, P. Petit, J. Robert, C. Xu, H. Lee, S. G. Kim, D. T. Colbert, G. Scuseria, D. Tomanek, J. E. Fischer and R. E. Smalley, Crystalline ropes of metallic carbon nanotubes, *Science* **273**, 483 (1996).

89. P. Petit, E. Jouguelet, J. E. Fischer, A. Thess and R. E. Smalley, Electron spin resonance and microwave resistivity of single-wall carbon nanotubes, *Phys. Rev. B* **56**, 9275 (1997).

90. A. S. Claye, N. M. Nemes, A. Janossy and J. E. Fischer, Structure and electronic properties of potassium-doped single-wall carbon nanotubes, *Phys. Rev. B* **62**, R4845 (2000).

91. A. A. Maarouf, C. L. Kane, and E. J. Mele, Electronic structure of carbon nanotube ropes, *Phys. Rev. B* **61**, 156 (2000).

92. O. Cespedes, M. S. Ferreira, S Sanvito, M. Kociak and J. M. D. Coey, Contact-induced magnetism in carbon nanotubes, *J. Phys.: Condens. Matter* **16**, L155–L161 (2004).

93. R. Saito, G. Dresselhaus and M. S. Dresselhaus, Physical properties of carbon nanotubes, Imperial College Press, London, 1999

94. A. Yu. Kasumov, R. Deblock, M. Kociak, B. Reulet, H. Bouchiat, I. I. Khodos, Yu. B. Gorbatov, V. T. Volkov, C. Journet and M. Burghard, Supercurrents through single-walled carbon nanotubes, *Science* **284**, 1508–1511 (1999).

95. Z. K. Tang, L. Zhang, N. Wang, X. X. Zhang, G. H. Wen, G. D. Li, J. N. Wang, C. T. Chan and P. Sheng, Superconductivity in 4 Å single-walled carbon nanotubes, *Science* **292**, 2462–2465 (2001).

2 Chemistry of Carbon Nanotubes

Eduard G. Rakov
D.I. Mendeleev University of Chemical Technology,
Moscow, Russia

CONTENTS

Abstract 38
2.1 Introduction...38
2.2 Carbon Nanotube Morphology and Structure 39
2.3 Synthesis of Carbon Nanotubes 40
2.4 Opening of Carbon Nanotubes 41
2.5 Functionalization of Carbon Nanotubes..42
    2.5.1 Attachment of Oxidic Groups 43
    2.5.2 Reactions of Carboxylic Groups Attached to Nanotubes....................43
    2.5.3 Fluorination..47
    2.5.4 Amidation 48
    2.5.5 Other Types of Covalent Bonding 50
    2.5.6 Noncovalent Bonding 53
    2.5.7 Dispersions in Oleum 56
    2.5.8 Self-Assembly, Film, and Fiber Formation 56
2.6 Filling the Inner Cavity of Carbon Nanotubes 59
    2.6.1 *In Situ* Filling 60
    2.6.2 Post-Processing Filling 61
        2.6.2.1 Filling from Liquid Media 61
        2.6.2.2 Filling from Gas Phase...63
    2.6.3 Reactions inside Nanotube 64
    2.6.4 The Structure of Crystals inside Nanotubes..65
2.7 Adsorption and Storage of Gases...66
    2.7.1 Hydrogen Problem 67
    2.7.2 Carbon Nanotube Gas Sensors 69
2.8 Attachment of Biomolecules 70
    2.8.1 Biosensors..70
    2.8.2 Others Fields of Application...72
2.9 Nanotubes as Templates...72
    2.9.1 Substitution of the Carbon Atoms of Nanotubes 72
    2.9.2 Decoration of Carbon Nanotubes 73
2.10 Intercalation of "Guest" Moieties..75
2.11 Summary and Conclusions 77
Acknowledgments 77
References 77

ABSTRACT

The main trends and recent achievements in carbon nanotube chemistry are reviewed. Apart from "traditional" subjects such as opening, filling, and decoration of nanotubes, some new subjects have also been discussed. A special emphasis has been placed on the functionalization and solubilization of carbon nanotubes, their self-assembly, film and fiber formation, and sensor and biosensor preparation. Some basics on carbon nanotubes are introduced.

2.1 INTRODUCTION

Owing to their electronic, mechanical, optical, and chemical characteristics, carbon nanotubes (NTs) attract a good deal of attention from physicists, chemists, biologists, and scientists from several other fields.[1–5] Possible applications in the fields of molecular electronics, nanomechanic devices, information display, sensing, energy storage, and composite materials are of interest for industry.[6]

The perspectives of NT application are greatly dependent on NT chemistry. Chemical behavior of NTs is very diverse and the processes of NT synthesis, purification, modification, and solubilization[7,8] all contribute to this diversity.

There are different ways to modify NTs:

- Partial oxidation and decapping (opening) of NTs
- Attachment of functional groups to the open ends of NTs
- Attachment of functional groups to the sidewalls of NTs
- Chemical reactions of functionalities attached to NTs
- Filling of inner cavities of the NTs with different substances (gaseous, liquid, or solid) and carrying out a chemical reaction inside the NTs
- Replacement of the carbon atoms of NTs by atoms of other chemical elements or groups
- Intercalation (insertion) of "guest" atoms or molecules into the intertubular space of single-wall NTs (SWNTs) bundles or in between the walls of multiwall NTs (MWNTs)
- Decoration of outer walls of NTs and using NTs as templates
- Adsorption of gases

NTs having very large molecular weights cannot form true solutions. Therefore, solubilization of NT means the formation of their colloid solutions (dispersions). The solubilization can be subdivided into the formation of aqueous, organic, or polymeric dispersions.

The geometry and size of NTs allows them to take part in self-assembly and aligning processes. The directional deposition on certain surfaces, structuring by Langmuir–Blodgett films and by liquid crystals, and formation of colloidal systems are mentioned here.

In its chemical behavior, NTs partially resemble graphite and fullerenes; however, there are also noticeable differences from both graphite and fullerenes.[8,9] Graphite represents a typical layered polymeric crystal and each fullerene can be considered as a molecule which can form molecular crystals (fullerites). However, a great many NTs cannot be classified as either usual molecules or crystals. SWNTs are similar to polymeric molecules of simple substances and MWNTs resemble structured nanoparticles. The individual NT can be also assumed as 1-D crystal and well-ordered NT bundles as 2-D crystals.

Graphite has planar structure corresponding to sp^2 for σ bonds and p for π bond. Fullerenes and NTs have hybrid bonding orbitals between sp^2 and sp^3. NTs having low percentage of sp^3 bonds are in this regard nearer to graphite. All these bonds in straight NTs are concentrated in half-spherical or conical caps, and in curved NT they are concentrated in the bends. The difference in chemical activity between NT caps and sidewalls as well as between straight and curved NTs is determined by the $sp^2:sp^3$ bonds ratio.

Specific chemical properties of graphite, fullerenes, and NTs as well as of fullerenes or NTs of different diameters are also different owing to dissimilar surface curvatures.[10]

The chemistry of new carbon allotropes has been described with a pyramidalization angle θ_p formalism.[11] For graphite, $\theta_p = 0°$. All carbon atoms in C_{60} have $\theta_p = 11.6°$. Pyramidalization angle of SWNTs (n, n) of various n ($n = 2$ to 10) is calculated to vary between $14–17°$ and $2°$.[12] Pyramidalization changes the hybridization of atomic orbitals at the C atom so that the π orbital contains different portions of s and p orbitals, leading to different chemical reactivity. Because of this hybridization, fullerenes and NTs are known to be more reactive than graphite. The deformation energy of sp^2 bond is inversely proportional to the diameter of NT, and therefore tubes with smaller diameter have greater reactivity. The enthalpy of the reaction decreases with increase in the diameter. The enthalpy of homolytic reactions of H atoms and methyl radicals with atoms of curved carbon plane is linearly dependent on the pyramidalization angle and can differ by 1 eV.[13]

The cups of NT, which are often similar to fullerene molecule halves, contain more reactive C atoms, than the sidewalls. Experimental evidence of the higher chemical reactivity of conformationally strained carbon sites in MWNTs has been reported.[14]

NTs exhibit more differences owing to various structure and morphology (kinked, branched, conical NTs, etc.).[15] The NTs differ from fullerenes by the larger size of internal cavity, and from graphite by the greater share of accessible surface sites.

Fullerites, graphite, and NTs contain van der Waals gaps, which on being filled can form intercalates, or "guest–host" compounds. The dimensionality of such compounds is different: 0-D for fullerites, 2-D for graphite, and 1-D for NTs. NTs can form "guest–host" compounds of different types: for instance, the "guests" can exist in the inner cavity of SWNT or MWNT, in the intertube space of the bundled NTs, or in between the walls of MWNTs having scroll structure.

The development of the chemistry of NTs will determine the fields of their practical application. Primarily, NT chemistry is involved with the rational routes of NT purification and sorting. Chemical modification of NTs can open the way to modify the properties of these materials.

Unlike physical investigation of the solid-state properties of NTs, the study of their chemical reactivity is still in its infancy. At the same time, there have been very impressive achievements in the study of the chemistry of NTs in recent years. There are a large number of original publications and some reviews on the subject.[8,10,16–21] The present review pertains mainly to works in the last five years.

2.2   CARBON NANOTUBE MORPHOLOGY AND STRUCTURE

Dozens of different morphological varieties of filamentous carbon nanoparticles were revealed in a remarkably short period of time after the discovery of "classical" NTs. As was mentioned by Hilding et al.,[10] NTs can differ by "aspect ratio, NT diameters, surface structures, defect densities, and physical entanglements."

The NT of the "classical" type represents a cylindrical particle formed from graphene — a flat carbon net with the atoms located in the corners of joined hexagons. They may be single-walled with a diameter between ~0.3 and ~5.0 nm (typical SWNTs have diameters of 1.0 to 1.4 nm, and length up to 50 to 100 μm) or multi-walled. SWNTs as a rule contain less topological defects and possess better mechanical and electrophysical properties. The specific surface area of SWNTs is independent of their diameter and is equal to 1315 m^2/g (for outer surface).

MWNTs consist of several (from two to tens) coaxial tubes, with outer diameter of ~1.4 to 100 nm. Their specific surface area depends on the number of walls, and to a lesser extent on the diameter of the inner tube. The theoretical surface area of double-wall NTs is between 700 and 800 m^2/g, and of ten-wall NTs is about 200 m^2/g. Nanofibers (NFs) have a diameter from tens to hundreds of nanometers.

Some MWNTs and NFs have internal cross connections formed by curved graphenes. A moderate relative number (density) of cross connections correspond to a "bamboo-like" structure of

MWNT, while a high relative number of connections may be attributed to conical structures such as "herringbone-like" or "cup-stacked" types. The angle of conicity in the conical structures varies from 15 to 85°. Grafene layers in NFs may also be located perpendicular to the fiber axis. Nanofibers having regularly changing diameter ("nanobeads") have been synthesized.[22] And finally, there are amorphous NTs or NFs.

SWNTs, MWNTs, and NFs can spontaneously form "secondary" structures. A classification of carbon nanotubular particles into primary, secondary, and tertiary forms was first introduced by De Jong and Geus.[23] The most popular secondary form of the SWNT is a bundle ("rope"), which consists of tubes in a 2-D triangular lattice, with a lattice constant of 1.7 nm. The bundles usually have greater lengths as compared to individual NTs. For example, 1-m-long bundles of NTs have been produced. They can combine to form a structure of higher order. Tangled SWNTs or MWNTs can form "bucky paper" ("nanopaper" and "nanomats").

Single-walled nanohorns (nanocones) quite often combine to produce flower- or bud-like structures.

"Secondary" structures formed by MWNTs are more diverse. The formation of bundles of MWNT is less typical. More often, kinked (L-shaped) and branched (Y-, T-, H-like, "octopus," tree-like) forms were synthesized. Tangled MWNTs form aggregates up to 3 mm in diameter ("worms" and "boiled spaghetti"). The aggregates along with conical "ceder forest," and "bucky pearl" some others are the members of "tertiary" structure.

A pyrolytic synthesis sometimes yields coiled NTs, helixes, double helixes, helixes inside NTs, and even more complex structures, such as brushes, entangled worm-like features, and foil-like structures at the nanometer scale. Transitional forms, such as nanopeas (fullerenes inside SWNTs), are also among the structures formed.

The chemistry of different structures must be notably diverse. This diversity has not been studied yet, and the overwhelming majority of published results pertain to SWNTs, SWNT bundles, and MWNTs.

2.3 SYNTHESIS OF CARBON NANOTUBES

There are two main methods or groups of processes of NT synthesis: sublimation of graphite with subsequent desublimation, and decomposition of carbon-containing compounds.[5,7,24] The first group of processes is associated with high temperatures (up to 4000°C), which can be obtained in electric arcs (see, for example, Refs. 25–29), by the process of laser ablation,[29–34] by focused solar radiation,[34,35] or by resistive heating of graphite.

The arc process is remarkable for the larger number of versions, among which some versions allow the realization of a semicontinuous process in automated facilities. Arching process in liquid nitrogen,[36,37] in water,[36] and in aqueous solutions[38,39] have also been developed.

Arching processes in gaseous[40] or liquid[41,42] hydrocarbons are classified as combined methods, where the pyrolysis of hydrocarbons takes place along with the sublimation–desublimation of graphite.

The arc-discharge technique is a popular method. The main disadvantages of the method lies in the difficulty of organizing a continuous process, the concurrent formation of amorphous carbon, metal clusters coated with carbon and in some cases fullerenes along with NTs. The total yield of SWNT, as a rule, does not exceed 20 to 40%.

The second group of methods has its own variations: pyrolysis of gases (chemical vapor deposition, CVD process), solids (e.g., pyrolysis of polymers), aqueous solutions (hydrothermal synthesis),[43,44] or organic solutions (supercritical toluene).[45]

The CVD method can produce NTs in large quantities and can be realized at temperatures of 500 to 1300°C. In accordance with the chemical composition of the carbon source, the method can be subdivided on the disproportionation of CO,[29,33,46] the pyrolysis of hydrocarbons (CH_4, C_2H_2, C_6H_6, etc., including polymers),[23,47–49] pyrolysis of CH_xO_y compounds (for example, alcohols[50]), and pyrolysis of heteroatomic $CH_xA_yB_z$ compounds (A, B = N, O, S, Cl, ..., e.g., amines).

It is also possible to synthesize NTs via a template method, applying, for example, porous anodic alumina membrane. CVD using hydrocarbons as NT precursors over patterned catalyst arrays leads to the formation of different complex structures and opens up exciting opportunities in nanotechnology.[48]

The thermal disproportionation of CO is realized in two main variants: HiPco[2,5,46] and CoMoCAT.[5,51] These methods are considered by American researchers as the most promising for the commercial production of SWNTs.

High-value NTs, viz. SWNTs, are produced at temperatures of 900 to 1200°C, mainly using CO or CH_4 as a precursor. The availability of precursors, such as C_2H_2, methanol, and ethanol, for SWNT synthesis is also shown. SWNTs in all instances are produced only by catalytic reactions.

Methods other than the ones mentioned above, for example, thermal decomposition of metal carbides[7,52,53] or carbonitrides,[54] chlorination of carbides,[7] electrolysis of molten salts,[7] interaction of cesium metal with microporous carbon,[7] defoliation of graphite by forming and subsequent transformation of "guest-host" intercalated compound, transformation of fullerenes C_{60} and C_{70},[55] "rapid thermal processing" of amorphous carbon film containing iron,[56] sonochemical production,[57] AC plasma processing,[58] etc., are not widely used.

2.4 OPENING OF CARBON NANOTUBES

Independent of their synthesis method, pristine SWNTs have large aspect ratio (10^3 to 10^4) and closed ends. Open-ended NTs offer unique possibilities as conduits for flow of low-surface-tension fluids through their cylindrical pores. Applications of NTs in molecule separation devices, in biocatalysis, in molecule detection, and as encapsulation media have been proposed. The end cups on NTs must be destroyed to make the inner cavity available for filling. The simplest method to open NTs is their oxidative treatment. The oxidation tends to initiate on the end cups, thus providing a mechanism for opening the NT. The treatment is accompanied by functionalization of NTs with oxygen-containing groups (see below).

Liquid or gaseous chemical agents are used as oxidants. Refluxing, sonication, or microwave (MW) digestion in concentrated acids, such as HNO_3, H_2SO_4 and their mixtures, are the most popular methods of oxidation. Refluxing of MWNTs in concentrated HNO_3 leads not only to oxidative opening, but also in the reduction of the NT length and diameter as well as the breaking of entangled MWNTs.[59] Treatment of SWNTs results in tube cutting and narrowing of diameter distribution.[60,61] The bundles of SWNTs become disordered and partly exfoliated, when they are immersed in 70% HNO_3 for a long period of time.[62] Nitric acid can be intercalated into SWNT bundles, in that way disintegrate the tube walls into graphitic flakes and then reforms them into various multi-shell phases such as MWNTs, cone-shaped phases, and onion-like phases.[63] A prolonged treatment of SWNTs leads to amorphization and complete destruction of the tubes.

Bundles in raw SWNT material are significantly thinner (<10 NTs) than those found in sonicated and HNO_3-oxidized samples (>30 NTs).[64,65] It has been suggested that this thickening is promoted by the H-bonding between –COOH side groups formed during oxidation.

The concentrated H_2SO_4/HNO_3 mixture (3:1) is a better agent for cutting SWNTs[60,66] and for enrichment of large-diameter SWNTs.[67,68]

Oxidation of SWNTs with H_2O_2 is a common tube-opening procedure.[69–71] SWNTs react exothermically with the H_2SO_4/H_2O_2 mixture.[60] Oxidation in solution ($HNO_3/H_2O_2/H_2SO_4$) has been found to be effective for opening SWNTs. Aqueous solutions of OsO_4, $OsO_4/NaIO_3$,[72] RuO_4, $KMnO_4$, $H_2SO_4/KMnO_4$, $H_2SO_4/(NH_4)_2S_2O_8$[73] have also been successively used to open NTs. Superacids such as $HF–BF_3$, have been shown to act as an etchant of NTs at room temperature.

The opening of MWNTs can be realized at room temperature electrochemically in H_2SO_4[74] or in dilute aqueous KCl solutions.[75] Supercritical water acts on MWNT as an opening and thinning agent.[76]

Gaseous thermal oxidation is more effective than acid treatment. Thinner SWNTs burn more quickly than thicker SWNTs during oxidation by oxygen gas.[77,78] Fixed ambient air,[79,80] fixed air activated by microwave irradiation,[81] air flow,[82–84] 5% O_2/Ar mixture,[85,86] reduced O_2 atmosphere,[61] O_2/H_2S mixture,[87,88] O_2, and H_2O plasma[89] used for oxidation. It has been shown that the oxidative stability of SWNT is higher than that of amorphous carbon but lower than the oxidative stability of graphitic carbon.

The pore structure and specific surface area of SWNT aggregates are changed by air oxidation.[82] However, unlike oxidation with H_2SO_4/HNO_3 (3:1) solution, the air oxidation process preferentially oxidizes SWNTs without introducing sidewall defects.[90] The air oxidation rate of SWNTs is clearly correlated to the amount of metal impurity. Ultrafine gold particles catalyze the oxidation.[91]

A mechanism for oxidative etching by O_2 includes adsorption of O_2 molecules on the tube cup or wall, successive transformation of adsorbed molecules, and tube cup being etched away.[92] Defective sites on the ends and the walls of MWNTs facilitate the thermal oxidative destruction of the tubes.[93] The kinetics of oxidation in an air flow has been studied at 400 to 450°C.[94] The apparent activation energy of oxidation has been found to be equal to 150 kJ/mol and corresponds to the data for oxidation of carbon soot.[95] A kinetic model of the process has been proposed.[84]

Ozone at reduced or room temperature[96–101] and CO_2/Ar (2:1) mixture at 600°C[102] are suitable for the oxidation of NTs.

As a rule, the oxidation procedure is used for the purification of crude SWNTs and MWNTs containing amorphous carbon, catalyst, and graphitic nanoparticles. Acid reflux followed by thermal oxidation or reciprocal manner of treatment are common.[5,9,10,79,83,85,86,103–105] Acid treatment of SWNTs in combination with tangential filtration[106] or centrifugation[104] have been tested. Microwave acid digestion allows a reduction in the operational time.[107] In some cases, HCl[71,108] or HF[109,110] is used for the dissolution of metal impurities. Multistep purification procedures including acid treatment[80,111] or air oxidation[112,113] have been developed. Hydrogen peroxide has been shown to be an effective agent in the process of carbon nanostructure purification from amorphous carbon impurities.[114] The methods that are usually used to remove impurities from the as-prepared SWNT material lead to hole-doped purified SWNTs.[115]

2.5 FUNCTIONALIZATION OF CARBON NANOTUBES

Functionalization allows the segregation of entangled or bundled NTs for their subsequent alignment. It is widely used for solubilization of NTs and for purification and classification of NTs in solutions. The surface modification of NTs plays an important role in their use in composites, providing strong fiber–matrix bonding and thus improving the mechanical properties of the material. The integration of NTs into integrated circuits and working devices, such as sensors and actuators requires robust, well-defined connections, for which few processes are better than covalent functionalization.

All the existing methods of chemical derivatization of NTs are divided into two groups, depending on whether attached moieties are introduced onto the NT tips or sidewalls. The use of the latter offers wider opportunities to change the original NT properties, since it allows high coverage with attached groups. The attachment can be realized either by covalent bond formation, or by simple adsorption via noncovalent interactions (hydrophobic, π stacking, etc.).

The covalent bonding can be realized via chemical or electrochemical reactions. The chemical functionalization involves oxidation, fluorination, amidation, and other reactions. Two main paths are usually followed for the functionalization of NTs: attachment of organic moieties either to carboxylic groups that are formed by oxidation of NTs with strong acid, or by direct bonding to the surface double bonds.[116]

Using NTs as either anode or cathode in an electrochemical cell enables oxidation or reduction of small molecules on the surface of the NT, leading to the formation of radical species which can be covalently bonded.

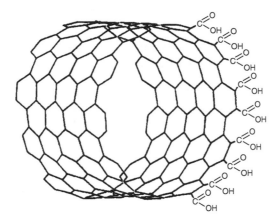

FIGURE 2.1   Structure of (10,10) SWNT–COOH.

2.5.1   ATTACHMENT OF OXIDIC GROUPS

By analogy with other carbonaceous materials, concentrated HNO_3 and mixtures of H_2SO_4 with HNO_3, H_2O_2, or $KMnO_4$ have been widely used for attaching acidic functionalities to NTs. First, acidic groups are attached to the open ends of SWNTs (Figure 2.1).

Refluxing NTs in a H_2SO_4/HNO_3 mixture results in a clear, colorless solution, which on evaporation of the solvent and removal of excess acid, gives a white solid containing functionalized NTs.[72] Neutralization of the acidic solution by alkali results in precipitation of a brown solid containing nanotubes.

The main acidic functionalities comprise –COOH, –C=O, and –OH groups[117] approximately in the proportion of 4:2:1.[118] The concentration of surface acid groups in the NTs treated by different oxidants varies in the range of 2×10^{20} to 10×10^{20} sites/g.[72] On a molar basis, the concentration of acid groups is equal to 5.5 to 6.7%,[119] ~6%,[120] ~5%[121] for shortened SWNTs or ~4% for full-length SWNTs.[122] Simple acid–base titration method shows that three different samples of purified SWNTs had about 1 to 3% of acidic sites and about 1 to 2% of –COOH functionalities.[123] The functional group concentration is time-dependent.

Treatment of SWNTs with concentrated H_2SO_4 containing $(NH_4)_2S_2O_7$ and P_2O_5, followed by treatment with H_2SO_4 and $KMnO_4$, results in the formation of material containing C/O/H in the atomic ratio of 2.7:1.0:1.2.[73]

A "one-pot" oxidative method via ozonolysis of the NT sidewall has been developed.[99] The ozonized NTs can react with several types of reagents, thus providing control over functional groups (Table 2.1).

Along with functionalization with carboxylic, alcoholic, aldehydic, and ketonic groups, acidic treatment leads to sizeable attaching of protons. The MWNTs after acidic purification contain 76.6% CH_x, 13.0% C–O, 4.2% C=O, and 6.2% N–C=O and O–C=O groups[124] –CSO_3H groups are also attached using sulfuric acid.

Acid-functionalized, purified, and shortened SWNTs can be dispersed in water by sonication.[125] No tube precipitation was observed with solutions containing 0.03 to 0.15 g/L after a month. The solubility and stability of the solution are pH-dependent.

2.5.2   REACTIONS OF CARBOXYLIC GROUPS ATTACHED TO NANOTUBES

The carboxylic groups at the SWNT tips can chemically react in organic solutions to form closed rings (Figure 2.2).[126] The average diameter of the rings is 540 nm with a narrow size distribution.

The most important aspect for further covalent or ionic functionalization is the possibility of exploiting carboxylic groups at the tube ends or walls. Amines are among the reagents that have drawn

TABLE 2.1

Relative Amounts of Different Surface Oxygenated Groups (%) on HiPco SWNTs Subjected to Ozonolysis at –78°C in Methanol Followed by Selective Chemical Treatment

Sample	C–OH	C=O	COOH, O–C=O
Ozonated	13.3	50.8	35.9
Treated with H_2O_2	37.0	9.4	53.6
Treated with DMS	28.7	41.1	30.2
Treated with $NaBH_4$	29.1	36.3	34.6

Source: From Banerjee, S. and Wong, S.S., *J. Chem. Phys.*, **B 106**, 12144–12151, 2002. With permission.

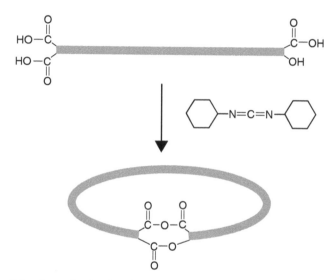

FIGURE 2.2   A possible scheme for the ring-closure reaction with 1,3- dicyclohexylcarbodiimide. (Reprinted with permission from Sano, M. et al., *Science*, **293**, 1299–1301, 2001.)

the greatest attention. There are three types of carboxylic group reactions with amines: (1) amidation, (2) acid–base interaction, and (3) condensation. Besides, amines can be physisorbed on NT walls.

Haddon and coworkers[118,120,127,128] pioneered the approach of functionalizing the carboxylic groups of shortened oxidized SWNTs through amidation with amines bearing long alkyl chains. To modify SWNTs with the amide functionality, the reactions shown in Figure 2.3 were used.

Shortened SWNTs were stirred in $SOCl_2$ containing dimethylformamide (DMF) at 70°C for 24 h, and after centrifugation, decantation, washing, and drying, the residual solid was mixed with octadecylamine (ODA) and heated at 90 to 100°C for 96 h. During this process, the volume of the SWNTs expanded several times.

The second and the third routes to attach amines are the direct reactions of carboxylic groups with amine (see Section 2.5.4).

The concentration of functional groups bound in SWNTs functionalized with $SOCl_2$ seems to be sensitive to gamma irradiation.[129] SWNTs and MWNTs containing acyl chloride groups are solubilized via poly(propionylethylenimine-*co*-ethylenimine) attachment.[130–132] Reaction with polyethyleneimine caused the formation of a product, which is soluble in chloroform.[133] MWNTs functionalized with –COCl groups can covalently attach polythiophene.[134]

The refluxing of functionalized NTs with an excess of $NaBH_4$ in absolute ethanol leads to the reduction of the carboxylic acid groups into hydroxyl groups.

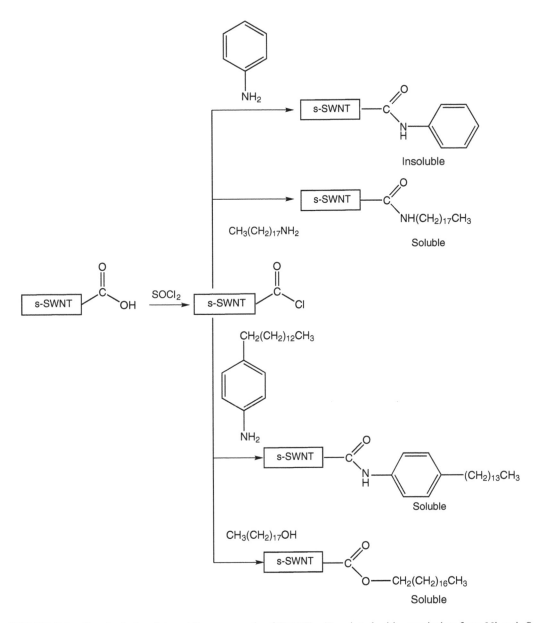

FIGURE 2.3 Covalent chemistry at the open ends of SWNTs. (Reprinted with permission from Niyogi, S. et al., *Acc. Chem. Res.*, **35**, 1105–1113, 2002. Copyright 2002. American Chemical Society.)

Esterification of the NT ends or sidewalls after their carboxylation[130,135–141] differs from many other methods of functionalization in the simplicity of defunctionalization.[136] The attached groups can be easily removed by a hydrolysis reaction, catalyzed by acids or bases. The NTs evolve from the solution after hydrolysis.

Oxidized carbon atoms can act as specific sites for adsorption of metal ions.[142,143] The simplest reaction may be expressed by the equation

$$NT{-}COOH + M^+X^- \rightarrow NT{-}COOM + HX$$

Individual Pb^{2+}, Cu^{2+}, and Cd^{2+} ion-adsorption capacities are equal to 97, 28, and 11 mg/g, respectively.[143] Hg(II) ions form groups of two types: $(-COO)_2Hg$ and $(-O)_2Hg$, in the ratio (%) of

30:70.[142] Ultrasonication of a dispersion of MWNTs in water–isopropanol solution containing RuCl$_3$·3H$_2$O leads to Ru attachment.[144] The surface carboxylic groups are used to attach the relatively bulky metal complexes such as Vaska's complex (*trans*-IrCl(CO)(PPh$_3$)$_2$),[145] Wilkinson's complex (RhCl(PPh$_3$)$_3$),[146] and also TiO$_2$ or CdSe nanoparticles.[147,148] It has been shown that Ir coordinates to the NTs by two distinctive pathways. With raw tubes, the metal attaches as if the tubes were electron-deficient alkenes. With oxidized tubes, oxygen atoms form a hexacoordinate around the Ir atom. The Rh atom similarly coordinates to these NTs through the increased number of oxygenated species. The functionalization reaction, in general, appears to increase significantly oxidized NT solubility in DMF in the case of Vasca's and in dimethyl sulfoxide (DMSO) in the case of Wilkinson's.

Among the range of reagents tested, the most effective for MWNTs silylation were *N*-(*tert*-butyldimethylsilyl)-*N*-methyltrifluoroacetamide and 1-(*tert*-butyldimethylsilyl)imidazole.[149]

The oxidized groups present on SWNTs allow the formation of polymer/NT films by the alternate adsorption of the polyelectrolyte and SWNTs onto substrates.[150] Such groups on MWNT walls can react with 3-mercaptopropyl trimethoxysilane.[151] Alkoxysilane-terminated amide acid oligomers are used to disperse NTs.[152] Alkoxysilane functional ends on the oligomer, once hydrolyzed, react with functionalities on the ends of the purified SWNTs, thus leading to polymer formation.

The formation of NT arrays by self-assembling COOH-terminated NTs onto certain metal oxide substrates (e.g., Ag, Cu, Al) has been demonstrated.[153] In such reactions, the ability of carboxylic groups to deprotonate in contact with metal oxides is utilized.[20] An assembling acid-functionalized SWNTs on patterned gold surface has been developed.[154,155] The reaction mechanism is presumed to include an ester intermediate formation.

The carboxylated tips of NWNT are used to force titrations by atomic force microscope (AFM).[156] The ability of carboxylic groups at the tips of NT to be readily derivatized by a variety of reactions allows the preparation of a wide range of molecular probes for AFM.[157,158]

Air heating of derivatized NTs at controlled temperatures and for controlled periods leads to the decomposition of carboxylic groups and to the formation of hydroxyl groups.[159] The carboxylic groups could be removed by thermal vacuum decarboxylation without damaging the electron system of the NTs, but defects remain on the tube walls.[65] It is generally accepted that carboxylic groups decomposed on heating to CO$_2$ gas and carbonyl groups, desorbed in the form of CO (Figure 2.4).[160,161] The CH$_x$ groups decompos giving CH$_4$ and H$_2$.[161]

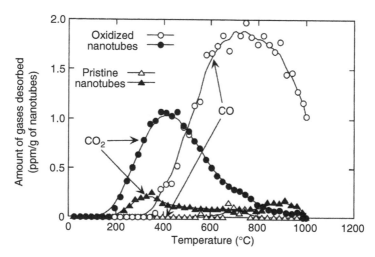

FIGURE 2.4 CO$_2$ and CO temperature programmed desorption patterns of oxidized and pristine NTs. (Reprinted from Kyotani, S. et al., *Carbon*, **39**, 771–785, 2001. With permision from Elsevier.)

2.5.3 FLUORINATION

Fluorination plays an important role in the chemistry of NTs because of the simplicity in acheiving a high degree of functionalization, the very high stability of fluorinated NTs, and the possibility to change the attached fluorine atoms to other functional groups. The fluorination reaction can easily be scaled-up.

The first work devoted to fluorination of fibrous carbon material was published by Nakajima et al.[162] before NTs were discovered. This work showed that the reaction starts at room temperature. The composition of the products prepared at fluorine pressure of 1 atm corresponds to $C_{8-12}F$.

After a decade, MWNTs were fluorinated at different temperatures and the formation of $(CF)_n$ at 500°C was documented.[163] A year later, French specialists, using a mixture of F_2–HF–IF_5 for fluorination of MWNTs, observed a modification of NT structure after reaction at high temperatures, and studied the electrochemical behavior of the fluorinated NT ("fluorotubes") as electrode material in a lithium cell.[164,165] The fluorination of MWNTs by vapor over a solution of BrF_3 in liquid Br_2 at room temperature revealed a decrease of cage nanoparticles in the fluorinated material relative to the pristine sample, which was connected with unrolling the NTs during fluorination.[166] Insufficient purity of the samples used in these works makes the full interpretation of the results difficult.

The amount of doped fluorine increases with increasing doping temperature. Doping at lower temperatures resembles the intercalation of graphite with fluorine and leads to the buckling of the outer MWNT walls.[167]

The fluorination of the internal surfaces of NTs, prepared by a template carbonization technique and which are less crystalline than those synthesized by arc discharge or laser methods, by elemental fluorine at 200°C shows that the resulting compound corresponds to $CF_{1.42}$.[168]

Fluorination of purified SWNTs in the form of "bucky paper," by flow of fluorine gas diluted by helium at a reaction time of 5 h demonstrated that the composition of fluorinated NTs varied from $CF_{0.1}$ at 150°C to $CF_{1.0}$ at 600°C.[169] It appeared that fluorination at 400°C and higher temperatures leads to destruction (e.g., "unzipping") of SWNTs, to the formation of structures resembling MWNTs, and to the evolution of gaseous products such as CF_4, C_2F_4, C_2F_6. Once fluorinated at temperatures up to 325°C, which corresponds to the formation of C_2F, SWNTs were defluorinated with anhydrous hydrazine and were rejuvenated. Partial or complete elimination of fluorine can be done by $LiBH_4$/$LiAlH_4$ treatment.[170]

In their subsequent works, Peng et al.[171] realized the fluorination of purified SWNTs with F_2/HF mixture at 250°C (HF acts as a catalyst).

Heat annealing of fluorinated SWNTs having C/F ratios between 2.0 and 2.3, in a flow of noble gas, indicates that NTs could be recovered at 100°C.[172] Tubes fluorinated at 250°C to a $CF_{0.43}$ stoichiometry lose fluorine on annealing under flowing helium gradually with increasing temperature.[173] Upon heating, the largest fluorine loss occurred between 200 and 300°C.

The fluorination of purified HiPco tubes to a stoichiometry CF_x ($x \leq 0.2$) followed by pyrolysis of partially fluorinated material up to 1000°C was found to have "cut" the NTs.[174] In an argon atmosphere, the fluorine was driven off the NT structure in the form of gaseous CF_4 and COF_2. Short bundles comprising strongly interacting individual NTs were found in the cut nanotube sample.

As a result of band-gap enlargement, the resistivity of fluorinated SWNT mat increases with increasing fluorination temperature, i.e., fluorine content.[175] The electronic properties are also altered by fluorination. As they are fluorinated, NTs reduce their tendency to self-agglomerate.

The most important property of fluorotubes is their ability to form soluble derivatives (Figure 2.5).[176–179]

Sidewall-alkylated NTs are obtained by interaction of fluorinated NTs with alkyl magnesium bromides in a Grignard synthesis or by reaction with alkyllithium precursors. The alkylated NTs are soluble in various organic solvents, including chloroform, methylene chloride, and tetrahydrofuran. For example, the solubility of hexyl-solubilized SWNTs in chloroform is up to ~0.6 g/L,

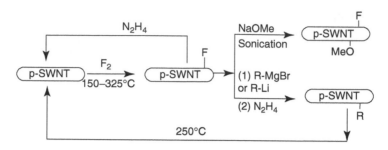

FIGURE 2.5 Sidewall fluorination of SWNTs and fluorine substitution reactions. (Reprinted with permission from Niyogi, S. et al., *Acc. Chem. Res.*, **35**, 1105–1113, 2002. Copyright 2002. American Chemical Society.)

in tetrahydrofuran to ~0.4 g/L, in methylene chloride to ~0.3 g/L, as compared with maximum concentration of 0.1 g/L of pristine NTs in DMF.

Sonication of SWNTs in some solvents for ~5 min also results in the selective solubilization of highly fluorinated (isopropanol) or sparcely fluorinated (DMF) samples. Fluorotubes can be solvated in alcohols yielding metastable solutions. Of the solvent used, 2-propanol and 2-butanol seemed to be the best, reaching SWNT concentration of 1 g/L. A probable mechanism of such solvation would be hydrogen bonding between the hydroxyl hydrogen atom in alcohol and NT-bound fluorine. Water, diethylamine, perfluorinated solvents, or acetic acid do not solvate NTs.

Fluorotubes dissolved in alcohols can react with alkoxides or terminal diamines such as $H_2N(CH_2)_nNH_2$ ($n = 2, 3, 4, 6$). They are capable of reacting with hydrogen peroxide, organic peroxides (e.g., lauroyl, benzoyl, *tert*-butyl), and with a number of solid inorganic compounds, such as alkali halides, Li_2S, ZnS, Li_2O_2, and AlP.

The atomic and electronic structures of fluorinated SWNTs have been examined in a few experimental and theoretical works.[175,180–188] As a result of fluorination, a significant charge transfer occurs from the NT wall to the fluorine atoms, resulting in partially ionic bonds. This transforms the non-polar SWNT to the polar one.

X-ray photoelectron spectroscopy can identify the type of bonding within CF_x compounds. The spectra of fluorinated samples give peaks appearing at 287 eV (semi-ionic C–F), at 288 to 299 eV (nearly covalent C–F), and at 292.0 to 294.05 eV (covalent CF_2 and CF_3).[175]

Fluorinated SWNTs are used to form composites with poly(ethylene oxide).[189]

2.5.4 AMIDATION

Amines, particularly ODA, have attracted special attention in the studies of functionalization of CNs. The SWNT–COOH product treated with oxalyl chloride at 0°C and then heated with ODA at 100°C, after purification contains 4 mol% of amine.[190] Shortened MWNTs attach considerably larger amounts of ODA, up to 41.7 wt%, after 96-h functionalization.[191]

Acid-chloride-functionalized SWNTs are used to attach glucosamine,[192] didecylamine,[193,194] 4-dodecyl-aniline, and $4\text{-}CH_3(CH_2)_{13}C_6H_4NH_2$.[118] NTs functionalized with aniline prove to be soluble only in aniline, whereas NTs derivatized with tetradecylaniline are soluble in CS_2 and aromatic solvents. The anilination reaction solubilizes SWNTs and allows their purification chromatographically using excess of adsorbed aniline. A product with the ratio of NT carbons to aniline sites of 360:1 has been prepared by refluxing oxidatively end-cut SWNTs.[195] The functionalized group of MWNTs modified with aniline was determined to be $C_6H_6N^-$.[196] The terminal chlorinated carboxylic groups were used to append pyrenyl subunits[197] and *n*-pentyl ethers.[141]

SWNTs absorb MW radiation, and thus tubes can be rapidly heated by radiated. A procedure based on MW heating, which allow the attachment of monoamine-terminated poly(ethylene glycol) molecules to shortened SWNTs–COCl, has been developed.[198]

The acid-chloride-functionalized SWNTs were attached in pyridine suspension to chemically functionalized Si surfaces.[199]

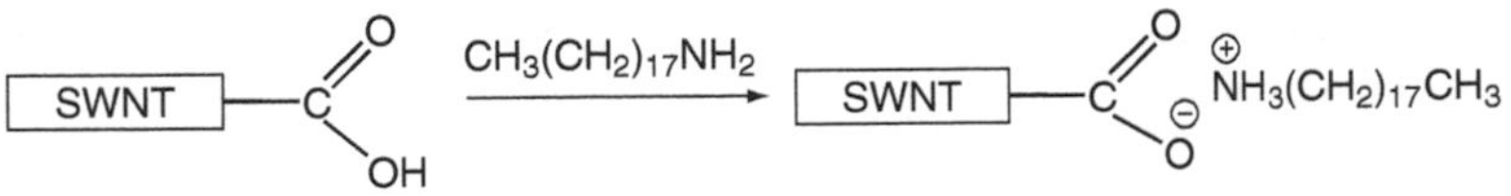

FIGURE 2.6 Zwitter-ionic functionalization of SWNTs. (Reprinted with permission from Niyogi, S. et al., *Acc. Chem. Res.*, **35**, 1105–1113, 2002. Copyright 2002. American Chemical Society.)

It is supposed that SWNT in DMF can form covalent bonds with tenth generation poly(amidoamine) starburst dendrimer.[200] A "grafting-from" method has been developed for production of hyperbranched poly(amidoamine)-modified NTs.[201]

Solution of SWNTs in DMF derivatized with ODA is used for chromatographic purification procedure.[202] The NT end-to-side or end-to-end junctions are created by the reaction between modified NTs and diamines.[203,204]

The CdSe quantum dots were coupled to individual acid-chloride-modified SWNTs via amide-bond formation.[205]

Oxygen-containing groups that are present on NT tips can condense with alkoxysilane groups-terminated amide acid polymers and facilitate the formation of NT/polymer films for electrostatic charge mitigation.[152,206]

The simplest possible route to the solubilization of SWNTs is direct reaction of the molten amine with the shortened SWNT–COOH. Thus a simple acid–base reaction is realized and zwitterions are formed (Figure 2.6).[118] The products of the reaction are found to be soluble in tetrahydrofuran (THF) and CH_2Cl_2. Zwitterion-functionalized shortened SWNTs are used for length separation via gel permeation chromatography[207] and for selective precipitation of metallic SWNTs upon solvent evaporation.[208] The method is used to solubilize full-length SWNTs.[122] The derivatization of the oxidized MWNTs with triethylenetetramine leads to subsequent covalent bonding with the epoxy resin used as a matrix for MWNTs.[209] $^+NH_3(CH_2)_{17}CH_3$ ions, as has been stated, can readily exchange with other ions, e.g., metal ions.[11] Cysteamin allows the realization of thiolization reaction of carboxyl-terminated SWNTs and deposition of the NTs onto a gold surface.[210] A version of region-specific NT deposition onto prepatterned surface via amidation of acid-functionalized NTs is described.[211,212]

Sun et al.[213] studied the reaction and dispersal of noncarboxylated SWNTs by refluxing in aniline. Dark red complexes are formed in the process. The solubility of SWNTs in aniline is up to 8 g/L. This aniline–NT solution can be readily diluted with other organic solvents such as acetone, THF, and DMF. As evidenced by their relatively high solubility in aniline, NTs may form donor–acceptor interactions with aniline.[213] Complexing with aromatic amines makes both single- and multiwall tubes dispersable in organic solvents.

The condensation reaction of acid-functionalized SWNTs with 2-aminoethanesulfonic acid allows to supply the end of SWNT with sulfonic groups and enhance its solubility in water.[214] This technique has been used for attachment of aminopolymers.[132] The amidation of NT-bound carboxylic acids can be accomplished in diimide-activated reactions. Functionalization with poly(propionylethylenimine-*co*-ethylenimine) in the presence of 1-ethyl-3-(3-dimethylaminopropyl)carbodiimide has been found to be significantly improved in both efficiency and yield by sonication under ambient conditions.[215]

The attachment of poly(styrene-*co*-aminomethylstyrene) is possible under amidation reaction.[139] The ball-milling process in ammonia atmosphere allows the introduction of amine and amide groups onto MWNTs.[216]

Basiuk et al.[217] attempted to simplify convenient solution method, and to apply gas-phase derivatization of oxidized NTs containing carboxylic groups on their tips according to the equation

$$\text{SWNT–COOH} + \text{HNR}^1\text{R}^2 \rightarrow \text{SWNT–CO–NR}^1\text{R}^2 + \text{H}_2\text{O}$$

Nonylamine, dipentilamine, ethylenediamine, and propylenediamine[217] have been used as test compounds. This procedure consists of treating SWNTs with amine vapors under reduced pressure and at a temperature of 160 to 170°C. Amine molecules not only formed derivatives with SWNT tips but physisorbed inside SWNTs. The content of physisorbed nonylamine is about one order of magnitude higher than the amide content.

Theoretical consideration of the amidation reaction with methylamine shows that the formation of amide derivatives on carboxylated armchair SWNT tips is more energetically preferable than that on the zigzag NTs.[218]

The physisorption on metallic SWNTs causes no significant change in the electrical conductance, whereas adsorption of amines (such as butylamine and propylamine) on partial length of semiconducting NTs causes modulated chemical gating.[219]

Other works on amidation have been published; among them is the amidation of HiPco SWNTs,[220] and the functionalization of SWNTs with phthalocyanine molecules through amide bonds.[221]

2.5.5 OTHER TYPES OF COVALENT BONDING

Direct covalent functionalization of NT can be realized via addition of carbenes,[127,222–226] nitrenes,[223,227–229] 1,3-dipoles,[230–232] aryl cations,[19,233,234] and radicals (Figure 2.7).[19,233,235–241]

For direct functionalization, one can use processes such as ultrasonication in organic media,[10] plasma treatment, UV irradiation, or irradiation with energetic particles.

Carbenes have the general formula CRR′, where R, R′ = H, halogen, organic residuum, etc., and represent unstable compounds of bivalent carbon. Dichlorocarbene is an electrophilic reagent that adds to deactivated double bonds, but not to benzene. It is capable of attacking C=C bonds, replacing them by CCl_2 bridges. The addition of dichlorocarbene took place at the sidewall of both insoluble SWNT[222] and shortened SWNTs (s-SWNTs).[127] It was reacted with NTs in a refluxing chloroform/water suspension. Around 5% of chlorine was incorporated into or onto the SWNTs.[222]

Hu et al.[225] used dichlorobenzene solution of $PhCCl_2HgBr$ and showed that the addition of dichlorocarbene converts metallic SWNTs to semiconducting SWNTs. Thermal treatment of (s-SWNT)CCl_2 above 300°C results in the breakage of C–Cl bonds, but the electronic structure of the SWNTs was not recovered. Monthioux[224] published a method for dichlorocarbene formation and attachment to the SWNT by the decomposition of chloroform under UV irradiation. The C=Cl$_2$ bridges are assumed to be removed under UV treatment.

The two-level Our owN n-layered Integrated molecular Orbital + molecular mechanics Method (ONIOM) technique has been employed to study the [2+1] cycloadditions of dichlorocarbene, silylene, germilene, and oxycarbonitrene onto the sidewall of SWNT.[226] Results showed that the reactions are site-selective and yield three-membered ring species. The thermal stability of the SWNT derivatives follows the order oxycarbonitrene >> dichlorocarbene > silylene > germilene. The derivatives can be good starting points for further functionalization.

Nitrenes are analogs of carbenes; they represent unstable compounds of monovalent nitrogen and have general formula RN, where R = alkyl, aryl, getaryl, NR′$_2$, CN, etc. Among the methods of nitrene generation, thermal and photochemical decomposition of azides and other compounds should be mentioned. The addition of (R-)-oxycarbonyl nitrenes allows the bonding of a variety of different groups such as alkyl chains, aromatic groups, dendrimers, crown ethers, and oligoethylene glycol units.[227]

For functionalization based on the 1,3-dipolar cycloaddition of azomethine ylides,[230–232] the heterogeneous reaction mixture of SWNTs suspended in DMF together with excess aldehyde and modified glycine was heated at 130°C for 5 days. The modified NTs are remarkably soluble in most organic solvents ($CHCl_3$, CH_2Cl_2, acetone, methanol and ethanol) and even in water. The solubility of SWNTs in $CHCl_3$ is close to 50 g/L without sonication. The reactions were successful with the use of either short-oxidized or long-nonoxidized SWNTs, without notable differences in their solubility. The functionalized NTs are less soluble in toluene and THF, and practically insoluble in less polar solvents including diethyl ether and hexane.

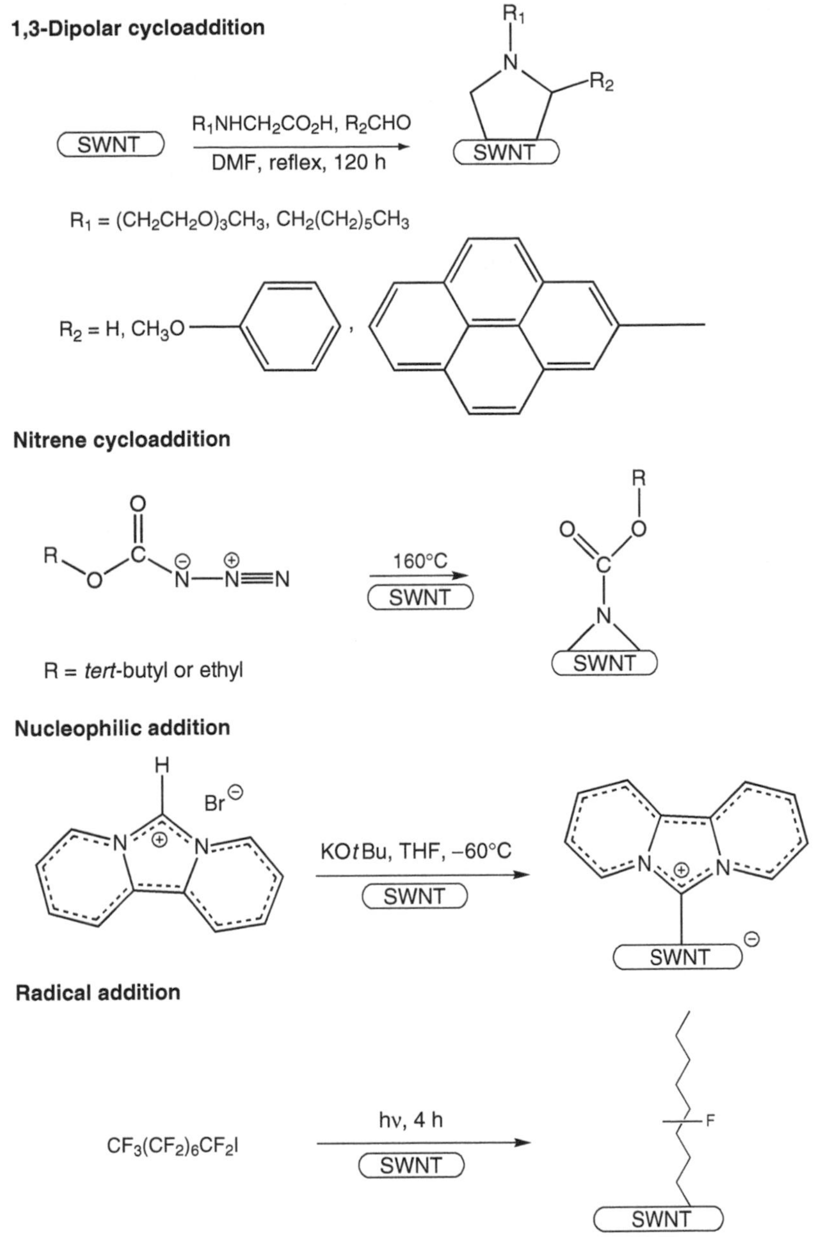
1,3-Dipolar cycloaddition
R1
N
R2
SWNT
R1NHCH2CO2H, R2CHO
DMF, reflex, 120 h
SWNT
R1 = (CH2CH2O)3CH3, CH2(CH2)5CH3
R2 = H, CH3O
Nitrene cycloaddition
R
O
C
N
N
160°C
SWNT
R = tert-butyl or ethyl
SWNT
Nucleophilic addition
H
Br
KOtBu, THF, −60°C
SWNT
SWNT
Radical addition
hv, 4 h
CF3(CF2)6CF2I
SWNT
F
SWNT

FIGURE 2.7 Sidewall covalent chemistry on SWNTs. (Reprinted with permission from Niyogi, S. et al., *Acc. Chem. Res.*, **35**, 1105–1113, 2002. Copyright 2002. American Chemical Society.)

It was reported that sonication and homogenization of a mixture of SWNTs and a monochlorobenzene solution of poly(methylmetacrylate) increased the ratio of shorter and thinner SWNTs[242–244] and led to the chemical modification of SWNTs.[245] Organic molecules decompose at the hot spots, and reactive species react with damaged SWNT sidewalls. FT-IR spectra show the formation of C–H and C=O groups. The sonication of purified SWNTs in monochlorobenzene solution leads to the formation of two kinds of modified SWNTs.[243,244] Sonochemical decomposition of *o*-dichlorobenzene[246] and 1,2-dichrorobenzene[247] also leads to the attachment of decomposition products to SWNTs and to the stabilization of SWNT dispersions. Other organic solvents formed NT dispersions under sonication (see the next section) most likely by an analogous mechanism.

For the purpose of SWNT sidewalls functionalization using organic radicals, Ying et al.[239] decomposed benzoyl peroxide in the presence of alkyl iodides and obtained phenyl radicals. These radicals reacted with alkyliodides, which generated iodobenzene and alkyl radicals. The procedure allowed them to attach long-chain alkanes, alkyl halides, amides, nitriles, and ethers to the sidewalls of the NTs. Methyl radicals can also bond to the sidewalls, but the resulting NTs are generally insoluble in organic solvents.

Water-soluble diazonium salts can react with NTs.[233,235,237] A reactive radical can be produced by electrochemical reduction of different aryl diazonium salts using a bucky-paper electrode.[235] The estimated degree of functionalization is up to 5% of carbon atoms. Along with reductive coupling, it is possible to provide oxidative coupling (Figure 2.8).[238,248] The derivatization with aryl diazonium salts is not limited to the electrochemically induced reaction.[237]

NTs derivatized with a 4-*tert*-butylbenzene moiety exhibit the highest solubility in organic solvents. Solvent-free functionalization has short reaction times.[233]

Addition of diazonium salts to NTs suspended in aqueous solution opens a way to select chemically and separate NTs based on their electronic structure. Metallic NTs under certain controlled conditions give up electrons more readily than semiconducting NTs, a factor that the diazonium reagent can respond to.[240] The chemistry is reversible. Heating of the functionalized NTs in inert media at 300 to 400°C stimulates pyrolysis of arene groups and leads to restoration of the pristine electronic structure of NTs. This work also proves that the assumption that NT chemistry is controlled solely by their diameter (with smaller-diameter NTs being less stable) is in fact not always true.

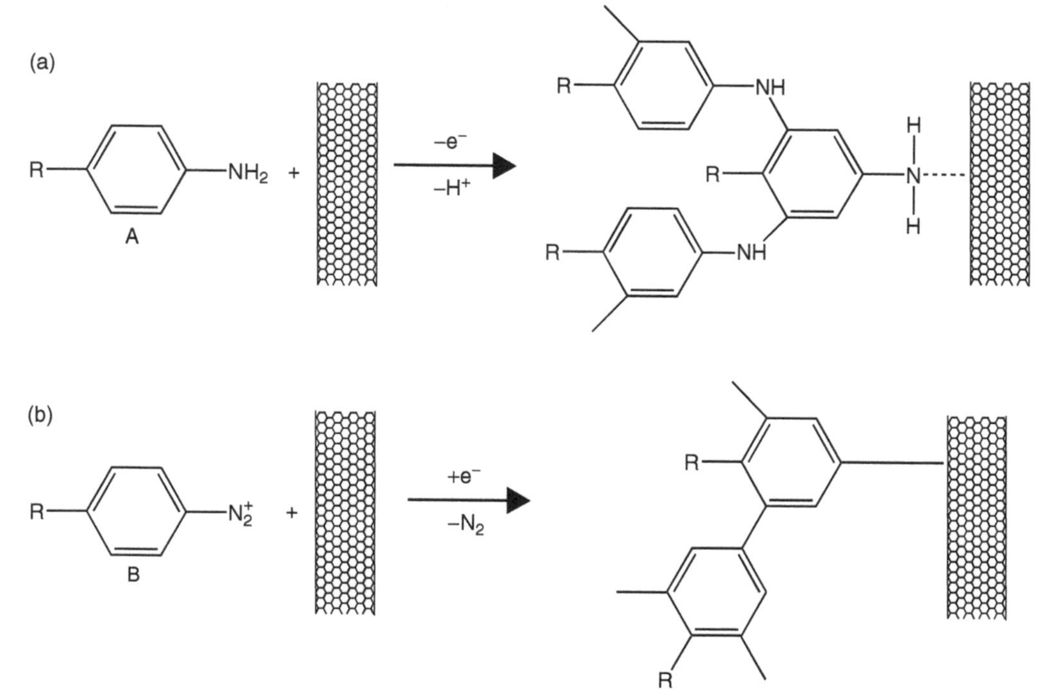

FIGURE 2.8 Oxidative (a) and reductive (b) electrochemical modification of SWNTs. The broken line in (a) indicates the formation of electro-polymerized layers of A on the SWNT, without the creation of chemical bonds. (Reprinted with permission from K. Balasubramanian, et al., *Adv. Mater.*, **15**, 1517–1518, 2003).

Since organic thiol derivatives are generally well known to interact strongly with noble metal surfaces, therefore, the selective thiolation may be used to make a good electrical junction between a NT and a metal electrode, or to position the NT relative to a metal surface. The first thiolated NTs, $NT-(CH_2)_{11}-SH$, with long alkyl chains, were synthesized by Smalley and coworkers.[66] Because of the long and flexible alkyl chain, the latter compounds do not anchor on a metal surface in a specific orientation and give rise to a large contact resistivity. To overcome these problems, another type of compound, $NT-CONH-(CH_2)_2SH$, with a shorter alkyl chain, was synthesized.[210] The compound, however, contains an amide bond that tends to react easily in an acid or basic environment.

The new form of thiolated NT, which contains thiol groups almost directly linked to the body of NT, was synthesized by Lim et al.[249] The formation of $NT-CH_2-SH$ is achieved via successive carboxylation, reduction, chlorination, and thiolation of the open ends of NTs.

Time-dependent plasma etching and irradiation with energetic particles can provide controlled introduction of defects and functionalities into NTs. Irradiation creates links between NTs, and leads to coalescence and welding of NTs (see, e.g., Refs. 250 and 251). Argon ion irradiation enhances the field emission of NTs.[252]

It has been shown that H_2O plasma can be used to open end-caps selectively of perpendicularly aligned NTs without any structural change.[253] The treatment of NTs with oxygen plasma at a low pressure for some minutes results in an oxygen concentration up to 14% and formation of outer layers consisting of hydroxide, carbonyl, and carboxyl groups.[254]

Interaction of SWNTs with high-energy protons at low irradiation doses causes the formation of wall defects.[255] The NTs curve at higher doses (>0.1 mC) and degrade into amorphous material at even higher doses (approaching 1 mC). The hydrogenation of NTs can be achieved in a cold MW plasma at low pressure.[256] A 30 sec exposure to a plasma of H_2 generated by glow-discharge results in near-saturation coverage of SWNT with atomic hydrogen.[257]

Calculations using molecular dynamics show that CH_3 radicals with energies of higher than 19 eV can attach to NT sidewalls.[258,259] The heat of the attachment reaction changes from ~0.8 eV for graphite to ~1.6 eV for fullerene C_{60}, and has intermediate, close to figures for graphite, values for NTs.[13]

The experiments show evidence of chemical functionalization of MWNTs by attachment of CF_3^+ ions at incident energies of 10 and 45 eV.[259] The exposure of SWNTs to CF_4 gaseous plasma leads to the formation of semi-ionic and covalent C–F bonds.[260,261] The ion bombardment does not result in loss of NT structure.[261]

Ion beams of certain energies can be used to create nanotube-based composites with improved adhesion between the filler and polymer matrixes as well as to create covalent cross-links between NTs and the C_{60} molecules.[262]

The modification of NTs is possible with acetaldehyde plasma activation.[212] Plasma-modified NTs improve the properties of nanotube-epoxy composites.[263] The hydrogen plasma treatment enhances field emission of NTs.[264] By using hydrothermal synthesis it is possible to produce hydrophilic SWNTs or MWNTs that are wetted by water and water solutions, because their outer and inner surfaces are terminated with OH groups.

An electrochemical derivatization method can also be used to attach carboxylate groups to NT walls (see Ref. 265 and references therein).

2.5.6 NONCOVALENT BONDING

Carbon NTs have been solubilized in water with the aid of surfactants, which can deposit on the NT surface and help to form stable colloidal dispersion. The repulsive force introduced by the surfactant overcomes the van der Waals attractive force between the carbon surfaces. However, there is a problem when the surfactant is removed from the NT surface.

Sodium dodecyl sulfate (NaDDS, $CH_3(CH_2)_{11}OSO_3Na$),[266–276] lithium dodecyl sulfate (LiDDS, $CH_3(CH_2)_{11}OSO_3Li$),[277–279] and sodium dodecylbenzene sulfonate (NaDDBS, $C_{12}H_{25}C_6H_4SO_3Na$)[280–284] are among the simplest and most popular surfactants used for NT solubilization.

At low NaDDS concentration, large and dense clusters of the initial NTs were still found after sonication. At higher surfactant concentrations, black and apparently homogeneous solutions, stable over several weeks, were obtained.[269] The phase diagram of the NaDDS–SWNT–water system is not a simple one.[271] The domain of homogeneously dispersed NTs is limited and has an optimum (good NT solubility and system stability) at ~0.35 wt% in NTs and 1.0 wt% in NaDDS.

Suspensions of MWNTs or SWNTs in water stabilized by 0.25% NaDDS solution have been used for purification and size separation of tubes.[266–268] An individual SWNT encased in close-packed columnar NaDDS micelle has a specific gravity of ~1.0, whereas that of an NaDDS-coated bundle has a specific gravity of ~1.2 or more.[270] Therefore, NaDDS suspensions (2 g/L) prepared by sonication of raw, solid SWNTs in 0.5% NaDDS solution are capable of separating bundled SWNTs from isolated individuals.[272] The MWNT dispersion stabilized by NaDDS allows the production of MWNTs/hydroxyapatite composites.[276]

Non-specific physical adsorption of NaDDBS allows the solubilization of lightweight fraction SWNTs in water.[280,282] The NT stabilization depends on the structure of surfactant molecules that lie on the tube, parallel to the cylindrical axis (Figure 2.9). It was possible to achieve relatively high SWNT concentration (up to 10 g/L as a mixture of isolated and small bundles of SWNTs) without nematic ordering in suspension. The optimum NT/surfactant ratio was found to be 1:10 (by weight). The properties of a dispersion depend on the sonication technique used (high-power or mild mode of operation, tip or bath sonicator). The mechanism of NT solubilization determines the hydrophobic forces between the surfactant tail and the NT surface. Each NT is covered by a monolayer of NaDDBS molecules, in which the heads form a compact outer surface of a cylindrical micelle.[283] The aqueous (D_2O) suspension in the presence of NaDDBS surfactant exhibits the presence of SWNT aggregates, but not rigid rods.[284]

The dispersing power of Triton X-100 (TX-100, $C_8H_{17}C_6H_4(OCH_2CH_2)_nOH$; $n \sim 9.5$),[60,66,131,280,285–288] sodium octylbenzene sulfonate ($C_8H_{17}C_6H_4SO_3Na$),[280] sodium butylbenzene sulfonate ($C_4H_9C_6H_4SO_3Na$),[280] sodium benzoate ($C_6H_5CO_2Na$),[280] dodecyltrimethylammonium bromide (DTAB, $CH_3(CH_2)_{11}N(CH_3)_3Br$),[280,289,290] cetyltrimethylammonium bromide (C_{16}TMAB, $CH_3(CH_2)_{14}CH_2N(CH_3)_3Br$),[291] cetyltrimethylammonium chloride ($CH_3(CH_2)_{14}CH_2N(CH_3)_3Cl$),[289] cetyl alcohol derivative ($CH_3(CH_2)_{14}CH_2(OC_2H_5)_{10}OH$),[291] pentaoxoethylenedodecyl ether ($C_{12}E_5$),[289] and hexadecyltrimethylammonium bromide ($CH_3(CH_2)_{16}N(CH_3)_3Br$)[292] have been studied. Both NaDDBS and TX-100 dispersed the NTs better than NaDDS, because of their benzene rings; NaDDBS dispersed better than TX-100 because of its head groups and slightly longer alkyl chain.[280] DTAB and $C_{12}E_5$ solutions, at concentrations ranging from 0.05% to a few percent, do not stabilize the NTs.[289]

Ultrasonication of a mixture of distilled water and MWNTs in the presence of 5 vol% TX-100, followed by centrifugation to remove unsuspended material allows the production of a suspension of

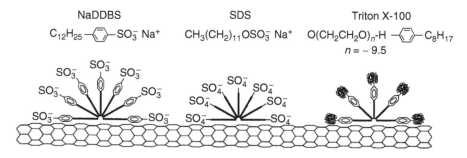

FIGURE 2.9 A representation of surfactant molecules adsorbed onto NT surface. (Reprinted with permission from Islam, M.F. et al., *Nano Lett.*, **3**, 269–273, 2003. Copyright 2003. American Chemical Society.)

concentration of 0.1 g/L.[287] An aqueous (or alcoholic) solution of TX-100 has been used to prepare SWNT dispersion, followed by alignment under AC electric field.[286] Such dispersions are suitable to prepare thin film coatings on flexible plastic substrates.[288,291]

Spectral study reveals that the most essential spectral shift of lines compared with the spectrum of SWNT in KBr pellet is observed for NT aqueous solutions with the surfactants containing charged groups.[293]

Acidification of a solution of surfactant-dispersed SWNTs in water in the pH region of 6.0 to 2.5 results in the reversible and selective reaction of protons at the sidewall of SWNTs.[290] The equilibrium constants are dependent on the NT band gap, and metallic NTs appear more sensitive to acidity of the solution. A crucial role is played by adsorbed O_2, which controls both the rate and equilibrium extent of the reaction. The results of this investigation hold promise for chemical separation and sorting of NTs having different electronic structures.

C_{16}TMAB or other surfactants are used to prepare a SiO_2/NT composite.[291,294]

SWNTs can be solubilized in water at g/L concentrations by non-covalent wrapping them with water-soluble linear polymers, most successfully with polyvinyl pyrrolidone[295,296] and sodium polystyrene sulfonate.[295,297,298] Polymetacrylic acid,[298] polypyrrole,[299] poly(phenylacetylene),[300] poly(diallyldimethylammonium chloride)[301,302] have also been tested. The solubilization of SWNTs by polymer wrapping might provide a series of useful techniques, such as purification, fractionation, and manipulation of the SWNTs.

For many applications, bio-compatible water-soluble derivatives of NTs are desirable. For this reason, the solubilization of NTs in cyclodextrins,[303–305] polysaccharides and natural mixtures of polysaccharides such as gelatine,[306,307] Gum Arabic,[275,289] and starch[308] has been studied.

Nanotubes are not soluble in aqueous solutions of starch but they are soluble in a starch–iodine complex. The starch, wrapped helically around small molecules, will transport NTs into aqueous solutions.[308] The process is reversible at high temperatures, which permits the separation of NTs in their starch-wrapped form. The addition of glucosidases to these starched NTs results in the precipitation of the NTs from the solution. Readily available starch complexes can be used to purify NTs. An effective process to produce colloidal solution of SWNT–amylose complexes is elaborated.[309] The solubility of sonicated NTs improves by dilution of water with DMSO (10 to 25 vol%). Some natural polysaccharides wrap SWNTs, forming helical suprastructures.[310]

An amphiphilic α-helical peptide specifically designed not only to coat and solubilize NTs, but also to control the assembly of the peptide-coated NTs into macromolecular structures, is described.[311] The NTs can be recovered from their polymeric wrapping by changing their solvent system.

As for the solubility of pure SWNTs in organic solvents, these solvents are divided into three groups.[285] The "best" solvents are N-methylpirrolidone (NMP), DMF, hexamethylphosphoramide, cyclopentanone, tetramethylene sulfoxide, and ε-caprolactone, which readily disperse SWNTs, forming light gray, slightly scattering liquid phases. All of these solvents are nonhydrogen-bonding Lewis bases. Group 2 includes DMSO, acrylonitrile, 4-chloroanisole, and ethylisothyocyanate. The third group includes 1,2-dichlorobenzene, 1,2-dimethylbenzene, bromobenzene, iodobenzene, and toluene.

Using solvochromic and thermochemical parameters of different solvents Torrens also categorized them into three groups.[312] The first group include the "best" solvents mentioned earlier. In the group of "good" solvents, he includes toluene, 1,2-dimethylbenzene, CS_2, 1-methylnaphthalene, iodobenzene, chloroform, bromobenzene, and o-dichlorobenzene. Group 3 are the "bad" solvents, n-hexane, ethyl isothyocyanate, acrylonitrile, DMSO, water, and 4-chloroanisole.

As reported earlier, the best solvents for generating SWNT dispersions in organic solvents are amides, particularly DMF and NMP.[176] The solubilities of SWNTs in 1,2-dichlorobenzene, chloroform, 1-methylnaphthalene, and 1-bromo-2-methylnaphthalene are equal to 95, 31, 25, and 23 mg/L,[247] respectively. Solubilities of purified and functionalized SWNTs in ethanol, acetone, and DMF is 0.5, 1.06, and 2.0 mg/L,[313] respectively. According to Ref. 247, the solubilities are <1, <1, and 7.2 mg/L.

MWNTs cannot be dispersed in toluene into the level of single tubes even when diluted to a concentration of ~10^{-3} g/L.[314] It has been found that the aggregation decreases with increasing temperature.

A procedure for the quantitative evaluation of the purity of bulk quantities of SWNT soot on the basis of near infrared (NIR) spectroscopy of a sample dispersed in DMF is described.[315]

Organic solutions of NTs in poly(*p*-phenylenevinylene-*co*-2,5-dioctoxy-*m*-phenylenevinylene),[316] poly(*m*-phenylenevinylene-*co*-2,5-dioctoxy-*p*-phenylenevinylene),[317] poly(2,6-pyridinylenevinilene-*co*-2,5-doictyloxy-*p*-phenylenevinylene),[317] a family of poly(*m*-phenylenevinylene-*co*-*p*-phenylene-vinylene)s,[318] can be formed due to the physical adsorption of polymers. Certain polymers such as polyphenylenevinylene derivatives, and vinyl-based polymers such as polyvinylalcohol and polyvinylpyrrolidone tend to disperse NTs, while rejecting other carbon-based impurities (see Ref. 319).

The solubilization of small-diameter NTs is possible using rigid side-chain poly(aryleneethyl-ene).[320] The method includes the dissolution of SWNTs in methylene chloride and polymer under vigorous stirring or sonication, and yields a solubility as high as 2.2 g/L. Researchers believe that the most probable mechanism is a π stacking which stabilizes the polymer–NT interaction.

Noncovalent solubilization of NTs can be realized by encapsulation of SWNTs by metallo-macrocyclic rings.[216,321] A DMF solution of poly(vinylidene fluoride) can be used for size fractionation of MWNTs.[322]

2.5.7 DISPERSIONS IN OLEUM

Successive dispersion of SWNTs in oleum at concentrations up to 4 wt% was first achieved at Rice University.[323] It was shown that at very low concentrations of SWNTs (<0.25 wt%), a single phase with uniformly dispersed tubes was formed. The SWNTs in the dilute system behave as Brownian (noninteracting) rods. The SWNT concentration in the dispersion can be increased up to 10 wt%; the dispersion process is promoted by the protonation of the SWNT sidewall, and the tubes are stabilized against aggregation due to the formation of electrostatic double layer of protons and negatively charged counterions.[324] Increasing the concentration of SWNTs in the acid leads to the formation of a highly unusual nematic phase of spaghetti-like, self-assembled supra-molecular strands of SWNTs. As concentration increases (to 4 vol% in 102% H_2SO_4), the strands self-assemble into a single-phase nematic liquid crystal. If a small amount of water is added, the liquid crystal separates into ~20 μm long, needle-like strands of highly aligned SWNTs, termed "alewives." This phase can be processed, under anhydrous conditions, into highly aligned fibers of pure SWNTs with a typical alignment ratio of 20:1 to 30:1. A syringe pump was used to extrude the mixture through a 0.15-mm internal diameter needle, followed by spinning the neat SWNT fibers up to 1 m long. The detailed structure and properties of the fiber have been studied.[325] High-temperature annealing of the fibers does not affect the SWNT alignment.

A solution of purified SWNTs in oleum (H_2SO_4/30% SO_3) was used to cast optically isotropic film exhibiting fibrillar morphology.[326] The electrical conductivity of this film (1×10^5 S/m) is about an order of magnitude higher than that for the SWNT bucky paper.

Chlorosulfonic acid, triflic acid, and anhydrous HF–BF_3 solution can also be used for the solubilization of NTs.

2.5.8 SELF-ASSEMBLY, FILM, AND FIBER FORMATION

A promising feature of NTs in nanotechnology is that they are potentially amenable to a "bottom-up," self-assembly-based manufacturing approach. It is essential to fabricate well-aligned structures of NTs for various electric and optical applications. The necessity for aligned and micropatterned NTs has been elaborated in many articles (see, e.g., review 20).

There are two general strategies to align NTs: (1) *in situ* synthesis and (2) postsynthesis ordering. The first approach includes CVD processes on a prepatterned (with catalyst) surfaces or a template-based synthesis.[20,48,327] This approach is not considered here.

The second approach includes a variety of chemical or electrochemical processes[9] such as different functionality reactions and noncovalent (e.g., electrostatic) interactions between the NTs and surface-bound moieties.

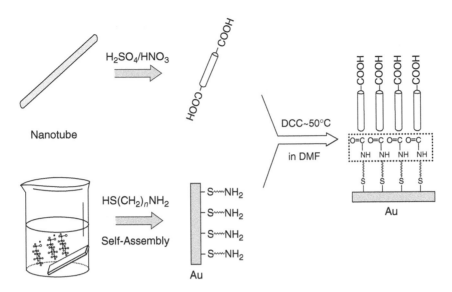

FIGURE 2.10 Schematic illustration of the formation of highly aligned SWNTs on gold surface. (Reprinted from Nan, X. et al., *J. Colloid Interf. Sci.*, **245**, 311–318, 2002. Copyright 2002. With permission from Elsevier.)

To organize NTs on gold or silver substrates, the carboxylic groups at the end of NTs were thiol-functionalized to form Au–S or Ag–S chemical bonds (Figure 2.10).[153–155,210–212] Flexible SWNTs with many thiol groups at their ends are more likely to bend on metal surface to form the "bow-type" structure, while a more rigid form of SWNT or MWNT with less thiol groups, will stand upright on the surface, forming a rod-like structure.[249] The monolayer of randomly tangled SWNTs is attached to a gold surface containing $HS-(CH_2)_{10}-COOH$.[328]

It is possible to use a patterned self-assembled monolayer, which can either enhance or deter NT adherence.[329–330] Silicon wafers can be coated with either nonpolar methyl groups or with polar carboxyl and amino groups.[199,331] When the substrate is placed in a suspension of SWNTs, the tubes are attracted toward the polar regions and self-assemble to organize pre-designed structures. To form self-assembled monolayers with amino-terminated surface, silicon wafers were silanized using 3-aminopropyltriethoxysilane.[332] The octadecyltrichlorosilane is used to attach methyl-terminated groups.

Polyelectrolyte layers on a silicon substrate have been used to align MWNTs.[333] The carboxylate anion groups of MWNTs bind on the oppositely charged polycationic poly(diallyldimethylammonium chloride) (PDSC). The possibility of forming multilayer assemblies such as Si/PDSC/PSS/PDSC/MNT (where PSS stands for poly(sodium 4-styrenesulfonate)) using coulombic interaction has been demonstrated.

An original method to produce a hollow spherical cage of nested SWNTs using self-assembly technique consists of attaching of NTs to amidated silica gel spheres and subsequent drying and dissolution of the template.[334]

The aligning of NTs can be realized under the influence of the capillary force and the tensile force that appear in the process of solvent evaporation.[335–339] The vacuum evaporation of concentrated (20 to 50 g/L) aqueous dispersion of purified MWNTs at 100°C yields long ribbons of aligned NTs, self-assembled on the wall of the container.[337] The ribbons form in one of the two orthogonal orientations to the bottom of the container (glass beaker): perpendicular when vacuum is applied, and parallel when no vacuum is applied. The ribbons are 50 to 100 µm wide, 4 to 12 µm thick, and 100 mm long. Presumably, the key factor is the rate of evaporation.

Using a dispersion of shortened purified SWNTs in de-ionized water, Shimoda et al.[338,339] formed a thin film on the surface of a soaked glass substrate with natural vaporization of water. The

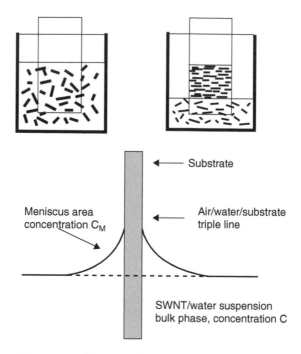

FIGURE 2.11 Self-assembly process of shortened SWNTs onto a hydrophilic glass slide.

SWNT bundles were uniaxially aligned parallel to the bottom of the container (Figure 2.11). In this process, SWNT bundles are first dispersed at a concentration of 0.5 to 1.0 g/L. Then a clean hydrophilic glass sheet is immersed vertically into the suspension. As the water evaporates, the NTs deposit near the triple borderline air/water/NTs and the deposit progresses downward. By means of patterned hydrophilic regions, patterned deposits were produced in the form of squares and strips, 100 μm in width.

Capillary forces arising during the evaporation of liquids from dense NT arrays were used to reassemble the NTs into 2-D contiguous cellular foams.[340]

Natural self-assembly and cooperative mechanisms of liquid crystals can be employed to manipulate the alignment of NTs.[341] Thermotropic liquid crystals used as a solvent provide a tool for aligning SWNTs and MWNTs.[342] The broad range of possibilities of liquid crystals has been demonstrated.

An alignment of NTs can be realized under the electric field with both DC and AC voltage between the electrodes.[343] This method allows the orientation and spatial positioning of the SWNTs[344] to be controlled. The room-temperature method, called "minimal-lithography" technique, has been used to prepare crossbars of SWNT ropes and deterministic wiring networks from SWNTs. Electrophoretic deposition of NT films[345] and dielectrophoretic formation of fibrils[346] has been demonstrated.

Multilayer polymer/SWNT films can be formed by electrostatic assembly.[150] The Langmuir–Blodgett method is used to deposit thin uniform films of SWNTs onto substrates.[277,347,348] Thin films have been made with NTs embedded in a surfactant matrix suspended on top of an aqueous subphase and then pulling the substrate through the surface. The Langmuir–Blodgett method has been used to prepare a monolayer of crown ether-modified full-length MWNTs and SWNTs.[349] A method of laying down thin uniform films of NTs on substrates of arbitrary composition resembling the Langmuir–Blodgett deposition technique has been developed.[350] The composite Langmuir–Schaefer conducting organic/MWNT films with useful optical and electrochemical properties have been studied.[351]

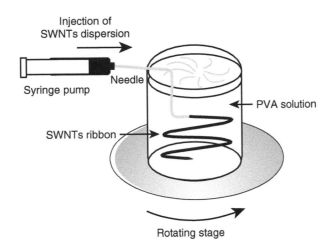

FIGURE 2.12 Simplified drawing of the experimental setup used to make NT ribbons. (From Poulin, P. et al., *Carbon*, **40**, 1741–1749, 2002. With permission.)

Fukushima et al.[352] have found a way to distribute NTs evenly through a gel, to form an electrically versatile material. The "bucky gel" materials were produced by grinding suspensions of SWNTs in imidasolium cation-based ionic liquids in an agate mortar. The gel can be printed using inkjet printer or polymerized. Lowering the temperature of the gels results in long-range ordering of the ionic liquid molecules and formation of crystal-like materials.

Yodh and his colleagues embedded isolated SWNTs coated with NaDDBS into a cross-linked polymer matrix, an *N*-isopropyl acrylamide gel.[281] The volume of the gel is highly temperature-dependent, and a change in temperature results in volume-compression transition. The condensed gel thus creates concentrations of isolated, aligned NTs that cannot be achieved when they are suspended in water.

Re-condensing of surfactant-stabilized NT solutions is used for the formation of aligned NT fibers.[269,353] In this method, NTs are sonicated in an aqueous solution of NaDDS. The dispersion is injected into a co-flowing stream of poly(vinyl alcohol) (PVA) via capillary tube. This principle was modified to produce long aligned fibers of NTs.[278] The process consists of introducing SWNT dispersion into a co-flowing stream of PVA in a cylindrical pipe, thereby causing the agglomeration of the SWNTs into a ribbon. The fibers are then unwound and passed through a series of washing stages to remove the excess PVA.

The extrusion of aqueous dispersions of NTs into rotating viscous solution of PVA leads to aggregation of NTs into narrow strips (Figure 2.12).[269] These strips, a few micrometers thick and a few millimeters wide, contract when dried in air forming dense fibers.

Electro-spinning allows the fabrication of an oriented poly(ethylene oxide) NFs in which MWNTs are embedded mostly aligned along the fiber axis (Figure 2.13).[275] The feasibility of this electrostatically induced self-assembly process for incorporation of NTs into NFs, production of membranes, and nanofiber yarn have been demonstrated.[354,355]

Macroscopic fibers have been produced from NT dispersions in oleum by spinning technique.[324,325]

2.6 FILLING THE INNER CAVITY OF CARBON NANOTUBES

Numerous attempts to fill the nanoscale cavities of NTs have been made following the discovery of these materials. The filling was attempted to achieve one of the two goals. First, being a kind of template synthesis, filling allows the preparation of nanostructured materials with controlled size, shape, and purity. Secondly, doping can modify the electronic properties of NTs. The thus prepared

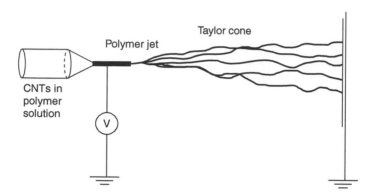

FIGURE 2.13   Schematic of the electrospinning process used to form SWNT-filled composites.

compounds are also interesting as nanosized objects to investigate size–crystal structure relations and size effects.

The filling of SWNTs attracts more attention than filling of MWNTs due to the smaller diameter of SWNT inner cavities, better stability, and more perfect structure of SWNTs. The afterward removal of SWNT used as nanomolds is a more easy procedure than removing the MWNT shield.

The filler can exist either in solid, liquid, or gaseous state. As far as solid materials are concerened, the inner cavity of NTs can be filled with single crystalline nanorods, polycrystalline nanorods, amorphous nanorods, or discrete nanoparticles.

Many solid substances can fill the inner cavities of NTs. The list of fillers includes metals (Cs, Cu, Ag, Au, Sn, Fe, Co, Ni, Pd, Rh, etc.), alloys (Fe–Ni, Fe–Pt, Pt–Ru, $Nd_2Fe_{14}B$), nonmetals (Ge, S, Se, Te, I_2, etc.), oxides (SnO, Sb_2O_3, NiO, UO_{2-x}), hydroxides ($Ni(OH)_2$), halides (KI, $LaCl_3$, $ZrCl_4$), salts ($AgNO_3$), carbides (B_4C, LaC_x, NbC_x, FeC_x), sulfides (AuS_x, CdS, CoS_x), nitrides (BN, GaN), organic substances ($CHCl_3$), acids (HNO_3), polymers (polystyrene), complex inorganic compounds and eutectic mixtures ($FeBiO_3$, $CoFe_2O_4$, AgCl–AgBr, KCl–UCl_4), fullerene and endofullerene molecules (C_{60}, $Gd@C_{82}$), complex hybrid materials ($FeCl_3$–C_{60}, K–C_{60}, Pt–WO_3).[7,8,224,356–358] Quantum chemical simulations predicted the stability of alkali metal compounds (Na@SWNT) and metallo-carbohedrene derivatives (Ti_8C_{12}@SWNT).

Water, inorganic acids, aqueous solutions, $CHCl_3$ solutions, and molten salts are among liquid substances suitable to fill NTs.

The solid and liquid substances can fill the cavity entirely or partially. Materials produced by filling of NT inner cavities can be used as magnetic media, catalysts, sorbents, quantum wires, field emitters, electromagnetic shielding, etc.

The results of theoretical calculations show that the radius as well as the helicity of the most stably doped SWNT are different for different kinds of impurity atoms.[359]

There are two basic strategies of filling: *in situ* synthesis of filled NTs or post-production methods that require an opening of the tubes.

2.6.1   *In Situ* Filling

All synthesis strategies may be accompanied by filling of NTs produced. In an arching process, the filler precursor can be introduced either by graphite anode doping (the most commonly used technique) or by dissolution in a liquid medium (if the process takes place in liquid environment). In the first stage of NT study, most information was obtained by using arc-discharge method in an inert gas flow. For filling of arc-produced NTs, a variety of metals, oxides, or salts have been used to dope the anode. With a few exclusions (Co, Cu, Pd), the encapsulated materials were always carbides.

Close-capped NTs can be filled *in situ* with metallic Co, S, and CoS_x from aqueous solution of $CoSO_4$ by arching.[38] A simplified arc-discharge in aqueous solution of $PdCl_2$ yields Pd-nanoparticles-filled NTs.[360]

All types of pyrolytic synthesis of NTs (using supported, dissolved, or floating catalysts, different physical activation methods) are inevitably accompanied by capture of some part of the catalyst and encapsulation of catalyst particles. Consequent purification with boiling HNO_3 or other oxidants cannot eliminate the metals completely.[361] The filling can be controlled to some extent by varying the process parameters, but sometimes, relative amount of incorporated material reach substantial values.

For example, the HiPco technique results in SWNTs partially filled with Fe (total Fe content in crude product is about 20 to 30%). The oxidation treatment of $LaNi_2$ alloy followed by CVD process using a CH_4/Ar mixture at 550°C leads to the formation of MWNTs filled with single-crystal Ni nanowires.[362] The synthesis of NTs filled with Ni by CVD over the Raney-Ni catalyst gives straight and two types of bamboo-shaped NTs.[363] The synthesis of Fe-, Ni-, and Co-filled NTs by using the pyrolysis of metallocenes (cyclopentadienyles) has been performed at 900 to 1150°C.[364,365] Invar ($Fe_{65}Ni_{35}$) has been introduced into NTs by pyrolyzing an atomized solution of $NiCp_2/FeCp_2$ in C_6H_6 at 800°C (Cp stands for cyclopentadiene).[366] The pyrolysis of methane over Fe_2O_3/Al_2O_3 binary aerogel at 880°C yields multi-wall nanohorns filled with Fe nanoparticles.[367] The decomposition of gaseous $Fe(CO)_5$ in a mixture with CO or C_6H_6 yields NTs partially filled with Fe.[368] MWNT-encapsulated Co particles have been produced by the catalytic method using water-soluble NaCl/NaF mixture as a support for the metal.[369]

Plasma-enhanced CVD on Si wafers allows the production of NTs-containing magnetic Fe, $Nd_2Fe_{14}B$, or Fe–Pt nanoparticles.[358] The formation of simple and branched Cu-filled NTs has been observed in the plasma-activated CVD process.[370] The Cu electrodes serve as a metal source in this process. Microwave-plasma-enhanced CVD process yields almost 100% GaN-filled NTs.[371]

"Double template" synthesis demands exploration of a material with aligned micropores. For example, NTs filled with Co have been synthesized using the CVD method and molecular sieve AlPO-5 and AlPO-31 as a primary template to formate NTs.[372]

A high-temperature process interaction of pulverized $Fe(NO_3)_3$ solution with carbon black and boron precursor results in the formation of Fe nanowire encapsulated in the inner cavity of carbon NT having an inner layer of BN.[373] The mechanism of the phase separation between C and BN is not clear.

An original, but complicated, method to produce relatively long cobalt nanowires filling the NT consists of a reaction of $Co(CO)_3NO$ with magnesium in closed vessel cell.[374,375]

During a hydrothermal synthesis of MWNTs, some gases, particularly CO_2, H_2O, and CH_4 can be trapped in the inner cavity of the tube.[376] Theoretical analysis of phase equilibria in such systems reveals an enhanced layering effect in the liquid phase.[377]

2.6.2 POST-PROCESSING FILLING

Nanotubes of two types are used in the filling process: NT synthesized by an usual method and NTs prepared by a template method in pores of Al_2O_3, AlPO-5, AlPO-31, or other suitable membranes. The usual methods lead to the formation of NTs of different diameters, whereas membrane synthesis ("double template" or "second-order" template synthesis) allows for the preparation of NTs of similar diameter and therefore in the production of encapsulated materials of uniform size. The second method is more complex and less productive.

2.6.2.1 Filling from Liquid Media

Filling by capillarity is possible only if the NTs are opened. The classification of liquid fillers or precursors includes:

- water and aqueous solutions
- liquid organic compounds

- hydrothermal solutions
- supercritical solutions
- molten metals, salts, and molten eutectic mixtures

The wetting properties of different carbons have been found to depend on the surface tension of the fluid, with a threshold value of 100 to 200 mN/m. Therefore, water having surface tension of 72 mN/m should wet NTs.[378,379] The boundary value is dependent on the inner diameter of NT.

The study of the filling of NT with water is of considerable importance in many biological systems and for the development of molecular devices. The behavior of water in the inner channel of NT has been discussed in a set of works.[380–387] Hummer et al.[380] showed, by molecular dynamics simulations that water molecules enter NTs of diameter greater than 0.81 nm even though carbon is hydrophobic. The possibility of transport of an electrolytic solution (KCl) through a carbon NT by a molecular dynamics simulation has also been explored.[388]

Certain metal oxides (V, Bi, Mo, Mn, Fe) can be doped by refluxing closed NTs with HNO_3 in the presence of metal nitrate and by subsequent calcination of metal nitrates inside NTs.[389,390] Filling with Ag (presumably silver oxide) has been achieved using concentrated aqueous solution of $AgNO_3$.[391] Aligned, open MWNTs can also be filled with Ag nanorods of up to 9 μm in length, using aqueous nitrate solution.[89] The reduction to metal Ag was achieved with CH_3COOH at 300°C. Interaction of the opened MWNTs with inner channel diameter of 60 nm with aqueous nitrate solution and subsequent heat treatment under inert atmosphere at 100°C, leads to the formation of spinel $CoFe_2O_4$ in the form of nanowire several micrometers long.[392–394] Immersion of a sample of pre-treated SWNT in a saturated aqueous solution of $RuCl_3$ leads to filling with this salt.[395]

Platinum–ruthenium nanoparticle (1.6 nm) filled NTs have been prepared by immersing the carbon/alumina template composite in an aqueous solution of H_2PtCl_2 and $RuCl_3$ for 5 h, and subsequently drying at room temperature, reducing in H_2 at 550°C, and dissolution of the membrane.[396,397] This method was used to load Pt, Pt–Ru, and Pt–WO_3 nanoparticles inside NTs.[398,399] A second-order template method was used for electrochemical Ni deposition inside NTs.[400] Highly crystalline nanorods of α-Fe_2O_3 were synthesized in the cavity of carbon NTs by hydrothermal treatment with NaOH aqueous solution.[401]

A solution of chromium(III) oxide in hydrochloric acid is used to fill NTs with the oxide at room temperature.[391,402] Wet chemical techniques to produce CuO nanoparticles, 20 to 90 nm in diameter and 250 to 700 nm in length, using MWNTs as templates has been developed.[403]

Gold nanoparticles can be introduced into the inner cavity of MWNT after its heat treatment in NH_3.[404]

The rapid filling of metals (Pd, Ni, and Cu) into MWNTs can be achieved using super-critical CO_2 as the reaction medium.[405] Britz et al.[406] used a metal-β-diketone complex M(hfa)$_2$ (hfa stands for hexafluoroacetylacetonate), which was reduced by H_2 dissolved in super-critical CO_2. Super-critical CO_2 solutions were used to fill SWNTs with C_{60} and $C_{61}(COOEt)_2$ molecules.

The solvent method of peapod preparation based on the refluxing of n-hexane suspension makes it possible to get a high yield of C_{60}@SWNT, C_{70}@SWNT and to synthesize N@C_{60}:C_{60}@SWNT.[407] Two others methods of nanopeapod formation consist of establishing a contact of SWNT and fullerene ethanol or toluol solutions/suspensions (Figure 2.14).[408]

A synthetic method of integrating organic molecules, such as *tetrakis*(dimethyl-amino)ethylene and tetracyanoquinodimethane into SWNTs[409] has strong potential for molecular electronics. Molecular dynamic simulations have indicated that even DNA could be encapsulated inside NTs in a water solute environment via an extremely rapid dynamic interaction process, provided that the tube size exceeds a certain critical value.[410]

Capillary filling with molten oxides,[411,412] molten metal salts, such as chlorides,[395,413–415] and some molten metals[416,417] is also widely used. Metal nanowires encapsulated in NTs have been obtained by treating SWNTs with metal salts at melting temperatures in vacuum-sealed quartz tubes, followed by reduction with hydrogen.

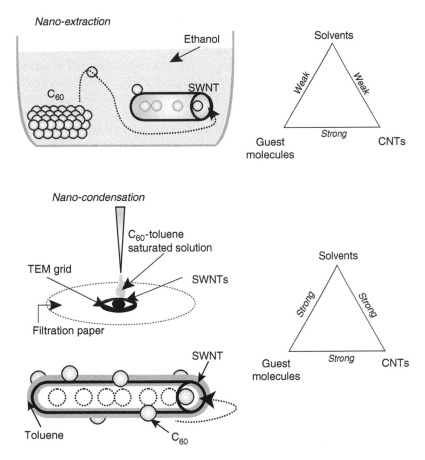

FIGURE 2.14 Schemes of nanoextraction and nanocondensation methods. The difference in relative affinities among the three components are shown in the triangles. (From Yudasaka, M. et al., *Chem. Phys. Lett.*, **380**, 42–46, 2003. With permission.)

The doping of SWNT bundles by immersion in molten iodine gives quite different results compared to using iodine vapors. This is due to the dissociative character of the melt.[418] Iodine atoms form charged polyiodide chains inside MWNTs of different inner diameter.[419]

A carbon nanotube 75 nm in diameter and 10 μm long, filled with liquid Ga, acts as a thermometer as claimed by Gao and Bando.[420] The height of the continuous, unidirectional column of Ga inside the nanotube varies linearly with temperatures ranging from 50 to 500°C.

2.6.2.2 Filling from Gas Phase

In SWNTs with open ends, nitrogen and oxygen are adsorbed first in the inside of the tubes, and next in the interstitial channels of bundles (see Section 2.7).[421] In each site, gases can be adsorbed with the stoichiometry of $C_{20}N_2$ or $C_{20}O_2$, as a monolayer.

Theoretical analysis of the behavior of some of the simplest molecules (methane, ethane, ethylene) in the NT cavity predicts normal-mode molecular diffusion for methane, and intermediate between molecular diffusion and single-file diffusion for ethane and ethylene at room temperature.[422,423]

The filling of NT with vapors of halcogenes (Se),[390] hydrides (SiH$_4$),[424] metal halogenides (ZrCl$_4$),[425] fullerenes (C$_{60}$, Sc$_2$@C$_{84}$, La@C$_{82}$),[426–429] and metallocenes (Fe(Cp)$_2$, and Ni(Cp)$_2$)[429–432] has been studied. Volatile metal carbonyls, metal diketonates, and metal phthalocyanines can also be used.

Reaction with Se in vapor phase using the two-bulb method allows to reach a filling of 50 to 80% of NTs.[390] The thermal decomposition of SiH_4 inside NTs by double-template technique leads to the formation of NTs/Si core-sheath composite.[424] Filling of SWNTs by $ZrCl_4$ is possible by heating of the mixture of components with mass ratio of 1:1, in a sealed ampoule at 623 K.[425] It is possible that under the experimental conditions (under pressure), $ZrCl_4$ formed melt and the NT filling is induced by capillary effect. Filling of opened NTs with fullerene molecules is possible by heating the mixture of components at 400 to 650°C in a sealed quartz or glass tube.[426,428] Mechanochemical activation under nitrogen atmosphere of solid-phase mixture of SWNTs and C_{60} fullerene leading to fullerene-modified NTs[433] presumably involved evaporation of fullerene.

An unusual process of Cs encapsulation inside SWNT via plasma-ion irradiation has also been demonstrated.[434]

2.6.3 REACTIONS INSIDE NANOTUBE

Many interesting chemical reactions can be performed inside NT cavities. Among these are:

- thermal decomposition of salts (e.g., metal nitrates)
- thermal decomposition and pyrolysis of volatile compounds (e.g., silane, metallocenes)
- chemical reduction of salts (e.g., metal halides or nitrates)
- air oxidation (e.g., metal halides)
- hydrolysis and pyrohydrolysis (metal halogenides, trimethylaluminum)
- polymerization (styrene, C_{60})
- formation of complex salts ($CoFe_2O_4$)
- alloy formation ($Nd_2Fe_{14}B$)
- sorption of vapors
- photolytic reduction

Some processes are complex and involve reactions of different type, e.g., thermal decomposition, reduction, and alloying. It is also possible to modify the inserted solid materials by *in situ* electron irradiation at intermediate accelerating voltages (100 to 400 kV). Electron[435] or carbon-ion[436] beam irradiation can form a connection between NTs. The structural change observed in NTs with encapsulated Ni particles is quite different as compared to empty samples, and leads to the formation of short-range fibrous or amorphous structures instead of a layered structure.[437] The crystalline–amorphous phase transformation has been observed under irradiation of encapsulated Fe.[438]

The first example of a chemical reaction inside SWNT is the H_2 reduction of $RuCl_3$ yielding Ru metal.[395] Some other examples of reactions (thermal decomposition of $AgNO_3$ to produce Ag nanowire, reduction of the salts, and CVD processes) have been mentioned earlier. One-dimensional crystals may be reduced to form metallic wires or templated 0-D nanocrystals of regulated sizes.

The reaction of trimethylaluminum and water vapor at 300°C using atomic-layer deposition technique yields Al_2O_3 nanorods inside MWNTs.[439]

Bando and coworkers[440] revealed that MWNT filled with Ga at 800°C can effectively absorb copper vapor, forming Ga–Cu alloy. The absorption rate for sealed tubes was equal to or even greater than the rate for open-ended tubes. This is presumably due to the fact that the Ga vapor pressure within a closed NT is much higher than that in an open one. and the Cu penetrates inside via defects.

Polymerization of C_{60} and generation of double-wall NTs takes place under heat treatment (the temperature must be above ~800°C) or irradiation.[116,441–443] Photopolymerization of C_{60} or C_{70} inside NT under blue-laser irradiation has also been detected.[444] A Raman investigation revealed that inner SWNTs produced by polymerization are remarkably defect-free.[443] It is interesting that in the polymerization of C_{60}, short inner NTs with diameters of ~0.7 nm are preferentially formed first. Such short NTs merge together and lengthen with diameter increasing in the course of polymerization.[445,446]

The polymerization of C_{60} fullerenes takes place also after heavy doping of the peapod system with an alkali metal.[116] A polymeric phase of K_6C_{60} has not been observed so far outside the NTs. The coalescence of Sm@C_{82} has been observed under electron irradiation.[447]

In contrast to fullerenes, in the case of $LnCl_3$ and $ZrCl_4$, the formation of clusters has been observed.[414,425]

MWNTs can be hydrothermally synthesized from different precursors in the presence of metal powder catalyst. One of the important features of some MWNTs produced in hydrothermal conditions is their large inner diameter, well suited for inner-tubular chemical reactions.[43] Since a part of these MWNTs have closed tips and contain encapsulated water and gases (supercritical mixtures of CO, CO_2, H_2O, H_2, and CH_4),[43,44,448] it was important to investigate interaction between the fluid and inner carbon walls.[376] It has been shown that strong interaction between the tube wall and aqueous mixture results in penetration of liquid between the graphitic layers, swelling of the inner wall layers, bending of graphite layers toward the tube center, and inter-calation of several inner layers with O–H species, and interlayer spacing increasing to 0.61 nm. The chemical interaction leads to dissolution of the tube wall and increase in the internal diameter of the tube in the vicinity of the liquid inclusions.

Recently, NTs with an outer diameter of ~90 nm and tube walls inclined with respect to the tube axis were found to form at 770°C and 50 to 90 MPa.[449] The conical-scroll structure of these tubes, with a high density of graphene edges at both inner and outer surfaces, enables functionalization of both surfaces. The outer wall of these NTs is shown to be covered by a hair-like layer of functional groups containing oxygen and carbon, thus making the hydrothermal NTs hydrophilic.

It is expected that nanopeapods could be used as chemical nanoreactors.[445,447]

Selectively opened NTs with 2.0 nm opening would allow many organic molecules to enter and leave. As pointed by Green,[450] rather than having a "ship-in-the-bottle" situation, one would have what could be described as a "fleet-in-the-harbor," and the idea of NT catalysis is very attractive. Encapsulation of polystyrene within MWNTs can be realized using solutions of monomer styrene and benzoyl peroxide (initiator) in super-critical CO_2.[451]

Some inner-tubular chemical and physical processes of NTs have been reviewed by Yang et al.[452]

2.6.4 THE STRUCTURE OF CRYSTALS INSIDE NANOTUBES

The restricted diameter range of NT confines the crystals formed inside the NT cavity. As nanocrystal size decreases, surface energy becomes an increasing factor in structural stability. For a given size, surface energy also influences the structural phase diagram in the temperature–pressure plane. Simple thermodynamic models, which predict a lowering of the melting point if the surface energy of the solid is higher than that of the liquid, describe the data semi-quantitatively.

The size of an NT inner cavity is usually small enough to show the dimension effect on embedded particles. The effect consists, for example, of size dependency of melting temperature of encapsulated nanocrystals.[453] But, besides enhanced surface energy, there are other factors affecting the crystal structure of material inside NT; e.g., van der Waals interaction and interaction with the NT wall.

The structure of the encapsulated materials significantly deviates from the structure of bulk materials, with lower surface coordination and substantial lattice distortion. The most noticeable studies in these fields were carried out at Oxford University and Cambridge University (see Refs. 414, 454, 455 and references therein).

The "Feynman crystal's" growth inside SWNTs is atomically regulated. One-dimensional KI having nine atoms in cross section (inside an NT of 1.6 nm diameter) and four atoms (inside NT of 1.4 nm diameter) were among the first such crystals studied.[453,456] In both cases, the crystal structure peculiar to the bulk crystal is distorted due to interaction with NT walls and change in the coordination number (CN). Especially, noticeable contraction was observed in 4 × 4 crystals, in which crystal parameters were about 4/9 of the standard ones for the bulk crystal. A novel, low-dimensional, crystal structures such as "twisted" crystal, was found.[457,458]

TABLE 2.2

Packing Behavior Observed in SWNTs as a Function of Bulk Structure Type

Halide Filling	Common Bulk Structure Type	CN in Bulk	Structure Inside SWNTs	CNs Observed or Predicted within SWNTs[a]
KI	3-D rock salt	6	Rock salt	**4, 5, 6**
AgX[b]	3-D rock salt	6	Rock salt	**4, 5, 6**
	Wurzite	4	Wurzite	**3, 4**
SrI$_2$	3-D network	7	1-D PHC[c]	**4, 6**
BaI$_2$	3-D network	7 + 2	1-D PHC	**5, 6**
PbI$_2$	2-D layered	6	1-D PHC	**5, 6**
LnCl$_3$[d]	2-D layered	6	1-D PHC	**5, 6**

[a] CNs in bold have been observed experimentally. Bulk coordination is predicted for polyhedra in the center of wide capillary SWNTs.

[b] X includes Cl or Br and I.

[c] 1-D PHC, 1-D polyhedral chain.

[d] Ln = La to Tb (UCl$_3$-type) only, Ln = Tb to Lu (PuBr$_3$-type) only.

Source: From Sloan, J. et al., *Chem. Commun.*, 1319–1332, 2002. With permission.

It was stated later that the form of the structures varies depending on whether the bulk material is a single packed structure (e.g., rocksalt or wurzite) or represents 1-D chain structure of simple coordination polyhedra (Table 2.2).

The chains of lanthanoide halides[415] and zirconium tetrachloride[425] consist of edge-shared MCl$_6$ polyhedra. "Twisted" CdCl$_2$ structure closely resembles helical iodine chains.[418] Double-helix iodine chains with the period of ~5 nm correspond to a composition of IC$_{20}$.

In the case of Sb$_2$O$_3$ filling, the material is identified as a single crystal in the valentinite form (i.e., the high-pressure modification as opposed to its senarmontite form). In comparison to the bulk structure of valentinite, the encapsulated crystal shows a longitudinal contraction of ~13% and significant lattice distortions.

The fullerenes inside NTs of appropriate diameter (close to that of (10,10) NT for C$_{60}$) form 1-D molecular chains. Two different domains, where the dominant alignment of C$_{70}$ molecular long axis is standing or lying with respect to the NT axis, have been observed.[459] The ratio of the standing to the lying C$_{70}$ domains is roughly 7:3.

Molecular dynamic simulations show that water molecules inside NTs at 300 K and 1 atm tend to organize themselves into a solid-like wrapped-around ice sheets.[384] A variety of new ice phases that are not seen in bulk ice, and a solid–liquid critical point where the distinction between solid and liquid phases disappears, are observed.[383] The water has been observed to exhibit first-order freezing transition to hexagonal and heptagonal ice nanotubes, as also a continuous phase transformation into solid-like square or pentagonal ice nanotubes.

2.7 ADSORPTION AND STORAGE OF GASES

Owing to the specific structure and morphology of NTs, there are a variety of possible sites available for adsorption. In ideal, defect-free, and bundled SWNTs, the sites having different energy of gas adsorption include (1) the internal cavity of opened SWNTs, (2) the interstitial channels between NTs in bundles, and (3) the grooves at the periphery of the bundles (Figure 2.15). Adsorptive capacity of single-wall nanohorns (nanocones) differs from the capacity of SWNTs. The surface structure of MWNTs may be more complex due to different topological defects, such as

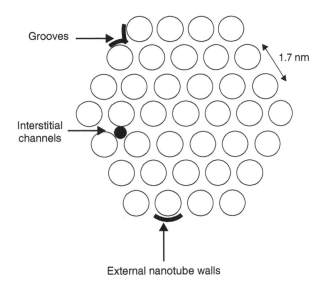

FIGURE 2.15   The three possible adsorption sites on a bundle of SWNTs with closed ends. (From Babaa, M.R. et al., *Carbon*, **42**, 1549–1554, 2004. With permission.)

pentagons and heptagons, holes and dislocations in sidewalls, etc. The structure of MWNT can be scroll-like, branched, or coiled. The tubes can form second- and third-order structures (see Section 2.2) and aligned arrays. Therefore the adsorption behavior of MWNTs is more complicated than that of SWNTs. Admixtures such as amorphous carbon, graphitic nanoparticles, or catalyst residues, attached to tube walls functionalities, exert influence upon the adsorption characteristics.

2.7.1 Hydrogen Problem

Hydrogen is an ideal fuel for the future. If hydrogen fuel cells were used to power automobiles and other vehicles, air pollution would be reduced. But the problem is not only to develop effective fuel cells or cost-effective method of hydrogen production. It is necessary to supply every vehicle with a safe, cheap, compact on-board hydrogen accumulator. The nearest goal is to achieve an adsorption capacity of 6.5 wt% or 62 kg H_2 m^{-3} at room temperature.

There are two methods to store hydrogen in NTs: one is to store it under high gas pressure, and the other is to employ an electrochemical charge–discharge cycling method.[5]

The first article on the possibility of hydrogen storage in NTs was published in 1997.[460] Dillon et al.[460] estimated that SWNTs can adsorb from 5 to 10 wt% of hydrogen under ambient conditions. In 1998, Rodriguez and coworkers[461] claimed that certain graphite NFs can store more than 50 wt% at room temperature. In 1999, Chen et al.[462] reported hydrogen uptake at 300 K and a pressure of 0.1 MPa up to 20 wt% by alkali metal-doped MWNTs. Aligned MWNTs with diameters of 50 to 100 nm show a reversible hydrogen storage capacity of 5 to 7 wt% at room temperature under 1 MPa pressure.[463] Pre-treatments of MWNTs increase the storage capacity up to 13 wt%.

Subsequent studies could not confirm these results. Other experimental data are given in Table 2.3.

Most of the work given in Table 2.3 has been done with minor quantities of highly impure and badly characterized samples of NT materials, and only some of the latest publications seemed to be accurate. Therefore the measurements using samples up to 85 g show low adsorptive capacities of MWNTs.[486] The thermal desorption spectra show that hydrogen desorbs below 300 K and at 656 K.[495] The D_2 sorption at 77 K has been found to be 0.45 wt% in raw SWNTs and 0.67 wt% in potassium-doped SWNTs.[496]

Results indicate that hydrogen uptake by metal-containing NT bundles is substantially enhanced compared with uptake by pure tubes. MWNTs seems to be more promising materials than SWNTs.

TABLE 2.3
Hydrogen Adsorption Levels in NTs

NT Type	Adsorption Conditions	Adsorption Capacity (wt%)	Year	Reference
SWNTs	300 K, 0.1 MPa, E[a]	0.35	1999	464
SWNTs	300 K, 0.1 MPa, E[a]	2.9	2000	465
SWNTs	300 K, 0.1 MPa	0.04–1.5	2001	466
SWNTs	300 K, 0.1 MPa, E[a]	1.84	2002	467
SWNTs	300 K, 0.1 MPa, E[a]	6	2002	468
SWNTs	300 K, 0.1 MPa	1.8	2004	469
SWNTs	300 K, 0.1 MPa	0.01	2004	470
SWNTs	295 K, 0.1 MPa	0.93	2002	471
SWNTs	300 K, 9 MPa	0.3	2003	472
SWNTs	300 K, 9 MPa	0.3–0.4	2004	473
SWNTs	300 K, 10 MPa	3.5	2001	474
SWNTs	300 K, 10 MPa	<0.2	2002	475
SWNTs	300 K, 12 MPa	4.2	1999	476
SWNTs	253 K, 6 MPa	1	2004	477
SWNTs	77 K, 0.04 MPa	3	2004	478
SWNTs	77 K, 0.1 MPa	1	2004	470
SWNTs	77 K, 0.1 MPa	2.37	2002	471
SWNTs	77 K, 0.2 MPa	6	2002	479
SWNTs	77 K, 1 MPa	0.7	2002	475
SWNTs	77 K, 2 MPa	0.6	2003	480
SWNTs	77 K, 2.5 MPa	2.4	2003	474
SWNTs	77 K, 10 MPa	8.2	1999	481
MWNTs	300 K, 0.1 MPa	0.25	2000	482
MWNTs	300 K, 0.1 MPa, E[a]	0.7	2000	483
MWNTs	300 K, 0.1 MPa, E[a]	0.1–1.6	2001	484
MWNTs	300 K, 0.1 MPa, E[a]	1.6	2003	74
MWNTs	300 K, 0.1 MPa, E[a], Li	0.6–2.3	2001	484
MWNTs	300 K, 0.1 MPa, Li, K	1.8–2.5	2000	485
MWNTs	300 K, 0.1 MPa	0.30	2004	486
MWNTs	300 K, 3 MPa	<0.2	2002	475
MWNTs	300 K, 10 MPa	0.7–2.4	2001	487
MWNTs	300 K, 10 MPa	0.5–2	2001	488
MWNTs	300 K, 10 MPa	1.16–2.67	2000	489
MWNTs	290 K, 10 MPa	3.4	2001	490
MWNTs	300 K, 12 MPa, KNO_3	3.2	2002	491
MWNTs	300 K, 13.5 MPa	2.5–4.6	2003	492
MWNTs	77 K, 0.1 MPa	2.27	2004	486
MWNTs	77 K, 3 MPa	~0.3	2002	475
NFs	300 K, 0.1 MPa, KNO_3	5.1	2003	493
NFs	300 K, 12 MPa	6.5	2002	494

[a] E = electrochemical storage.

The results of theoretical calculations on hydrogen sorption in NTs[5,497–515] are also controversial. It is clear that purified SWNTs store hydrogen in the molecular form. The hydrogen uptake is due to physisorption, therefore the relative amount of adsorbed hydrogen is proportional to the specific surface area of the sorbent material (1.5 wt% per 1000 m^2/g at 77 K). It has been shown that different forms of carbon are essentially the same for hydrogen molecules.[516] Close-ended SWNTs and open SWNTs have a relatively low accessible surface areas due to bundling of the tubes.

MWNTs can be used as additives to metal adsorbents.[517] Mg–5 wt% MWNTs composite absorbed 4.86 wt% (80% maximum hydrogen storage capacity) within 1 min at 553 K, under 2.0 MPa.

The problem has been discussed in the reviews.[518–527] One can trace the change from rather optimistic estimations of the prospects to use NTs as effective storage media in early publications, to more cautious conclusions and even to a pessimistic point of view in the latest ones. Today, it is still unclear whether NTs will have real practical applications in the hydrogen storage area. Addition of some metals and salts to the NTs, chemical activation, and aligning of NTs leads to an increase in adsorption capacity. However, some other problems which need to be solved include achievement of fast kinetics of discharge, high stability during cycling, low self-discharge level, etc.

2.7.2 CARBON NANOTUBE GAS SENSORS

The adsorption of different gases and vapors in NT materials as reported in publications up to year 2000 has been briefly reviewed earlier.[8] Apart from hydrogen, adsorption of noble gases,[528–532] nitrogen,[531–534] oxygen,[531,532,534–536] carbon monoxide,[534] carbon dioxide,[531,533,537–539] nitrogen dioxide,[531,540] ammonia,[531,540] water vapor,[531,534] methane,[528,531,533,541–544] acetylene,[545] acetone,[546] carbon tetrachloride,[547] carbon tetrafluoride,[530] sulfur hexafluoride,[530] methanol,[548] ethanol,[548,549] linear-chain alkanes, and fatty acids[550] has been characterized in the last 5 years. The gas transport properties of NTs has been measured.[551,552]

NTs have been found to be an effective separation media for removing some admixtures from flue gases; they can store gases having relatively high molecular mass (e.g., methane). However, the most promising field of their application is the development of new sensing devices. Sensing gaseous molecules is important in environmental monitoring, control of chemical processes and agriculture, and biological and medical applications. NT-based sensors have huge prospects in outer planetary exploration and for incorporation into yarns and fabrics.

Chemical doping of NTs, particularly of SWNTs, changes their electronic properties and can induce strong changes in conductance[48,553,554] and in thermoelectric power.[555–557] Chemical sensors are measurement devices that convert a chemical or physical property of a specific analyte into a measurable signal whose magnitude is usually proportional to the concentration of the analyte. For sensing applications, NTs have advantages, such as small size with larger surface, high sensitivity, fast response, and good reversibility.

The first sensor based on SWNTs was reported by Dai and coworkers,[558] who demonstrated that small concentrations of NO_2 produced large changes in sensor conduction. To clarify the reason for the discrepancy of recovery time between theoretical results and experimental data, the formation of adsorbed NO_3 has been postulated.[559] Recent works[560,561] show the possibility of using MWNT thin films for measuring sub-ppm NO_2 concentrations (10 to 100 ppb) in dry air, with the maximum response at 165°C. The gas-sensing properties of NT thin films depend on the nature of the defect and concentration.[562]

Gas-sensing characteristics of MWNTs as applied to humidity, NH_3, CO, and CO_2 partial pressure have been studied.[563] A practical gas sensor for ammonia and water vapor, based on measuring the variation of the electrical conductivity of MWNT ropes, has been proposed.[564] In particular, the absorption of different gases in the $MWNT/SiO_2$ layer changes the permittivity and conductivity of the material, consequently altering the resonant frequency of the sensor, which is tracked remotely, using a loop antenna.[564] Detecting properties of MWNTs for NH_3 allow the exploitation of the sensor at room temperature.[565]

The construction of gas sensors based on MWNT thin films to measure NO_2, CO, NH_3, H_2O, and C_2H_5OH concentrations has been described.[561]

The thermoelectric response due to the interaction of adsorbed molecules with NT walls can be used to detect gases such as He, N_2, and H_2.[566]

The viability of using boron- or nitrogen-doped NT films as sensors has been suggested[567] and verified.[568] It was shown that sensors fabricated using individual films of CN_x NTs have a fast response for NH_3 and reach saturation within 2 to 3 sec. The experiments also indicate a great potential in the manufacture of solvent vapor sensors, and especially for ethanol, acetone, and chloroform detection. The simple casting of SWNTs on an inter-digitated electrode allows to the fabrication of a gas sensor for detection of NO_2 and nitrotoluene.[569]

Rational chemical modification of NTs leads to the fabrication of a sensor having good molecular selectivity. A selective molecular hydrogen sensor has been formed using SWNTs decorated with Pd nanoparticles.[570] Sensing properties of NTs can be amplified using conjugated polymer composites (see Ref. 571 and references therein).

2.8 ATTACHMENT OF BIOMOLECULES

NTs are considered to be biocompatible[572–576] and attract great interest as a material for biological applications. The main fields of such applications include: (1) fabrication of bioprobes and biosensors, (2) synthesis of molecular structures for the transport of vaccines and drugs, especially water-insoluble substances into the body or to a certain body organ, (3) formation of anti-fouling surfaces (enzyme-containing composites), (4) use of NTs as a template to grow cells, and (5) creation of implantable bioelectronic devices.

2.8.1 BIOSENSORS

The development of biosensors has taken greater importance in the past years of heightened security. Biosensors are defined as analytical devices incorporating biological material (e.g., DNA, enzymes, antibodies, microorganisms, etc.) associated with or integrated within a physico-chemical transducer. In contrast to gas sensors, biosensors act in the aqueous phase, particularly in physiological solutions. NTs have definite advantages in sensing applications: small size, large specific surface, high sensitivity, fast response, and good reversibility.[577]

Amperometric sensors are based on the reaction between a biomolecule immobilized on the sensor electrode with the analyte. The reaction produces an electric current, which is proportional to the analyte concentration and can be measured.

The NT film can be easily fabricated on a glassy carbon or edge-plane carbon surfaces to form an electrode.[576] Nonfunctionalized NT electrodes have been used in electrochemical oxidation of dopamine and norepinephrine[578,579] and to determine thyroxine.[580] The NT sidewalls can be functionalized for biocompatibility and to display selective binding sites for specific biological analytes.[573] Different biomolecules, such as proteins, enzymes, and DNAs, can be attached to the NTs. The attachment can be covalent or noncovalent. The noncovalent linking of DNA is a usual way to modify NTs.[573,581–584] The method elaborated by Chen et al.[583] involves a bi-functional molecule, 1-pyrenebutanoic acid, and succinimidyl ester, irreversibly adsorbed onto the inherently hydrophobic surfaces of SWNTs in an organic solvent. Succinimidyl ester groups are highly reactive to nucleophilic substitution by primary and secondary amines that exist in abundance on the surface of most proteins.

The covalent binding of DNA with NTs has been described.[585–587] A major area of interest for biosensor developers is the study of human DNA. The NT array electrode functionalized with DNA/PNA can detect the hybridization of targeted DNA/RNA from the sample. A nanoelectrode array based on vertically aligned MWNTs embedded in SiO_2 has been reported by Li et al.[588] and Koehne et al.[589] Oligonucleotide probes were selectively attached to the ends of the MWNTs. The hybridization of subattomole DNA targets was detected by combining the electrode array with ruthenium bipyridine-mediated guanine oxidation.

Such electro-chemical assay provides enhanced daunomycin signals.[590] A MWNTs–COOH-modified glassy carbon electrode was fabricated and oligonucleotides with 5′-amino groups were covalently bonded to the carboxyl group of NTs.

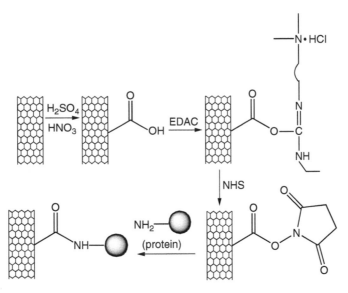

FIGURE 2.16 The attachment of proteins to NTs via a two-step process of diamide-activated amidation. (From Jiang, K. et al., *J. Mater. Chem.*, 14, 37–39, 2004. With permission.)

Trace levels of oligonucleotides and polynucleotides can be readily detected following short accumulation periods with detection limits of 25, 60, 126, and 219 µg/L for oligo(dG)(21), oligo(dG)(11), ss, and da calf thymus DNA, respectively.[591]

A novel glucose biosensor based on NT-modified electrode is also of great practical interest. Glucose oxidase (GOx) is a flavine enzyme used commercially on a massive scale to monitor blood glucose levels in diabetics. The GOx adsorbs spontaneously on NT sheets and can be reversibly oxidized without complete loss of enzymatic activity.[592] A vertical array of MWNTs with GOx molecules is adsorbed nonspecifically on the exposed ends.[593] The electrode exhibits an amperometric response in the presence of varying glucose concentrations. Different methods of GOx immobilization onto NTs have been described.[594–596]

SWNTs fabricated into field effect transistors have several advantages over conventional sensor systems, providing high sensitivity and allowing real-time detection.[597] This approach has been used to fabricate biosensors for streptavidin recognition.[598] NT coated with a polymer functional layer was functionalized with protein biotin, thus providing specific molecular detection.

The immobilization of different proteins and enzymes onto NTs for the development of new biosensors has been described.[599–606] Two different methods were employed for NTs functionalization with bioactive peptides:[607] (1) fragment condensation of fully protected peptides and (2) selective chemical ligation. The latter is shown in Figure 2.16.

Biosensors can be applied for detecting of diseases, particularly in cancer diagnostics.[608,609]

Detection of organophosphorus pesticides and nerve gases by biosensors is possible.[610] Detection of pathogens and toxins, and direct determination of total cholesterol in blood, has been studied. Bacterial binding has been used to create biosensors.[611] The capture of specific bacteria by particular antibodies can be detected by different sensor platforms, e.g., by acoustic wave sensors, using piezoelectric resonator. NTs can also be potentially used as ion channel biosensors.[388]

The study of the kinetics and thermodynamics of biological redox processes by biosensors is an important part of biochemistry.[576,612] As per theory, NIR-radiation-excited NT ropes could control the activity of proteins *in vivo*.[613]

2.8.2 OTHERS FIELDS OF APPLICATION

Enzyme-containing polymer/SWNT composites have been explored as unique biocatalytic materials.[614] NTs functionalized with bioactive peptides act as a medium to transfer peptides across the cell membranes.[615]

MWNTs functionalized with hemin are able to detect oxygen in solution.[616,617] Bucky paper can be used for transplanting cells into the retina. In experiments, the substrate worked as a scaffold for the growth of the retinal cells taken from white rabbits, for implanting in other rabbits. The NTs served as substrates for neuronal growth.[618,619]

Some biomolecules solubilize NTs. Lipid rings[620] and DNA[621] have been studied as solubilizers to dissolve NTs in aqueous solutions. The solubility of β-galactoside-modified SWNTs in water was shown to be increased by adsorption of lectin molecules.[622] NTs in the presence of galactose-specific lectins form micrometer-sized nanostructures.

Solubilization using single-stranded DNA allows the purification and separation of NTs.[623,624] By screening a library of oligonucleotides, scientists found that a certain sequence of single-stranded DNA self-assembles into a helical structure around individual NTs. Since NT/DNA hybrids have different electrostatic properties that depend on the NT diameter and electronic structure, they can be separated and sorted using anion-exchange chromatography. DNA-stabilized dispersions of NTs prepared by Barisci et al.[625] are concentrated (up to 1%, by mass) and are better suited for fiber spinning than conventional surfactants.

Selective localization of SWNTs on aligned DNA molecules on surfaces is an important tool in bottom-up bio-templated nanofabrication.[626] Self-assembly based on molecular recognition could be the best approach for constructing complex architectures (e.g., field-effect transistor[627]) for miniature biological electronics and optical devices. The attachment of DNA to oxidatively opened ends of MWNT arrays[587] and the DNA-guided assembly of NTs[628] have been studied.

2.9 NANOTUBES AS TEMPLATES

2.9.1 SUBSTITUTION OF THE CARBON ATOMS OF NANOTUBES

The incorporation of some chemical elements in carbon NTs leads to a new nanomaterial without a significant modification of the NT structure, but with possible improvement of their properties. Substitution of carbon atoms with nitrogen or boron atoms is most interesting from a practical standpoint. It allows the formation of hetero-junctions within a single NT, and opens a way to use such a tube in electronic devices. Besides, substitution with silicon has also been studied.

Tight-binding calculations show that the presence of nitrogen atoms is responsible for introducing donor states near Fermi level.[629,630] Nitrogen-doped carbon NTs have been found to be metals[630,631] (earlier it was stated that the modified tubes were metals or semiconductors depending on the relative position of nitrogen and carbon atoms[632]).

Arc-discharge process[633,634] and pyrolysis[629,635-646] are the main methods of nitrogen-substituted carbon NT synthesis. The third method is carbon-resistive heating under high isostatic nitrogen pressure.[647]

The incorporation of nitrogen into carbon NTs grown by arc-discharge method takes place in an helium–nitrogen atmosphere[633] or by introducing nitrogen-rich organic or inorganic precursors into the anode rods.[634] Metal phthalocyanines,[636,641] methane and nitrogen mixtures,[637-639,646] ferrocene and melamine mixtures,[629,640] acetylene/ammonia/iron carbonyl mixtures,[642] acetylene/ammonia/hydrogen mixtures,[643] and acetonitrile or acetonitrile/hydrogen mixtures[644,645] have been used in CVD processes.

The amount of incorporated nitrogen varies from some tenths of a percent to 9%[641] and even 13%.[647] The nitrogen incorporation enhances the growth of NTs by the CVD process and induces visible changes in the NT structure and morphology. Overall, the presence of nitrogen suppresses the formation of bundles, changes the tube conformation, and leads to the NT texturing.

All boron-doped carbon NTs exhibit highly metallic electronic character. Boron has been shown to play a key role at the open end of a growing carbon NT, thus increasing its overall length.

Boron-substituted NTs are formed by arc-discharge[634,648,649] or CVD[650,651] methods. Moreover, solid-state[652–655] and gas–solid[656,657] reactions are also employed to prepare the material.

It is possible to reach boron-substitutional level up to 20%[653,654] and synthesize carbon-free BN nanotubes.[656,658] The methods described allow the preparation of B–C–N NTs[635] and NFs.[634,658–661]

The determined properties of silicon-doped carbon NTs and a suggestion how such tubes can be synthesized have been published.[662,663] The interaction of carbon NTs with SiO_2 or Si yields SiC nanorods (see the next section). Some substitution reactions have been developed to produce hetero-structures of SWNTs and metal-carbide nanorods.[664]

2.9.2 DECORATION OF CARBON NANOTUBES

Surface-modified carbon NTs, inorganic NTs, and nanorods are of great significance in various practical applications, such as catalysis, electrocatalysis, photocatalysis, ion-exchange and gas sorption, electron emitter and nanodevice fabrication, and production of composites. "Decoration" is often referred to as a process of covering NTs with a substance that does not form strong chemical bonds with carbon atoms. At the same time, several decoration processes yield a covering substance linked by weak chemical or electrostatic forces to the NTs.

NTs can be decorated with metals (Cu, Ag, Au, Al, Ti, Ni, Pt, Pd), alloys (Co–B, Ni–P, Mo–Ge), nonmetals (Se), metal oxides (ZnO, CdO, Al_2O_3, CeO_2, SnO_2, SiO_2, TiO_2, V_2O_5, Sb_2O_5, MoO_2, MoO_3, WO_3, RuO_2, IrO_2), metal chalcogenides (ZnS, CdS, CdSe, CdTe), metal carbides (SiC), metal nitrides (SiN_x, AlN), and polymers (polyaniline, polypyrrole). The list of such substances can go on. It is possible to prepare the covering of NTs in the form of quantum dots and nanoparticles, to isolate nanotubules or nanorods and nanowires. The surface structure of NTs, and the strategy and extent of deposition greatly influence the morphology of a coating.

The approaches to decoration of NTs include physical evaporation, electroless deposition, electroplating, CVD, sol–gel process, chemical attachment of metal ions or complexes with subsequent reduction or thermal decomposition, solid-state reactions, solid–gas reactions, self-assembling, polymerization, and others chemical processes.

The decoration of NTs with nickel is achieved by electroless plating method,[665–668] by electrochemical deposition process,[669,670] by chemical reduction of $NiCl_2$ with hydrogen,[671] or by electron-beam evaporation of the metal.[672,673] Copper has been deposited onto NTs by electroless plating[667] or by reduction of $CuCl_2$ with hydrogen.[674] Cobalt has been plated onto NTs via electroless plating[675] or by reduction of $Na_2Co(OH)_4$ using KBH_4 solution.[676] Electroless plating usually demands preliminary chemical activation of a substrate.

Several procedures have been used to decorate acid-treated NTs with nanoscale clusters of Ag, Au, Pt, and Pd. For instance, Satishkumar et al.[677] carried out the refluxing with $HAuCl_4$ and HNO_3 or *tetrakis* hydroxymethyl phosphonium chloride, with H_2PtCl_6 and HNO_3 or ethylene glycol, and with $AgNO_3$ and HNO_3; Li et al.[678] refluxed NTs for 6 h with H_2PtCl_6 dissolved in ethylene glycol. The activity of Pt cathode catalyst for direct methanol fuel cells has been shown to be dependent on the ratio of water/ethylene glycol.[679] Pt/NTs composite prepared by reduction of H_2PtCl_6 with $Na_2S_2O_6$ in water/alcohol solution exhibited good catalytic properties.[69] It is possible to use sorption of Pd^{2+} ions from solution with subsequent hydrogen reduction.[680,681] Contact of an acetone solution of H_2PtCl_6 with NTs allows production of NTs decorated with platinum clusters without any surface functionalization.[682] Decomposition and reduction of H_2PtCl_6, $HPdCl_3$, $HAuCl_4$, or $AgNO_3$ dispersed on NTs at temperatures from 300 to 700°C gives metal nanoparticles with average size of 7 to 17 nm.[683] Nanoparticle/NT hybrid structures have been prepared by forming Au and Pd nanoparticles on the sidewalls of SWNTs using reducing reagents or catalyst-free electroless deposition, as a result of direct redox reaction between metal ions and NTs.[684] Because Au^{3+} and Pt^{2+} have much higher reduction potential than NTs, they are reduced spontaneously. Hydrogen reduction of a Pd(II)-β-diketone precursor in super-critical carbon dioxide produces Pd nanoparticles on MWNTs.[405,685]

As for gold and silver nanoparticles, the usual thiol-linking procedure can be used to anchor them to NTs.[686–688] Silver clusters on SWNT surface have been grown by the decomposition of (cycloocta-1,5-diene)(hexafluoroacetylacetonate)silver(I).[686] A highly selective electroless deposition of gold nanoparticles on SWNTs has been reported.[684] Sonication allows the deposition of gold onto MWNTs directly from gold colloid solution.[689] A method to fabricate gold nanowires using NTs as positive templates has been demonstrated.[690] The first step was to self-assemble gold nanocrystals along NTs. After thermal treatment, the nanocrystal assemblies were transformed into continuous polycrystalline gold nanowires. Monolayer-protected Au clusters strongly adsorb onto both end-opened and solubilized when refluxed in aniline SWNTs.[195]

Formation of different metal nanowires using NTs as positively charged template has been described.[691]

To form a SiO_2 or SiO_x coating on NTs as well as to prepare silica nanotubes, silica nanorods and silica/NT composites, sol–gel technique has been employed by many researchers.[692–696] Tetraethoxysilane (tetraethyl-o-silicate) was the usual precursor in this process. Silica-coated SWNTs have also been produced by adding silica/H_2SiF_6 solution to a surfactant-stabilized dispersion of SWNTs[697] or by hydrolysis of 3-aminopropyltriethoxysilane.[698]

NTs have been coated with titania by using titanium *bis*-ammonium lactato dihydroxide,[699] tetraethyl-o-titanate,[696,700] titanium tetraisopropoxide,[147,700] titanium oxysulfate,[700] and titanium tetrachloride[701] as precursors. Sol–gel method has been used to sheathe NT tips for field emission electron emitters (Figure 2.17).[702] The coating morphology depends on the precursor composition and method of deposition.[700] Covalent linking of TiO_2 nanocrystals to SWNTs by short-chain organic molecule linker is demonstrated.[147]

Coating of NTs with alumina[692,696,703] or $Al(OH)_3$[704] is possible by either thermal or chemical decomposition of aluminum sources, such as aluminum isopropoxide, aluminum trichloride, or aluminum nitrate. Oxidation of a powdered aluminum metal/MWNT mixture with O_2 or air leads to the formation of Al_2O_3 nanotubes or nanowires.[705] Treating the same mixture in NH_3 atmosphere at 300 to 500°C yields AlN nanowires or particles.[705]

SWNTs and MWNTs can be coated with a thin or thick film of SnO_2.[706–708] Heat treatment of MWNTs with zinc at various temperatures gives ultra-thin films of ZnO on the tubes, ZnO quantum dots, or nanowires.[709] Deposition of CeO_2 nanoparticles on the surface of MWNTs by hydrolysis of $CeCl_3$ in aqueous solution has been described.[710] Covering of NTs with oxides of different metals (V, Sb, Mo, Ir, Ru) has been characterized.[711,712]

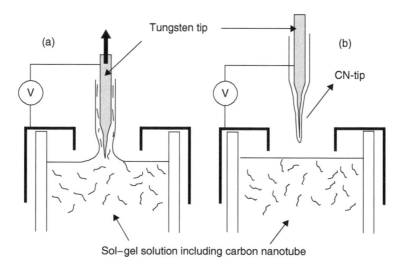

FIGURE 2.17 Schematic illustration of the sol–gel coating process under DC voltage. (From Brioude, A. et al., *Appl. Surf. Sci.*, **221**, 4–9, 2004. With permission.)

Semiconductor metal chalcogenide nanoparticles such as CdS,[713] CdSe,[147,205,714,205] CdTe,[148] and ZnS[715,714] have been bound to the surfaces of NTs. The example of an attachment process is presented in Figure 2.18.

Reaction of carbon NTs and carbon NFs with silicon[716,717] as well as with silica[718] allows the production of SiC coatings, NFs or nanorods. The plasma enhanced CVD (PECVD) method using mixtures of SiH_4 with C_2H_2 or with NH_3 introduced into the reaction chamber at 250°C has been used to coat MWNTs with amorphous films of SiC or SiN_x.[719] A process of simultaneous growth of carbon NTs and SiC nanorods has been realized by means of CVD.[720]

The aligned NTs can be used to make conducting polymer/NT coaxial nanowires by electrochemical deposition of a concentric layer of conducting polypyrrole.[721] Ultra-thin films of pyrrole were deposited on the surfaces of MWNTs using a plasma polymerization technique.[722] Polymer nanoshells on NTs can be formed via layer-by-layer deposition on NT template.[723]

Interesting and unusual nanowires of rare-earth phthalocyanine have been produced via the templated assembly method.[724] When the solvent is evaporated gradually, $RErPc_2$ nanoparticles and nanowires are found to self-assemble on the exterior walls of NTs.

2.10   INTERCALATION OF "GUEST" MOIETIES

Similar to graphite, NTs can form guest–host compounds with electron acceptors (e.g., bromine, iodine, interhalogens, $FeCl_3$, HNO_3) and electron donors (alkali metals).[8,16,18] The guest species can be inserted between the graphene shells of MWNTs or in the open inter-tubular spaces inside the bundles of SWNTs. Reversible inter-calation proceeds by both direct vapor/liquid contacting and electrochemical insertion of ions. The doping is accompanied by charge transfer and enhancement of NT conductivity.

The inter-calation of bromine into MWNTs at 50°C shows an unusual ordered accumulation along the direction perpendicular to the NT surface.[725] Iodine reversibly forms charged polyiodide chains into SWNT bundles,[726] which is similar to the chains in iodine-doped SWNTs[418] and MWNTs.[419]

The behavior of MWNTs depends on their structure. In addition to $FeCl_3$, other metal chlorides ($ZnCl_2$, $CdCl_2$, $AlCl_3$, and YCl_3) formed inter-calation compounds with MWNTs having scroll structure.[18]

The HNO_3 molecules intercalate into the bundles of SWNT,[727] and HNO_3, H_2SO_4, and HCl intercalate into the SWNT film ("bucky paper").[728] In all cases, an acceptor-type doping of the SWNTs has been observed.

Fluorine, as was mentioned earlier (see Section 2.5.3), forms strong semi-ionic or covalent bonds with carbon atoms of NTs and cannot be deintercalated in diatomic molecular form.

Much more information has been obtained on NT doping with alkali metals. Some theoretical models of inter-calation products have been offered.[16,18] Potassium and other heavy-weight alkali metals at saturation state form MC_8 compounds corresponding to the first-stage graphite intercalation compounds (GIC).[16,18] The doping of SWNTs is diameter-selective.[729,730] The dopant atoms initially deposited on the tips of aligned MWNTs diffuse toward their roots at extremely small rates.[731]

The rapid development of lithium-ion batteries, which use graphite or carbonaceous materials as one of the electrodes, gives rise to a great interest in NT material inter-calation with lithium.[5,522] A significant reversible capacity has been found.[732] SWNTs can be reversibly intercalated with lithium up to $Li_{1.7}C_6$ composition.[733] The reversible lithium storage capacity after etching of SWNTs increases to Li_2C_6.[734,735] Moreover, this saturation composition increases to $Li_{2.7}C_6$ by applying a suitable ball-milling treatment to the purified SWNTs.[736] It means that SWNTs can contain roughly twice the energy density of graphite. To explain the results of Refs. 732 and 736, a non-GIC mechanism of storage has been proposed.[737] In this mechanism, except for the lithium intercalated between graphene layers to form GIC, the higher charge capacity should be related to lithium doped into disordered graphitic structures, defects, microcavities, and edge of graphitic layers.

The reversible capacity increases as compared to that of raw NTs after treatment with lithium compounds such as $LiNO_2$.[738] The lithium diffusion coefficient decreases with an increase of the

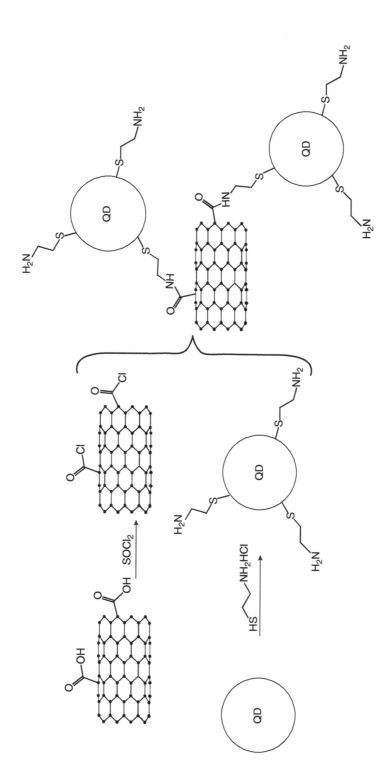

FIGURE 2.18 Quantum dots (QD) attachment at cut ends or at defects along NT sidewall. (From Haremza, J.M. et al., *Nano Lett.*, **2**, 1253–1258, 2002. With permission.)

open-circuit voltage. This phenomenon is presumably connected with a change of composition and properties of thin inorganic film covering the NTs.

2.11 SUMMARY AND CONCLUSIONS

Carbon NTs represent a set of materials including numerous variants with different structure, morphology, and properties. Some of these materials are now coming to the industrial arena as new attractive resources with many remarkable properties and dozens of promising applications. Several scaleable continuous chemical methods of NT production have been elaborated and scaled.

The practical significance of carbon NT chemistry and its achievements is sharply increased. Today, this branch of NT "general chemistry" subdivides into carbon NT inorganic, organic, bioorganic, colloidal, and polymeric chemistry. NT chemistry now occurs in solution, in such a way that it opens many new possibilities to purify, sort, and modify the material and to integrate it in electronic devices.

Carbon NTs are among the most important materials of modern nanoscience and nanotechnology, including molecular electronics, and the chemistry of NTs play a remarkable role in the development of this interdisciplinary area.

ACKNOWLEDGMENTS

The author is grateful to Mr. I. Anoshkin and Mr. Nguyen Tran Hung for their helpful assistance.

REFERENCES

1. J.-M. Bonard, H. Kind, T. Stöckli, and L.-O. Nilsson, Field emission from carbon nanotubes: the first five years, *Solid-State Electron.* **45**, 893–914 (2001).
2. R.H. Baughman, A.A. Zakhidov, and W.A. de Heer, Carbon nanotubes — the route toward applications, *Science* **297**, 787–792 (2002).
3. Ph. Avouris, Carbon nanotube electronics, *Chem. Phys.* **281**, 429–445 (2002).
4. T.W. Odom, J.-L. Huang, and C.M. Lieber, Single-walled carbon nanotubes: from fundamental studies to new device concepts, *Ann. N.Y. Acad. Sci.* **960**, 203–215 (2002).
5. M. Daenen, R.D. de Fouw, B. Hamers, P.G.A. Janssen, K. Schouteden, and M.A.J. Veld, *The Wondrous World of Carbon Nanotubes. A Review of Current Carbon Nanotube Technologies.* Eindhoven University of Technology, 2003, p. 93.
6. *Nanotubes: Technology and Directions.* Business Communication Co., Inc., 2003, p. 280 [http://www.marketresearch. com/map/prod/967076.html].
7. E.G. Rakov, Methods for preparation of carbon nanotubes, *Russ. Chem. Rev.* **69**, 35–52 (2000).
8. E.G. Rakov, The chemistry and application of carbon nanotubes, *Russ. Chem. Rev.* **70**, 827–863 (2001).
9. O. Zhou, H. Shimoda, B. Gao, S. Oh, L. Fleming, and G. Yue, Materials science of carbon nanotubes: fabrication, integration, and properties of macroscopic structures of carbon nanotubes, *Acc. Chem. Res.* **35**, 1045–1053 (2002).
10. J. Hilding, E.A. Grulke, Z.G. Zhang, and F. Lockwood, Dispersion of carbon nanotubes in liquids, *J. Disp. Sci. Technol.* **24**, 1–41 (2003).
11. S. Niyogi, M.A. Hamon, H. Hu, B. Zhao, P. Bhowmik, R. Sen, M.E. Itkis, and R.C. Haddon, Chemistry of single-walled carbon nanotubes, *Acc. Chem. Res.* **35**, 1105–1113 (2002).
12. Z. Chen, W. Thiel, and A. Hirsch, Reactivity of the convex and concave surfaces of single-walled carbon nanotubes (SWNTs) toward addition reactions: dependence on the carbon-atom piramidalization, *Chem. Phys. Chem.* **4**, 93–97 (2003).
13. T. Yu. Astakhova, G.A. Vinogradov, O.D. Gurin, and M. Menon, Effect of local strain on the reactivity of carbon nanotubes, *Russ. Chem. Bull.* **51**, 704–708 (2002).
14. K.D. Ausman, H.W. Rohrs, M. Yu, and R.S. Ruoff, Nanostressing and mechanochemistry, *Nanotechnology* **10**, 258–262 (1999).

15. D. Srivastava, D.W. Brenner, J.D. Schall, K.D. Ausman, M.F. Yu, and R.S. Ruoff, Predictions of enhanced chemical reactivity at regions of local conformational strain on carbon nanotubes: kinky chemistry, *J. Phys. Chem.* **B 103**, 4330–4337 (1999).

16. J.E. Fischer, Chemical doping of single-wall carbon nanotubes, *Acc. Chem. Res.* **35**, 1079–1086 (2002).

17. Y.-P. Sun, K. Fu, Y. Lin, and W. Huang, Functionalized carbon nanotubes: properties and applications, *Acc. Chem. Res.* **35**, 1096–1104 (2002).

18. L. Duclaux, Review of the doping of carbon nanotubes (multiwalled and single-walled), *Carbon* **40**, 1751–1764 (2002).

19. A. Hirsch, Functionalization of single-walled carbon nanotubes, *Angew. Chem. Int. Edit.* **41**, 1853–1859 (2002).

20. L. Dai, A. Patil, X. Gong, Z. Guo, L. Liu, Y. Liu, and D. Zhu, Aligned nanotubes, *Chem. Phys. Chem.* **4**, 1150–1169 (2003).

21. T. Lin, V. Bajpai, T. Ji, and L. Dai, Chemistry of carbon nanotube, *Aust. J. Chem.* **56**, 635–651 (2003).

22. J.-M. Ting and J.B.C. Lan, Beaded carbon tubes, *Appl. Phys. Lett.* **75**, 3309–3311 (1999).

23. K.De Jong and J.W. Geus, Carbon nanofibers: catalytic synthesis and applications, *Catal. Rev. Sci. Eng.* **42**, 481–510 (2000).

24. C.T. Kingston and B. Simard, Fabrication of carbon nanotubes, *Anal. Lett.* **36**, 3119–3145 (2003).

25. C. Journet, W.K. Maser, P. Bernier, A. Loiseau, M. Lamy de la Chapelle, S. Lefrant, P. Deniard, R. Lee, and J.E. Fischer, Large-scale production of single-walled carbon nanotubes by the electric-arc technique, *Nature* **388**, 756–758 (1997).

26. Z. Shi, Y. Lian, X. Zhou, Z. Gu, Y. Zhang, S. Iijima, L. Zhou, K.T. Yue, and S. Zhang, Mass-production of single wall carbon nanotubes by arc discharge method, *Carbon* **37**, 1449–1453 (1999).

27. M. Takizawa, S. Bandow, M. Yudasaka, Y. Ando, H. Shimoyama, and S. Iijima, Change of tube diameter distribution of single-wall carbon nanotubes induced by changing the bimetallic ratio of Ni and Y catalysts, *Chem. Phys. Lett.* **326**, 351–357 (2000).

28. M. Cadek, R. Murphy, B. McCarthy, A. Drury, B. Lahr, R.C. Barklie, M. in het Panhuis, J.N. Coleman, and W.J. Blau, Optimization of the arc-discharge production of multi-wall carbon nanotubes, *Carbon* **40**, 923–928 (2002).

29. A.V. Krestinin, N.A. Kiselev, A.V. Raevskii, A.G. Ryabenko, D.N. Zakharov, and G.I. Zvereva, Perspectives of single-wall carbon nanotube production in the arc discharge process, *Euras. Chem. Tech. J.* **5**, 7–18 (2003).

30. T. Guo, P. Nikolaev, A. Thess, D.T. Colbert, and R.E. Smalley, Catalytic growth of single-walled nanotubes by laser vaporization, *Chem. Phys. Lett.* **243**, 49–54 (1995).

31. M. Yudasaka, T. Komatsu, T. Ichihashi, and S. Iijima, Single-wall carbon nanotube formation by laser ablation using double-targets of carbon and metal, *Chem. Phys. Lett.* **278**, 102–106 (1997).

32. O. Jost, A.A. Gorbunov, J. Möller, W. Pompe, A. Graff, R. Friedlein, X. Liu, M.S. Golden, and J. Fink, Impact of catalyst coarsening on the formation of single-wall carbon nanotubes, *Chem. Phys. Lett.* **339**, 297–304 (2001).

33. D.T. Colbert and R.E. Smalley, Past, present and future of fullerene nanotubes: buckytubes. *Perspectives of Fullerene Nanotechnology*. Ed. E. Osawa. Kluwer Academic Publishers, 2002, pp. 3–10.

34. W.K. Maser, A.M. Benito, and M.T. Martínez, Production of carbon nanotubes: the light approach, *Carbon* **40**, 1685–1695 (2002).

35. D. Laplaze, P. Bernier, W.K. Maser, G. Flamant, T. Guillard, and A. Loiseau, Carbon nanotubes: the solar approach, *Carbon* **36**, 685–688 (1998).

36. M.V. Antisari, R. Marazzi, and R. Krsmanovic, Synthesis of multiwall carbon nanotubes by electric arc discharge in liquid environments, *Carbon* **41**, 2393–2401 (2003).

37. N. Sano, J. Nakano, and T. Kanki, Synthesis of single-walled carbon nanotubes with nanohorns by arc in liquid nitrogen, *Carbon* **42**, 686–688 (2004).

38. Y.L. Hsin, K.C. Hwang, F.-R. Chen, and J.-J. Kai, Production and in-situ metal filling of carbon nanotubes in water, *Adv. Mater.* **13**, 830–833 (2001).

39. H.W. Zhu, X.S. Li, B. Jiang, C.L. Xu, Y.F. Zhu, D.H. Wu, and X.H. Chen, Formation of carbon nanotubes in water by the electric-arc technique, *Chem. Phys. Lett.* **366**, 664–669 (2002).

40. K. Shimotani, K. Anazawa, H. Watanabe, and M. Shimizu, New synthesis of multi-walled carbon nanotubes using an arc discharge technique under organic molecular atmospheres, *Appl. Phys.* **73 A**, 451–454 (2001).

41. V.A. Ryzhkov, On producing carbon nanotubes by a self-regulated electric contact arc discharge in hydrocarbon liquids. *NANOTUBE'02 Workshop,* Abstr. P-103 [http://dielc.kaist.ac.kr/nt02/abstracts/P103.shtml].

42. V.A. Ryzhkov, Use of self-regulated arc discharge in hydrocarbon liquids for bulk production of carbon nanotubes, *Int. Conf. Sci. Applic. Nanotubes 2003,* July 7–11, 2003, Seoul, Korea [http://nt03.skku.ac.kr/abstract/view.html?idx= NTqRCh4KAnABY].

43. Y. Gogotsi, J.A. Libera, and M. Yoshimura, Hydrothermal synthesis of multiwall carbon nanotubes, *J. Mater. Res.* **15**, 2591–2594 (2000).

44. J.A. Libera and Y. Gogotsi, Hydrothermal syntyesis of graphite tubes using Ni catalyst, *Carbon* **39**, 1307–1318 (2001).

45. D.C. Lee, F.V. Mikulec, and B.A. Korgel, Carbon nanotube synthesis in supercritical toluene, *J. Am. Chem. Soc.* **126**, 4951–4957 (2004).

46. P. Nikolaev, M.J. Bronikowski, R.K. Bradley, F. Rohmumd, D.T. Colbert, K.A. Smith, and R.E. Smalley, Gas-phase catalytic growth of single-walled carbon nanotubes from carbon monoxide, *Chem. Phys. Lett.* **313**, 91–97 (1999).

47. R. Andrews, D. Jacques, D. Qian, and T. Rantell, Multiwall carbon nanotubes: synthesis and application, *Acc. Chem. Res.* **35**, 1008–1017 (2002).

48. H. Dai, Carbon nanotubes: synthesis, integration, and properties, *Acc. Chem. Res.* **35**, 1035–1044 (2002).

49. K.B.K. Teo, C. Singh, M. Chhowalla, and W.I. Milne, Catalytic synthesis of carbon nanofibers. *Encyclopedia of Nanoscience and Nanotechnology.* Ed. H.S. Nalwa. Vol. 10, American Sci. Pub. 2003, pp. 1–22.

50. S. Maruyama, R. Kojima, Y. Miyauchi, S. Chiashi, and M. Kohno, Low-temperature synthesis of high-purity single-walled carbon nanotubes from alcohol, *Chem. Phys. Lett.* **360**, 229–234 (2002).

51. CoMoCat Process (University of Oklahoma, School Chem. Eng. Mater. Sci.) [http://www.ou.edu/engineering/nanotube/comocat.html].

52. M. Kusunoki, T. Suzuki, C. Honjo, T. Hirayama, and N. Shibata, Selective synthesis of zigzag-type aligned carbon nanotubes on SiC (000-1) wafers, *Chem. Phys. Lett.* **366**, 458–462 (2002).

53. T. Nagano, Y. Ishikawa, and N. Shibata, Preparation of silicon-on-insulator substrate on large free-standing carbon nanotube film formation by surface decomposition of SiC film, *Jpn. J. Appl. Phys.* **42**, 1717–1721 (2003).

54. Y.L. Li, Y.D. Yu, and Y. Liang, A novel method for synthesis of carbon nanotubes: low temperature solid pyrolysis, *J. Mater. Res.* **12**, 1678–1680 (1997).

55. S. Maruyama, Y. Miyauchi, T. Edamura, Y. Igarashi, S. Chiashi, and Y. Murakami, Synthesis of single-walled carbon nanotubes with narrow diameter-distribution from fullerene, *Chem. Phys. Lett.* **375**, 553–559 (2003).

56. T.S. Wong, C.T. Wang, K.H. Chen, L.C. Chen, and K.J. Ma, Carbon nanotube growth by rapid thermal processing, *Diamond Rel. Mater.* **10**, 1810–1813 (2001).

57. R. Katoh, Y. Tasaka, E. Sekreta, M. Yumura, F. Ikazaki, Y. Kakudate, and S. Fujiwara, Sonochemical production of a carbon nanotube. *Ultrason. Sonochem.* **6**, 185–187 (1999).

58. F. Fabry, T.M. Gruenberger, J. Gonzalez-Aguilar, H. Okuno, E. Grivei, N. Probst, L. Fulcheri, G. Flamant, and J.-C. Charlier, Continuous mass production of carbon nanotubes by 3-phase AC plasma processing, *Nanotech 2004, Vol. 3, Techn. Proc. of the 2004 NSTI Nanotechnology Conf. and Trade Show,* 228–231 (2004). March 7–11, Boston Sheraton Hotel & Copley Convention Center, Boston, MA.

59. Z. Jia, Z. Wang, J. Liang, B. Wei, and D. Wu, Production of short multi-walled carbon nanotubes, *Carbon* **37**, 903–906 (1999).

60. E. Farkas, M.E. Anderson, Z. Chen, and A.G. Rinzler, Length sorting *cut* single wall carbon nanotubes by high performance liquid chromatography, *Chem. Phys. Lett.* **363**, 111–116 (2002).

61. E. Borowiak-Palen, T. Pichler, X. Liu, M. Knupfer, A. Graff, O. Jost, W. Pompe, R.J. Kalenczuk, and J. Fink, Reduced diameter distribution of single-wall carbon nanotubes by selective oxidation, *Chem. Phys. Lett.* **363**, 567–572 (2002).

62. C. Bower, A. Kleinhammes, Y. Wu, and O. Zhou, Intercalation and partial exfoliation of single-walled carbon nanotubes by nitric acid, *Chem. Phys. Lett.* **288**, 481–486 (1998).

63. K.H. An, K.K. Jeon, J.-M. Moon, S.J. Eum, C.W. Yang, G.-S. Park, C.Y. Park, and Y.H. Lee, Transformation of single-walled carbon nanotubes to multi-walled carbon nanotubes and onion-like structures by nitric acid treatment, *Synt. Met.* **140**, 1–8 (2004).

64. K.B. Shelimov, R.O. Esenaliev, A.G. Rinzler, C.B. Huffman, and R.E. Smalley, Purification of single-wall carbon nanotubes by ultrasonically assisted filtration, *Chem. Phys. Lett.* **282**, 429–434 (1998).

65. A. Kukovecz, Ch. Kramberger, M. Holzinger, H. Kuzmany, J. Schalko, M. Mannsberger, and A. Hirsch, On the stacking behavior of functionalized single-wall carbon nanotubes, *J. Phys. Chem.* **B** 106, 6374–6380 (2002).

66. J. Liu, A.G. Rinzler, H. Dai, J.H. Hafner, R.K. Bradley, P.J. Boul, A. Lu, T. Iverson, K. Shelimov, C.B. Huffman, F. Rodriguez-Macias, Y.-S. Shon, T.R. Lee, D.T. Colbert, and R.E. Smalley, Fullerene pipes, *Science* 280, 1253–1255 (1998).

67. Y. Yang, J. Zhang, X. Nan, and Z. Liu, Toward the chemistry of carboxylic single-walled carbon nanotubes by chemical force microscopy, *J. Phys. Chem.* **B** 106, 4139–4144 (2002).

68. Y. Yang, H. Zou, B. Wu, Q. Li, J. Zhang, Z. Liu. X. Guo, and Z. Du, Enrichment of large-diameter single-walled carbon nanotubes by oxidative acid treatment, *J. Phys. Chem.* **B** 106, 7160–7162 (2002).

69. J. Chen, C. Xu, Z. Mao, G. Chen, B. Wei, J. Liang, and D. Wu, Fabrication of Pt deposited on carbon nanotubes and performance of its polymer electrolyte membrane fuel cells, *Sci. China* 45, 82–86 (2002).

70. F. Simon, Á. Kukovecz, and H. Kuzmany, Controlled oxidation of single-wall carbon nanotubes: a Raman study, *AIP Conf. Proc.* 685, No. 1, 185–188 (2003).

71. Y. Feng, G. Zhou, G. Wang, M. Qu, and Z. Yu, Removal of some impurities from carbon nanotubes, *Chem. Phys. Lett.* 375, 645–648 (2003).

72. B.C. Satishkumar, A. Govindaraj, G. Mofokeng, G.N. Subbanna, and C.N.R. Rao. Novel experiments with carbon nanotubes: opening, filling, closing and functionalizing, *J. Phys.* **B** 29, 4925–4934 (1996).

73. N.I. Kovtyukhova, T.E. Mallouk, L. Pan, and E.C. Dickey, Individual single-walled nanotubes and hydrogels made by oxidative exfoliation of carbon nanotube ropes, *J. Am. Chem. Soc.* 125, 9761–9769 (2003).

74. J.M. Skowroński, P. Scharff, N. Pfänder, and S. Cui, Room temperature electrochemical opening of carbon nanotubes followed by hydrogen storage, *Adv. Mater.* 15, 55–57 (2003).

75. T. Ito, L. Sun, and R.M. Crooks, Electrochemical etching of individual multiwall carbon nanotubes, *Electrochem. Solid-State Lett.* 6, C4–C7 (2003).

76. J.-Y. Chang, A. Ghule, J.-J. Chang, S.-H. Tzing, and Y.-C. Ling, Opening and thinning of multiwall carbon nanotubes in supercritical water, *Chem. Phys. Lett.* 363, 583–590 (2002).

77. S. Nagasawa, M. Yudasaka, K. Hirahara, T. Ichihashi, and S. Iijima, Effect of oxidation on single-wall carbon nanotubes, *Chem. Phys. Lett.* 328, 374–380 (2000).

78. E. Borowiak-Palen, X. Liu, T. Pichler, M. Knupfer, J. Fink, and O. Jost, Diameter control of single-walled carbon nanotubes by selective oxidation, *NANOTUBE'02 Workshop,* Sa-RS2-Sy46, Log Number: P-53 [http://dielc.kaist.ac.kr/nt02/abstracts/P53.shtml].

79. J.-M. Moon, K.H. An, Y.H. Lee, Y.S. Park, D.J. Bae, and G.-S. Park, High-yield purification process of singlewalled carbon nanotubes, *J. Phys. Chem.* **B** 105, 5677–5681 (2001).

80. P.X. Hou, S. Bai, Q.H. Yang, C. Liu, and H.M. Cheng, Multi-step purification of carbon nanotubes, *Carbon* 40, 81–85 (2002).

81. E. Vázques, V. Georgakilas, and M. Prato, Microwave-assisted purification of HIPCO carbon nanotubes, *Chem. Commun.* 2308–2309 (2002).

82. C.-M. Yang, K. Kaneko, M. Yudasaka, and S. Iijima, Effect of purification on pore structure of HiPco single-walled carbon nanotube aggregates, *Nano Lett.* 2, 385–388 (2002).

83. C. Xu, E. Flahaut, S.R. Bailey, G. Brown, J. Sloan, K.S. Coleman, V.C. Williams, and M.L.H. Green, Purification of single-walled carbon nanotubes grown by a chemical vapour deposition (CVD) method, *Chem. Res. Chinese Univ.* 18, 130–132 (2002).

84. S. Gajewski, H.-E. Maneck, U. Knoll, D. Neubert, I. Dörfel, R. Mach, B. Strauß, and J.F. Friedrich, Purification of single walled carbon nanotubes by thermal gas phase oxidation, *Diamond Rel. Mater.* 12, 816–820 (2003).

85. I.W. Chiang, B.E. Brinson, R.E. Smalley, J.L. Margrave, and R.H. Hauge, Purification and characterization of single-wall carbon nanotubes, *J. Phys. Chem.* **B** 105, 1157–1161 (2001).

86. I.W. Chiang, B.E. Brinson, A.Y. Huang, P.A. Willis, M.J. Bronikowski, J.L. Margrave, R.E. Smalley, and R.H. Hauge, Purification and characterization of single-wall carbon nanotubes (SWNTs) obtained from gas-phase decomposition of CO (HiPco process), *J. Phys. Chem.* **B** 105, 8297–8301 (2001).

87. T. Jeong, W.-Y. Kim, and Y.-B. Hahn, A new purification method of single-wall carbon nanotubes using H_2S and O_2 mixture gas, *Chem. Phys. Lett.* 344, 18–22 (2001).

88. T. Jeong, T.H. Kim, W.-Y. Kim, K.-H. Lee, and Y.-B. Hahn, High yield purification of carbon nanotubes with H_2S–O_2 mixture, *Korean J. Chem. Eng.* 19, 519–523 (2002).

89. S. Huang and L. Dai, Plasma etching for purification and controlled opening of aligned carbon nanotubes, *J. Phys. Chem.* **B** 106, 3543–3545 (2002).

90. J.G. Wiltshire, A.N. Khlobystov, L.J. Li, S.G. Lyapin, G.A.D. Briggs, and R.J. Nicholas, Comparative studies on acid and thermal based selective purification of HiPCO produced single-walled carbon nanotubes, *Chem. Phys. Lett.* **386**, 239–243 (2004).

91. E. Mizoguti, F. Nihey, M. Yudasaka, S. Iijima, T. Ichihashi, and K. Nakamura, Purification of single-wall carbon nanotubes by using ultrafine gold particles, *Chem. Phys. Lett.* **321**, 297–301 (2000).

92. C.-Y. Moon, Y.-S. Kim, E.-C. Lee, Y.-G. Jin, and K.J. Chang, Mechanism for oxidative etching in carbon nanotubes, *Phys. Rev.* **B 65**, 155401.1–4 (2002).

93. D. Bom, R. Andrews, D. Jacques, J. Anthony, B. Chen, M.S. Meier, and J.P. Selegue, Thermogravimetric analysis of the oxidation of multiwalled carbon nanotubes: evidence for the role of defect sites in carbon nanotube chemistry, *Nano Lett.* **2**, 615–619 (2002).

94. E.G. Rakov, I.G. Ivanov, S.N. Blinov, N.V. Kazakov, V.V. Skudin, N.G. Digurov, and A.K. Bogdanovich, Kinetics of carbon multi-wall nanotube synthesis by catalytic pyrolysis of methane, *Fullerenes Nanotubes Carbon Nanostruct.* **12**, 29–32 (2004).

95. Summary of oxidation studies pertinent to soot and carbon solids [http://me.lsu,edu/~mechar/current/sootoxid.html].

96. D.B. Mawhinney, V. Naumenko, A. Kuznetsova, J.T. Yates, Jr., J. Liu, and R.E. Smalley, Infrared spectral evidence for the etching of carbon nanotubes: ozone oxidation at 298 K, *J. Am. Chem. Soc.* **122**, 2383–2384 (2000).

97. X. Lu, L. Zhang, X. Xu, N. Wang, and Q. Zhang, Can the sidewalls of single-wall carbon nanotubes be ozonized? *J. Phys. Chem.* **B 106**, 2136–2139 (2002).

98. K. Hernadi, A. Siska, L. Thiên-Nga, L. Forró, and I. Kiricsi, Reactivity of different kinds of carbon during oxidative purification of catalytically prepared carbon nanotubes, *Solid State Ionics* **141–142**, 203–209 (2001).

99. S. Banerjee and S.S. Wong, Rational sidewall functionalization and purification of single-walled carbon nanotubes by solution-phase ozonolysis, *J. Chem. Phys.* **B 106**, 12144–12151 (2002).

100. S. Banerjee, M.G.C. Kahn, and S.S. Wong, Rational chemical strategies for carbon nanotube functionalization, *Chem. Eur. J.* **9**, 1898–1908 (2003).

101. O. Byl, P. Kondratyuk, S.T. Forth, S.A. FitzGerald, L. Chen, J.K. Johnson, and J.T. Yates, Jr., Adsorption of CF_4 on the internal and external surfaces of opened single-walled carbon nanotubes: a vibrational spectroscopy study, *J. Am. Chem. Soc.* **125**, 5889–5896 (2003).

102. M.R. Smith, S.W. Hedges, R. LaCount, D. Kern, N. Shah, G.P. Huffman, and B. Bockrath, Selective oxidation of single-walled carbon nanotubes using carbon dioxide, *Carbon* **41**, 1221–1230 (2003).

103. O.P. Gorelik, P. Nikolaev, and S. Arepalli, Purification procedures for single-wall carbon nanotubes, *Report NASA/CR-2000-208926*, 2001, p. 64 [http://mmptdpublic.jsc.nasa.gov/hscnano/CR-2000-208926.pdf].

104. H. Huang, H. Kajiura, A. Yamada, and M. Ata, Purification and alignment of arc-synthesis single-walled carbon nanotube bundles, *Chem. Phys. Lett.* **356**, 567–572 (2002).

105. K.L. Strong, D.P. Anderson, K. Lafdi, and J.N. Kuhn, Purification process for single-wall carbon nanotubes, *Carbon* **41**, 1477–1488 (2003).

106. L. Vaccarini, C. Goze, R. Aznar, V. Micholet, C. Journet, and P. Bernier, Purification procedure of carbon nanotubes, *Synt. Met.* **103**, 2492–2493 (1999).

107. M.T. Martinez, M.A. Callejas, A.M. Benito, W.K. Maser, M. Cochet, J.M. Andrés, J. Schreiber, O. Chauvet, and J.L.G. Fierro, Microwave single walled carbon nanotubes purification, *Chem. Commun.* 1000–1001 (2002).

108. F. Li, H.M. Cheng, Y.T. Xing, P.H. Tan, and G. Su, Purification of single-walled carbon nanotubes synthesized by the catalytic decomposition of hydrocarbons, *Carbon* **38**, 2041–2045 (2000).

109. J.-F. Colomer, P. Piedigrosso, I. Willems, C. Journet, P. Bernier, G. Van Tendeloo, A. Fonseca, and J.B. Nagy, Purification of catalytically produced multi-wall carbon nanotubes, *J. Chem. Soc. Faraday Trans.* **94**, 3753–3758 (1998).

110. J.-F. Colomer, P. Piedigrosso, A. Fonseca, and J.B. Nagy, Different purification methods of carbon nanotubes produced by catalytic synthesis, *Synth. Met.* **103**, 2482–2483 (1999).

111. S.R.C. Vivekchand and A. Govindaraj, A new method of preparing single-walled carbon nanotubes, *Proc. Indian Acad. Sci.* **115**, 509–518 (2003).

112. Y. Sato, T. Ogawa, K. Motomiya, K. Shinoda, B. Jeyadevan, K. Tohji, A. Kasuya, and Y. Nishina, Purification of MWNTs combining wet grinding, hydrothermal treatment, and oxidation, *J. Phys. Chem.* **B 105**, 3387–3392 (2001).

113. H. Kajiura, S. Tsutsui, H. Huang, and Y. Murakami, High-quality single-walled carbon nanotubes from arc-produced soot, *Chem. Phys. Lett.* **364**, 586–592 (2002).

114. W.-K. Choi, S.-G. Park, H. Takahashi, and T.-H. Cho, Purification of carbon nanofibers with hydrogen peroxide, *Synth. Met.* **139**, 39–42 (2003).

115. M.E. Itkis, S. Niyogi, M.E. Meng, M.A. Hamon, H. Hu, and R.C. Haddon, Spectroscopic study of the Fermi level electronic structure of single-walled carbon nanotubes, *Nano Lett.* **2**, 155–159 (2002).

116. H. Kuzmany, A. Kukovecz, F. Simon, M. Holzweber, Ch. Kramberger, and T. Pichler, Functionalization of carbon nanotubes, *Synth. Met.* **141**, 113–122 (2004).

117. K. Esumi, M. Ishigami, A. Nakajima, K. Sawada, and H. Honda, Chemical treatment of carbon nanotubes, *Carbon* **34**, 279–281 (1996).

118. M.A. Hamon, J. Chen, H. Hu, Y. Chen, M.E. Itkis, A.M. Rao, P.C. Eklund, and R.C. Haddon, Dissolution of single-walled carbon nanotubes, *Adv. Mater.* **11**, 834–840 (1999).

119. A. Kuznetsova, I. Popova, J.T. Yates Jr., M.J. Bronikowski, C.B. Huffman, J. Liu, R.E. Smalley, H.H. Hwu, and J.G. Chen, Oxygen-containing functional groups on single-wall carbon nanotubes: NEXAFS and vibrational spectroscopic studies, *J. Am. Chem. Soc.* **123**, 10699–10704 (2001).

120. M.A. Hamon, H. Hu, P. Bhowmik, S. Niyogi, B. Zhao, M.E. Itkis, and R.C. Haddon, End-group and defect analysis of soluble single-walled carbon nanotubes, *Chem. Phys. Lett.* **347**, 8–12 (2001).

121. D.B. Mawhiney, V. Naumenko, A. Kuznetsova, J.T. Yates Jr., J. Liu, and R.E. Smalley, Surface defect site density on single walled carbon nanotubes by titration, *Chem. Phys. Lett.* **324**, 213–216 (2000).

122. J. Chen, A.M. Rao, S. Lyuksyutov, M.E. Itkis, M.A. Hamon, H. Hu, R.W. Cohn, P.C. Eklund, D.T. Colbert, R.E. Smalley, and R.C. Haddon, Dissolution of full-length single-walled carbon nanotubes, *J. Phys. Chem.* **B 105**, 2525–2528 (2001).

123. H. Hu, P. Bhowmik, B. Zhao, M.A. Hamon, M.E. Itkis, and R.C. Haddon, Determination of the acidic sites of purified single-walled carbon nanotubes by acid-base titration, *Chem. Phys. Lett.* **345**, 25–28 (2001).

124. L. Liu, Y. Qin, Z.-X. Guo, and D. Zhu, Reduction of solubilized multi-walled carbon nanotubes, *Carbon* **41**, 331–335 (2003).

125. W. Zhao, C. Song, and P.E. Pehrsson, Water-soluble and optically pH-sensitive single-walled carbon nanotubes from surface modification, *J. Am. Chem. Soc.* **124**, 12418–12419 (2002).

126. M. Sano, A. Kamino, J. Okamura, and S. Shinkai, Ring closure of carbon nanotubes, *Science* **293**, 1299–1301 (2001).

127. J. Chen, M.A. Hamon, H. Hu, Y. Chen, A.M. Rao, P.C. Eklund, and R.C. Haddon, Solution properties of single-walled carbon nanotubes, *Science* **282**, 95–98 (1998).

128. B. Zhao, H. Hu, S. Niyogi, M.E. Itkis, M.A. Hamon, P. Bhowmik, M.S. Meier, and R.C. Haddon, Chromatographic purification and properties of soluble single-walled carbon nanotubes, *J. Am. Chem. Soc.* **123**, 11673–11677 (2001).

129. V. Skakalova, U. Dettlaff-Weglikowska, and S. Roth, Gamma-irradiated and functionalized single wall nanotubes, *Diamond Rel. Mater.* **13**, 296–298 (2004).

130. J.E. Riggs, Z. Guo, D.L. Carroll, and Y.-P. Sun, Strong luminescence of solubilized carbon nanotubes. *J. Am. Chem. Soc.* **122**, 5879–5880 (2000).

131. J.E. Riggs, D.B. Walker, D.L. Carroll, and Y.-P. Sun, Optical limiting properties of suspended and solubilized carbon nanotubes, *J. Phys. Chem.* **B 104**, 7071–7076 (2000).

132. Y. Lin, A.M. Rao, B. Sadanadan, E.A. Kenik, and Y.-P. Sun, Functionalizing multiple-walled carbon nanotubes with aminopolymers, *J. Phys. Chem.* **B 106**, 1294–1298 (2002).

133. I. Yamaguchi and T. Yamamoto, Soluble self-doped single-walled carbon nanotube, *Mater. Lett.* **58**, 598–603 (2004).

134. B. Philip, J. Xie, A. Chandrasekhar, J. Abraham, and V.K. Varadan, A novel nanocomposite from multiwalled carbon nanotubes functionalized with a conducting polymer, *Smart Mater. Struct.* **13**, 295–298 (2004).

135. Y.-P. Sun, W. Huang, Y. Lin, K. Fu, A. Kitaygorodsky, L.A. Riddle, Y.J. Yu, and D.L. Caroll, Soluble dendron-functionalized carbon nanotubes: preparation, characterization and properties, *Chem. Mater.* **13**, 2864–2869 (2001).

136. K. Fu, W. Huang, Y. Lin, L.A. Riddle, D.L. Carroll, and Y.-P. Sun, Defunctionalization of functionalized carbon nanotubes, *Nano Lett.* **1**, 439–441 (2001).

137. Y.-P. Sun, B. Zhou, K. Henbest, K. Fu, W. Huang, Y. Lin, S. Taylor, and D.L. Carroll, Luminescence anisotropy of functionalized carbon nanotubes in solution, *Chem. Phys. Lett.* **351**, 349–353 (2002).

138. M.A. Hamon, H. Hui, P. Bhowmik, M.E. Itkis, and R.C. Haddon, Ester-functionalized soluble single-walled carbon nanotubes, *Appl. Phys.* **A 74**, 333–338 (2002).

139. D.E. Hill, Y. Lin, L.F. Allard, and Y.-P. Sun, Solubilization of carbon nanotubes via polymer attachment, *Int. J. Nanosci.* **1**, 213–221 (2002).

140. S. Qin, D. Qin, W.T. Ford, D.E. Resasco, and J.E. Herrera, Polymer brushes on single-walled carbon nanotubes by atom transfer radical polymerization of *n*-butyl methacrylate, *J. Am. Chem. Soc.* **126**, 170–176 (2004).

141. M. Alvaro, P. Atienzar, P. de la Cruz, J.L. Delgado, H. Garcia, and F. Langa, Synthesis and photochemistry of soluble, pentyl ester-modified single wall carbon nanotube, *Chem. Phys. Lett.* **386**, 342–345 (2004).

142. A.M. Bond, W. Miao, and C.L. Raston, Mercury(II) immobilized on carbon nanotubes: synthesis, characterization, and redox properties, *Langmuir* **16**, 6004–6012 (2000).

143. Y.-H. Li, J. Ding, Z. Luan, Z. Di, Y. Zhu, C. Xu, D. Wu, and B. Wei, Competitive adsorption of Pb^{2+}, Cu^{2+} and Cd^{2+} ions from aqueous solutions by multiwalled carbon nanotubes, *Carbon* **41**, 2787–2792 (2003).

144. G. Arabale, D. Wagh, M. Kulkarni, I.S. Mulla, S.P. Vernekar, K. Vijayamohanan, and A.M. Rao, Enhanced supercapacitance of multiwalled carbon nanotubes functionalized with ruthenium oxide, *Chem. Phys. Lett.* **376**, 207–213 (2003).

145. S. Banerjee and S.S. Wong, Functionalization of carbon nanotube with a metal-containing molecular complex, *Nano Lett.* **2**, 49–53 (2002).

146. S. Banerjee and S.S. Wong, Structural characterization, optical properties, and improved solubility of carbon nanotubes functionalized with Wilkinson's catalyst, *J. Am. Chem. Soc.* **124**, 8940–8948 (2002).

147. S. Banerjee and S.S. Wong, Synthesis and characterization of carbon nanotube-nanocrystal heterostructures, *Nano Lett.* **2**, 195–200 (2002).

148. S. Banerjee and S.S. Wong, In situ quantum dot growth on multiwalled carbon nanotubes, *J. Am. Chem. Soc.* **125**, 10342–10350 (2003).

149. M. Aizawa and M.S.P. Shaffer, Silylation of multi-walled carbon nanotubes, *Chem. Phys. Lett.* **368**, 121–124 (2003).

150. J.H. Rouse and P.T. Lillehei, Electrostatic assembly of polymer/single walled carbon nanotube multilayer films, *Nano Lett.* **3**, 59–62 (2003).

151. C. Velasco-Santos, A.L. Martínez-Hernández, M. Lozada-Cassou, A. Alvarez-Castillo, and V.M. Castaño, Chemical functionalization of carbon nanotubes through an organosilane, *Nanotechnology* **13**, 495–498 (2002).

152. J.G. Smith Jr., K.A. Watson, C.M. Thompson, and J.W. Connell, Carbon nanotube/space durable polymer nanocomposite films for electrostatic charge dissipation, p. 12, 2002 [http://techreport.larc.nasa.gov/ltrs/PDF/2002/mtg/NASA-2002-34sampe-jgs.pdf].

153. B. Wu, J. Zhang, Z. Wei, S. Cai, and Z. Liu, Chemical alignment of oxidatively shortened single-walled carbon nanotubes on silver surface, *J. Phys. Chem.* **B 105**, 5075–5078 (2002).

154. X.-L. Nan, J. Zhang, Z.-F. Liu, Z.-J. Shi, and Z.-N. Gu, Patterned assembly of shortened single-walled carbon nanotubes on gold surface, *Acta Phys. Chim. Sin.* **17** (5), 393–396 (2001).

155. X. Nan, Z. Gu, and Z. Liu, Immobilizing shortened single-walled carbon nanotubes (SWNTs) on gold using a surface condensation method, *J. Colloid Interf. Sci.* **245**, 311–318 (2002).

156. S.S. Wong, A.T. Woolley, E. Joselevich, and C.M. Lieber, Functionalization of carbon nanotube AFM probes using tip-activated gases, *Chem. Phys. Lett.* **306**, 219–225 (1999).

157. S.S. Wong, E. Joselevich, A.T. Woolley, C.L. Cheung, and C.M. Lieber, Covalently functionalized nanotubes as nanometre-sized probes in chemistry and biology, *Nature* **394**, 52–55 (1998).

158. S.S. Wong, A.T. Woolley, E. Joselevich, C.L. Cheung, and C.M. Lieber, Covalently-functionalized single-walled carbon nanotube probe tips for chemical force microscopy, *J. Am. Chem. Soc.* **120**, 8557–8558 (1998).

159. M. Sano, A. Kamino, and S. Shinkai, Activation of hydroxyl groups on carbon nanotubes by thermal theatment in air, *NANOTUBE '02 Workshops* Sa–P71–Sy17. Log Number: P160 [http://dielc.kaist.ac.kr/nt02/abstracts/P160.shtml].

160. T. Kyotani, S. Nakazaki, W.-H. Xu, and A. Tomita, Chemical modification of the inner walls of carbon nanotubes by HNO_3 oxidation, *Carbon* **39**, 771–785 (2001).

161. A. Kuznetsova, D.B. Mawhinney, V. Naumenko, J.T. Yates Jr., J. Liu, and R.E. Smalley. Enhancement of adsorption inside of single-walled nanotubes: opening the entry ports, *Chem. Phys. Lett.* **321**, 292–296 (2000).

162. T. Nakajima, N. Watanabe, I. Kameda, and M. Endo, Preparation and electrical conductivity of fluorine-graphite fiber intercalation compound, *Carbon* **24**, 343–351 (1986).
163. T. Nakajima, S. Kasamatsu, and Y. Matsuo, Synthesis and characterization of fluorinated carbon nanotubes, *Eur. J. Solid State Inorg. Chem.* **33**, 831–840 (1996).
164. A. Hamwi, H. Alvergnat, S. Bonnamy, and F. Béguin, Fluorination of carbon nanotubes, *Carbon* **35**, 723–728 (1997).
165. A. Hamwi, P. Gendrand, H. Gaucher, S. Bonnamy, and F. Beguin, Electrochemical properties of carbon nanotube fluorides in a lithium cell system, *Mol. Cryst. Liq. Cryst.* **310**, 185–190 (1998).
166. A.V. Okotrub, N.F. Yudanov, A.L. Chuvilin, I.P. Asanov, Yu.V. Shubin, L.G. Bulusheva, A.V. Gusel'nikov, and I.S. Fyodorov, Fluorinated cage multiwall carbon nanoparticles, *Chem. Phys. Lett.* **322**, 231–236 (2000).
167. H. Touhara and F. Okino, Property control of carbon material by fluorination, *Carbon* **38**, 241–267 (2000).
168. Y. Hattori, Y. Watanabe, S. Kawasaki, F. Okino, B.K. Pradhan, T. Kyotani, A. Tomita, and H. Touhara, Carbon-alloying of the rear surfaces of nanotubes by direct fluorination, *Carbon* **37**, 1033–1038 (1999).
169. E.T. Mickelson, C.B. Huffman, A.G. Rinzler, R.E. Smalley, R.H. Hauge, and J.L. Margrave, Fluorination of single-wall carbon nanotubes, *Chem. Phys. Lett.* **296**, 188–194 (1998).
170. I.W. Chiang, E.T. Mickelson, P.J. Boul, R.H. Hauge, R.E. Smalley, and J.L. Margrave, Fluorination, defluorination, and derivatization of single-wall carbon nanotubes, *Abstr. Pap. — Am. Chem. Soc. 2000, 220th*, IEC-153. Washington, DC, Aug. 20–24, 2000.
171. H. Peng, Z. Gu, J. Yang, J.L. Zimmerman, P.A. Willis, M.J. Bronikowski, R.E. Smalley, R.H. Hauge, and J.L. Margrave, Fluorotubes as cathodes in lithium electrochemical cells, *Nano Lett.* **1**, 625–629 (2001).
172. W. Zhao, C. Song, B. Zheng, J. Liu, and T. Viswanathan, Thermal recovery behavior of fluorinated single-walled carbon nanotubes, *J. Phys. Chem.* **B 106**, 293–296 (2002).
173. P.E. Pehrsson, W. Zhao, J.W. Baldwin, C. Song, J. Liu, S. Kooi, and B. Zheng, Thermal fluorination and ahhealing of single-wall carbon nanotubes, *J. Phys. Chem.* **B 107**, 5690–5695 (2003).
174. Z. Gu, H. Peng, R.H. Hauge, R.E. Smalley, and J.L. Margrave, Cutting single-wall carbon nanotubes through fluorination, *Nano Lett.* **2**, 1009–1013 (2002).
175. Y.S. Lee, T.H. Cho, B.K. Lee, J.S. Rho, K.H. An, and Y.H. Lee, Surface properties of fluorinated single-walled carbon nanotubes, *J. Fluor. Chem.* **120**, 99–104 (2003).
176. P.J. Boul, J. Liu, E.T. Mickelson, C.B. Huffman, L.M. Ericson, I.W. Chiang, K.A. Smith, D.T. Colbert, R.H. Hauge, J.L. Margrave, and R.E. Smalley, Reversible sidewall functionalization of buckytubes, *Chem. Phys. Lett.* **310**, 367–372 (1999).
177. K.F. Kelly, I.W. Chiang, E.T. Mickelson, R.H. Hauge, J.L. Margrave, X. Wang, G.E. Scuseria, C. Radloff, and N.J. Halas, Insight into the mechanism of sidewall functionalization of single-walled nanotubes: an STM study, *Chem. Phys. Lett.* **313**, 445–450 (1999).
178. E.T. Mickelson, I.W. Chiang, J.L. Zimmerman, P.J. Boul, J. Lozano, J. Liu, R.E. Smalley, R.H. Hauge, and J.L. Margrave, Solvation of fluorinated single-wall carbon nanotubes in alcohol solvents, *J. Phys. Chem.* **B 103**, 4318–4322 (1999).
179. V.N. Khabashesku, W.E. Billups, and J.L. Margrave, Fluorination of single-wall carbon nanotubes and subsequent derivatization reactions, *Acc. Chem. Res.* **35**, 1087–1095 (2002).
180. D.V. Kirin, N.N. Breslavskaya, and P.N. D'yachkov, Heterojunktions based on chemically modified carbon nanotubes, *Doklady Phys. Chem.* **374**, 161–166 (2000).
181. H.F. Bettinger, K.N. Kudin, and G.E. Scuseria, Thermochemistry of fluorinated single wall carbon nanotubes, *J. Am. Chem. Soc.* **123**, 12849–12856 (2001).
182. K.N. Kudin, H.F. Bettinger, and G.E. Scuseria, Fluorinated single-wall carbon nanotubes, *Phys. Rev.* **B 63**, 045413.1–8 (2001).
183. T. Hayashi, M. Terrones, C. Scheu, Y.A. Kim, M. Rühle, T. Nakajima, and M. Endo, NanoTeflons: structure and EELS characterization of fluorinated nanotubes and nanofibers, *Nano Lett.* **2**, 491–496 (2002).
184. P.R. Marcoux, J. Schreiber, P. Batail, S. Lefrant, J. Renouard, G. Jacob, D. Albertini, and J.-Y. Mevellec, A spectroscopic study of the fluorination and defluorination reactions on single-walled carbon nanotubes, *Phys. Chem. Phys.* **4**, 2278–2285 (2002).
185. H.F. Bettinger, Experimental and computational investigation of the properties of fluorinated single-walled carbon nanotubes, *Chem. Phys. Chem.* **4**, 1283–1289 (2003).
186. K.H. An, K.A. Park, J.G. Heo, J.Y. Lee, K.K. Jeon, S.C. Lim, C.W. Yang, Y.S. Lee, and Y.H. Lee, Structural transformation of fluorinated carbon nanotubes induced by in situ electron-beam irradiation, *J. Am. Chem. Soc.* **125**, 3057–3061 (2003).

187. K.A. Park, Y.S. Choi, and Y.H. Lee, Atomic and electronic structures of fluorinated single-walled carbon nanotubes, *Phys. Rev.* **B 68**, 045429.1–8 (2003).
188. R.L. Jaffe, Quantum chemistry study of fullerene and carbon nanotube fluorination, *J. Phys. Chem.* **B 107**, 10378–10388 (2003).
189. H. Geng, R. Rosen, B. Zheng, H. Shimoda, L. Fleming, J. Liu, and O. Zhou, Fabrication and properties of composites of poly(ethylene oxide) and functionalized carbon nanotubes, *Adv. Mater.* **14**, 1387–1390 (2002).
190. R.L. Jaffe, Quantum chemistry study of chemical functionalization reactions of fullerenes and carbon nanotubes, *Proc. Electrochem. Soc.* **12**, 153–162 (1999). Fullerenes Vol. 7: Recent Advances in the Chemistry and Physics of Fullerenes and Related Mater. (12th Int. Symp.), K.M. Kadish, P.V. Kamat, and D.M.Guldi, PV 99–12, Seattle, Washington, Spring 1999.
191. M. Xu, Q. Huang, Q. Chen, P. Guo, and Z. Sun, Synthesis and characterization of octadecylamine grafted multi-walled carbon nanotubes, *Chem. Phys. Lett.* **375**, 598–604 (2003).
192. F. Pompeo and D.E. Resasco, Water solubilization of single-walled carbon nanotubes by functionalization with glucosamine, *Nano Lett.* **2**, 369–373 (2002).
193. L. Liu, S. Zhang, T. Hu, Z.-X. Guo, C. Ye, L. Dai, and D. Zhu, Solubilized multi-walled carbon nanotubes with broadband optical limiting effect, *Chem. Phys. Lett.* **359**, 191–195 (2002).
194. W. Wu, J. Li, L. Liu, L. Yanga, Z.-X. Guo, L. Dai, and D. Zhu, The photoconductivity of PVK-carbon nanotube blends, *Chem. Phys. Lett.* **364**, 196–199 (2002).
195. J. Zhang, G. Wang, Y.-S. Shon, O. Zhou, R. Superfine, and R.W. Murray, Interactions of small molecules and Au nanoparticles with solubilized single-wall carbon nanotubes, *J. Phys. Chem.* **B 107**, 3726–3732 (2003).
196. W.-Y. Chen, C.-Y. Chen, K.-Y. Hsu, C.-C. Wang, and Y.-C. Ling, Reaction monitoring of polyaniline film formation on carbon nanotubes with TOF-SIMS, *Appl. Surf. Sci.* **231–232**, 845–849 (2004).
197. M. Álvaro, P. Atienzar, J.L. Bourdelande, and H. García, An organically modified single wall carbon nanotube containing a pyrene chromophore: fluorescence and diffuse reflectance laser flash photolysis study, *Chem. Phys. Lett.* **384**, 119–123 (2004).
198. F. Della Negra, M. Meneghetti, and E. Menna, Microwave-assisted synthesis of a soluble single wall carbon nanotube derivative, *Fullerenes Nanotubes Carbon Nanostruct.* **11**, 25–34 (2003).
199. H.-J. Lee, H. Park, S. Koo, and H. Lee, Vertical alignments of single-walled carbon nanotubes on chemically functionalized silicon substrates [http://otlf.hanyang.ac.kr/publications/pdf/2003/Vertical%20alignments%20of%20SWNTs%20on%20chemicall%20functionalized%20silicon%20substrates.pdf].
200. M. Sano, A. Kamino, and S. Shinkai, Construction of carbon nanotube "stars" with dendrimers, *Angew. Chem. Int. Edit.* **40**, 4661–4663 (2001).
201. L. Cao, W. Yang, J. Yang, C. Wang, and S. Fu, Hyperbranched poly(amidoamine)-modified multi-walled carbon nanotubes via grafting-from method, *Chem. Lett.* **33**, 490–491 (2004).
202. S. Niyogi, H. Hu, M.A. Hamon, P. Bhowmik, B. Zhao, S.M. Rozenzhak, J. Chen, M.E. Itkis, M.S. Meier, and R.C. Haddon, Chromatographic purification of soluble SWNTs, *J. Am. Chem. Soc.* **123**, 733–734 (2001).
203. P.W. Chiu, G.S. Duesberg, U. Dettlaff-Weglikowska, and S. Roth, Interconnection of carbon nanotubes by chemical functionalization, *Appl. Phys. Lett.* **80**, 3811–3813 (2002).
204. K. Niesz, Z. Kónya, A.A. Koós, L.P. Biró, Á. Kukovecz, and I. Kiricsi, Synthesis procedures for production of carbon nanotube junctions, *AIP Conf. Proc.* **685**, No. 1, 253–256 (2003).
205. J.M. Haremza, M.A. Hahn, T.D. Krauss, S. Chen, and J. Calcines, Attachment of single CdSe nanocrystals to individual single-walled carbon nanotubes, *Nano Lett.* **2**, 1253–1258 (2002).
206. J.G. Smith Jr., J.W. Connell, D.M. Delozier, P.T. Lillehei, K.A. Watson, Y. Lin, B. Zhou, and Y.-P. Sun, Space durable polymer/carbon nanotube films for electrostatic charge mitigation, *Polymer* **45**, 825–836 (2004).
207. D. Chattopadhyay, S. Lastella, S. Kim, and F. Papadimitrakopoulos, Length separation of zwitterion-functionalized single wall carbon nanotubes by GPC, *J. Am. Chem. Soc.* **124**, 728–729 (2002).
208. D. Chattopadhyay, I. Galeska, and F. Papadimitrakopoulos, A route to bulk separation of semiconducting from metallic single-wall carbon nanotubes, *J. Am. Chem. Soc.* **125**, 3370–3375 (2003).
209. F. Gojny, J. Nastalczyk, Z. Roslaniec, and K. Schulte, Surface modified multi-walled carbon nanotubes in CNT/epoxy composites, *Chem. Phys. Lett.* **370**, 820–824 (2003).

210. Z. Liu, Z. Shen, T. Zhu, S. Hou, L. Ying, Z. Shi, and Z. Gu, Organizing single-walled carbon nanotubes on gold using a wet chemical self-assembling technique, *Langmuir* **16**, 3569–3573 (2000).
211. Q. Chen and L. Dai, Plasma patterning of carbon nanotubes *Appl. Phys. Lett.* **76**, 2719–2721 (2000).
212. Q. Chen, L. Dai, M. Gao, S. Huang, and A. Mau, Plasma activation of carbon anotubes for chemical modification, *J. Phys. Chem.* **B 105**, 618–622 (2001).
213. Y. Sun, S.R. Wilson, and D.I. Schuster, High dissolution and strong light emission of carbon nanotubes in aromatic amine solvents, *J. Am. Chem. Soc.* **123**, 5348–5349 (2001).
214. B. Li, Z. Shi, Y. Lian, and Z. Gu, Aqueous soluble single-wall carbon nanotube, *Chem. Lett.* 2001, 598–599.
215. W. Huang, Y. Lin, S. Taylor, J. Gaillard, A.M. Rao, and Y.-P. Sun, Sonication-assisted functionalization and solubilization of carbon nanotubes, *Nano Lett.* **2**, 231–234 (2002).
216. Z. Kónya, I. Vesselenyi, K. Niesz, A. Kukovecz, A. Demortier, A. Fonseca, J. Delhalle, Z. Mekhalif, J.B. Nagy, A.A. Koós, Z. Osváth, A. Kocsonya, L.P. Biró, and I. Kiricsi, Large scale production of short functionalized carbon nanotubes, *Chem. Pgys. Lett.* **360**, 429–435 (2002).
217. E.V. Basiuk, V.A. Basiuk, J.-G. Bañuelos, J.-M. Saniger-Blesa, V.A. Pokrovskiy, T.Yu. Gromovoy, A.V. Mischanchuk, and B.G. Mischanchuk. Interaction of oxidized single-walled carbon nanotubes with vaporous aliphatic amines, *J. Phys. Chem.* **B 106**, 1588–1597 (2002).
218. V.A. Basiuk, ONIOM studies of chemical reactions on carbon nanotube tips: effects of the lower theoretical level and mutual orientation of the reactants, *J. Phys. Chem.* **B 107**, 8890–8897 (2003).
219. J. Kong and H. Dai, Full and modulated chemical gating of individual carbon nanotubes by organic amine compounds, *J. Phys. Chem.* **B 105**, 2890–2893 (2001).
220. F. Hennrich, M.M. Kappes, M.S. Strano, R.H. Hauge, and R.E. Smalley, Infrared analysis of amine treated single-walled carbon nanotubes produced by decomposition of CO, *AIP Conf. Proc.* **685**, No. 1, 197–201 (2003).
221. G. de la Torre, W. Blau, and T. Torres, A survey on the functionalization of single-walled nanotubes. The chemical attachment of phtalocyanine moieties, *Nanotechnology* **14**, 765–771 (2003).
222. Y. Chen, R.C. Haddon, S. Fang, A.M. Rao, P.C. Eklund, W.H. Lee, E.C. Dickey, E.A. Grulke, J.C. Pendergrass, A. Chavan, B.E. Haley, and R.E. Smalley, Chemical attachment of organic functional groups to single-walled carbon nanotube material, *J. Mater. Res.* **13**, 2423–2431 (1998).
223. M. Holzinger, O. Vostrowsky, A. Hirsch, F. Hennrich, M. Kappes, R. Weiss, and F. Jellen, Sidewall functionalization of carbon nanotubes, *Angew. Chem. Int. Edit.* **40**, 4002–4005 (2001).
224. M. Monthioux, Filling single-wall carbon nanotubes, *Carbon* **40**, 1809–1823 (2002).
225. H. Hu, B. Zhao, M.A. Hamon, K. Kamaras, M.E. Itkis, and R.C. Haddon, Sidewall functionalization of single-walled carbon nanotubes by addition of dichlorocarbene, *J. Am. Chem. Soc.* **125**, 14893–14900 (2003).
226. X. Lu, F. Tian, and Q. Zhang, The [2+1] cycloadditions of dichlorocarbene, silylene, germylene, and oxycarbonylnitrene onto the sidewall of armchair (5,5) single-wall carbon nanotube, *J. Phys. Chem.* **B 107**, 8388–8391 (2003).
227. M. Holzinger, J. Abraham, P. Whelan, R. Graupner, L. Ley, F. Hennrich, M. Kappes, and A. Hirsch, Functionalization of single-walled carbon nanotubes with (R-)oxycarbonyl nitrenes, *J. Am. Chem. Soc.* **125**, 8566–8580 (2003).
228. J. Abraham, P. Whelan, A. Hirsch, F. Hennrich, M. Kappes, D. Samaille, P. Bernier, A. Vencelová, R. Graupner, and L. Ley, Covalent functionalization of arc discharge, laser ablation and HiPCO single-walled carbon nanotubes, *AIP Conf. Proc.* **685**, No. 1, 291–296 (2003).
229. S. Qin, D. Qin, W.T. Ford, D.E. Resasco, and J.E. Herrera, Functionalization of single-walled carbon nanotubes with polystyrene via grafting to and grafting from methods, *Macromolecules* **37**, 752–757 (2004).
230. V. Georgakilas, K. Kordatos, M. Prato, D.M. Guldi, M. Holzinger, and A. Hirsch, Organic functionalization of carbon nanotubes, *J. Am. Chem. Soc.* **124**, 760–761 (2002).
231. V. Georgakilas, D. Voulgaris, E. Vázquez, M. Prato, D.M. Guldi, Kukovecz, and H. Kuzmany, Purification of HiPCO carbon nanotubes via organic functionalization. *J. Am. Chem. Soc.* **124**, 14318–14319 (2002).
232. N. Tagmatarchis and M. Prato, Functionalization of carbon nanotubes via 1,3-dipolar cycloadditions, *J. Mater. Chem.* **14**, 437–439 (2004).
233. C.A. Dyke and J.M. Tour, Solvent-free functionalization of carbon nanotubes, *J. Am. Chem. Soc.* **125**, 1156–1157 (2003).
234. C.A. Dyke and J.M. Tour, Overcoming the insolubility of carbon nanotubes through high degrees of sidewall functionalization, *Chem. Eur. J.* **10**, 812–817 (2004).

235. J.L. Bahr, J. Yang, D.V. Kosynkin, M.J. Bronikowski, R.E. Smalley, and J.M. Tour, Functionalization of carbon nanotubes by electrochemical reduction of aryl diazonium salts: a bucky paper electrode, *J. Am. Chem. Soc.* **123**, 6536–6542 (2001).

236. J.L. Bahr and J.L. Tour, Highly functionalized carbon nanotubes using in situ generated diazonium compounds, *Chem. Mater.* **13**, 3823–3824 (2001).

237. J.L. Bahr and J.M. Tour, Covalent chemistry of single-wall carbon nanotubes, *J. Mater. Chem.* **12**, 1952–1958 (2002).

238. S.E. Kooi, U. Schlecht, M. Burghard, and K. Kern, Electrochemical modification of single-carbon nanotubes, *Angew. Chem. Int. Ed.* **41**, 1353–1355 (2002).

239. Y. Ying, R.K. Saini, F. Liang, A.K. Sadana, and W.E. Billups, Functionalization of carbon nanotubes by free radicals, *Org. Lett.* **5**, 1471–1473 (2003).

240. M.S. Strano, C.A. Dyke, M.L. Usrey, P.W. Barone, M.J. Allen, H. Shan, C. Kittrell, R.H. Hauge, J.M. Tour, and R.E. Smalley, Electronic structure control of single-walled carbon nanotube functionalization, *Science* **301**, 1519–1522 (2003).

241. P. Umek, J.W. Seo, K. Hernadi, A. Mrzel, P. Pechy, D.D. Mihailovic, and L. Forro, Addition of carbon radicals generated from organic peroxides to single-wall carbon nanotubes, *Chem. Mater.* **15**, 4751–4755 (2003).

242. M. Yudasaka, M. Zhang, C. Jabs, and S. Iijima, Effect of an organic polymer in purification and cutting of single-wall carbon nanotubes, *Appl. Phys.* **A 71**, 449–451 (2000).

243. M. Zhang, M. Yudasaka, A. Koshio, and S. Iijima, Effect of polymer and solvent on purification and cutting of single-wall carbon nanotubes, *Chem. Phys. Lett.* **349**, 25–30 (2001).

244. M. Zhang, M. Yudasaka, A. Koshio, and S. Iijima, Thermogravimetric analysis of single-wall carbon nanotubes ultrasonicated in monochlorobenzene, *Chem. Phys. Lett.* **364**, 420–426 (2002).

245. A. Koshio, M. Yudasaka, M. Zhang, and S. Iijima, A simple way to chemically react single-wall carbon nanotubes with organic materials using ultrasonication, *Nano Lett.* **1**, 361–363 (2001).

246. S. Niyogi, M.A. Hamon, D.E. Perea, C.B. Kang, B. Zhao, S.K. Pal, A.E. Wyant, M.E. Itkis, and R.C. Haddon, Ultrasonic dispersions of single-walled carbon nanotubes, *J. Phys. Chem.* **B 107**, 8799–8804 (2003).

247. J.L. Bahr, E.T. Mickelson, M.J. Bronikowski, R.E. Smalley, and J.M. Tour, Dissolution of small diameter SWNTs in organic solvents? *Chem. Commun.* 193–194 (2001).

248. K. Balasubramanian, M. Friedrich, C. Jiang, Y. Fan, A. Mews, M. Burghard, and K. Kern, Electrical transport and confocal Raman studies of electrochemically modified individual carbon nanotubes, *Adv. Mater.* **15**, 1515–1518 (2003).

249. J.K. Lim, W.S. Yun, M.-H. Yoon, S.K. Lee, C.H. Kim, K. Kim, and S.K. Kim, Selective thiolation of single-walled carbon nanotubes, *Synt. Met.* **139**, 521–527 (2003).

250. A.V. Krasheninnikov and K. Nordlund, Irradiation effects in carbon nanotubes, *Nucl. Instr. Meth. Phys. Res.* **B 216**, 355–366 (2004).

251. A.V. Krasheninnikov and K. Nordlund, Signatures of irradiation-induced defects in scanning-tunneling microscopy images of carbon nanotubes, *Phys. Solid State* **44**, 470–472 (2002).

252. D.-H. Kim, H.-S. Jang, C.-D. Kim, D.-S. Cho, H.-D. Kang, and H.-R. Lee, Enhancement of the field emission of carbon nanotubes straightened by application of argon ion irradiation, *Chem. Phys. Lett.* **378**, 232–237 (2003).

253. L. Dai, H.J. Griesser, and A.W.H. Mau, Surface modification by plasma etching and plasma patterning, *J. Phys. Chem.* **B 101**, 9548–9554 (1997).

254. H. Bubert, S. Haiber, W. Brandl, G. Marginean, M. Heintze, and V. Brüser, Characterization of the uppermost layer of plasma-treated carbon nanotubes, *Diamond Rel. Mater.* **12**, 811–815 (2003).

255. V.A. Basiuk, K. Kobayashi, T. Kaneko, Y. Negishi, E.V. Basiuk, and J.-M. Saniger-Blesa, Irradiation of single-walled carbon nanotubes with high-energy protons, *Nano Lett.* **2**, 789–791 (2002).

256. B.N. Khare, M. Meyyappan, A.M. Cassell, C.V. Nguyen, and J. Han, Functionalization of carbon nanotubes using atomic hydrogen from a glow discharge, *Nano Lett.* **2**, 73–77 (2002).

257. B.N. Khare, M. Meyyappan, J. Kralj, P. Wilhite, M. Sisay, H. Imanaka, J. Koehne, and C.W. Baushchlicher Jr., A glow-discharge approach for functionalization of carbon nanotubes, *Appl. Phys. Lett.* **81**, 5237–5239 (2002).

258. B. Ni and S.B. Sinnott, Chemical functionalization of carbon nanotubes through energetic radical collisions, *Phys. Rev.* **B 61**, R16343–R16346 (2000).

259. B. Ni, R. Andrews, D. Jacques, D. Qian, M.B.J. Wijesundara, Y. Choi, L. Hanley, and S.B. Sinnott, A combined computational and experimental study of ion-beam modification of carbon nanotube bundles, *J. Phys. Chem.* **B 105**, 12719–12725 (2001).

260. N.O.V. Plank, L. Jiang, and R. Cheung, Fluorination of carbon nanotubes in CF_4 plasma, *Appl. Phys. Lett.* **83**, 2426–2428 (2003).

261. N.O.V. Plank and R. Cheung, Functionalization of carbon nanotubes for molecular electronics, *Microelectronic Eng.* **73–74**, 578–582 (2004).

262. Y. Hu and S. Sinnott, Nanometer-scale engineering of composites, *11th Foresight Conf. Molec. Nanotech.* [http://www.foresight.org/Conferences/MNT11/Abstracts/Hu/].

263. Y. Breton, S. Delpeux, R. Benoit, J.P. Salvetat, C. Sinturel, F. Beguin, S. Bonnamy, G. Desarmot, and L. Boufendi, Functionalization of multiwall carbon nanotubes: properties of nanotubes-epoxy composites, *Mol. Cryst. Liq. Cryst.* **387**, 135–140 (2002).

264. C.Y. Zhi, X.D. Bai, and E.G. Wang, Enhanced field emission from carbon nanotubes by hydrogen plasma treatment, *Appl. Phys. Lett.* **81**, 1690–1692 (2002).

265. S.A. Miller, V.Y. Young, and C.R. Martin, Electroosmotic flow in template-prepared carbon nanotube membranes, *J. Am. Chem. Soc.* **123**, 12335–12342 (2001).

266. G.S. Duesberg, J. Muster, V. Krstic, M. Burghard, and S. Roth, Chromatographic size separation of single-wall carbon nanotubes, *Appl. Phys.* **A 67**, 117–119 (1998).

267. G.S. Duesberg, M. Burghard, J. Muster, G. Philipp, and S. Roth, Separation of carbon nanotubes by size exclusion chromatography, *Chem. Commun.* 435–436 (1998).

268. G.S. Duesberg, W. Blau, H.J. Byrne, J. Muster, M. Burghard, and S. Roth, Chromatography of carbon nanotubes, *Synth. Met.* **103**, 2484–2485 (1999).

269. B. Vigolo, A. Pénicaud, C. Coulon, C. Sauder, R. Pailler, C. Journet, P. Bernier, and P. Poulin, Macroscopic fibers and ribbons of oriented carbon nanotubes, *Science* **290**, 1331–1334 (2000).

270. M.J. O'Connell, S.M. Bachilo, C.B. Huffman, V.C. Moore, M.S. Strano, E.H. Haroz, K.L. Rialon, P.J. Boul, W.H. Noon, C. Kittrell, J. Ma, R.H. Hauge, R.B. Weisman, and R.E. Smalley, Band gap fluorescence from individual single-walled carbon nanotubes, *Science* **297**, 593–596 (2002).

271. P. Poulin, B. Vigolo, and P. Lannois, Films and fibers of oriented single wall nanotubes, *Carbon* **40**, 1741–1749 (2002).

272. S.K. Doorn, M.S. Strano, M.J. O'Connell, E.H. Haroz, K.L. Rialon, R.H. Hauge, and R.E. Smalley, Capillary electrophoresis separation of bundled and individual carbon nanotubes, *J. Phys. Chem.* **B 107**, 6063–6069 (2003).

273. A.V. Neimark, S. Ruetsch, K.G. Kornev, P.I. Ravikovitch, P. Poulin, S. Badaire, and M. Maugey, Hierarchical pore structure and wetting properties of single-wall carbon nanotube fibers. *Nano Lett.* **3**, 419–423 (2003).

274. J. Wang, R.P. Deo, P. Poulin, and M. Mangey, Carbon nanotube fiber microelectrodes, *J. Am. Chem. Soc.* **125**, 14706–14707 (2003).

275. Y. Dror, W. Salalha, R.L. Khalfin, Y. Cohen, A.L. Yarin, and E. Zussman, Carbon nanotubes embedded in oriented polymer nanofibers by electrospinning, *Langmuir* **19**, 7012–7020 (2003).

276. L. Zhao and L. Gao, Novel in situ synthesis of MWNTs-hydroxyapatite composites, *Carbon* **42**, 423–426 (2004).

277. V. Krstic, G.S. Duesberg, J. Muster, M. Burghard, and S. Roth, Langmuir-Blodgett films of matrix-diluted single-walled carbon nanotubes, *Chem. Mater.* **10**, 2338–2340 (1998).

278. A.B. Dalton, S. Collins, E. Muñoz, J.M. Razal, Von H. Ebron, J.P. Ferraris, J.N. Coleman, B.G. Kim, and R. Baughman, Super tough carbon nanotube fibers, *Nature* **423**, 703 (2003).

279. A.B. Dalton, S. Collins, J. Razal, E. Munoz, Von H. Ebron, B.G. Kim, J.N. Coleman, J.P. Ferraris, and R. Baughman, Continuous carbon nanotube composite fibers: properties, potential applications, and problems, *J. Mater. Chem.* **14**, 1–3 (2004).

280. M.F. Islam, E. Rojas, D.M. Bergey, A.T. Johnson, and A.G. Yodh, High weight-fraction surfactant solubilization of single-wall carbon nanotubes in water, *Nano Lett.* **3**, 269–273 (2003).

281. M.F. Islam, A.M. Alsayed, Z. Dogic, J. Zhang, T.C. Lubensky, and A.G. Yodh, Nematic nanotube gels, *Phys. Rev. Lett.* **92**, 1–4 (2004).

282. J.I. Paredes and M. Burghard, Dispersions of individual single-walled carbon nanotubes of high length, *Langmuir*, **20**, 5149–5152 (2004).

283. O. Matarredona, H. Rhoads, Z. Li, J.H. Harwell, L. Balzano, and D.E. Resasco, Dispersion of single-walled carbon nanotubes in aqueous solutions of the anionic surfactant NaDDBS, *J. Phys. Chem.* **B 107**, 13357–13367 (2003).

284. W. Zhou, M.F. Islam, H. Wang, D.L. Ho, A.G. Yodh, K.I. Winey, and J.E. Fischer, Small angle neutron scattering from single-wall carbon nanotube suspensions: evidence for isolated rigid rods and rod networks, *Chem. Phys. Lett.* **384**, 185–189 (2004).

285. K.D. Ausman, R. Piner, O. Lourie, R.S. Ruoff, and M. Korobov, Organic solvent dispersions of SWNTs: toward solutions of pristine nanotubes, *J. Phys. Chem.* **B 104**, 8911–8915 (2000).

286. X. Liu, J.L. Spencer, A.B. Kaiser, and W.M. Arnold, Electric-field oriented carbon nanotubes in different dielectric solvents, *Curr. Appl. Phys.* **4**, 125–128 (2004).

287. Z. Hongbing, C. Wenzhe, W. Minquan, Zhengchan, and Z. Chunlin, Optical limiting effects of multi-walled carbon nanotubes suspension and silica xerogel composite, *Chem. Phys. Lett.* **382**, 313–317 (2003).

288. N. Saran, K. Parikh, D.-S. Suh, E. Muñoz, H. Kolla, and S.K. Manohar, Fabrication and characterization of thin films of single-walled carbon nanotube bundles on flexible plastic substrates, *J. Am. Chem. Soc.* **126**, 4462–4463 (2004).

289. R. Bandyopadhyaya, E. Nativ-Roth, O. Regev, and R. Yerushalmi-Rosen, Stabilization of individual carbon nanotubes in aqueous solutions, *Nano Lett.* **2**, 25–28 (2002).

290. M.S. Strano, C.B. Huffman, V.C. Moore, M.J. O'Connell, E.H. Haroz, J. Hubbard, M. Miller, K. Rialon, C. Kittrell, S. Ramesh, R.H. Hauge, and R.E. Smalley, Reversible, band-gap-selective protonation of single-walled carbon nanotubes in solution, *J. Phys. Chem.* **B 107**, 6979–6985 (2003).

291. J. Ning, J. Zhang, Y. Pan, and J. Guo, Surfactants assisted processing of carbon nanotube-reinforced SiO_2 matrix composites, *Ceramics Int.* **30**, 63–67 (2004).

292. Z. Jin, L. Huang, S.H. Goh, G. Xu, and W. Ji, Characterization and nonlinear optical properties of a poly(acrylic acid)–surfactant–multi-walled carbon nanotube complex, *Chem. Phys. Lett.* **332**, 461–466 (2000).

293. V.A. Karachevtsev, A.Yu. Glamazda, U. Dettlauff-Weglikowska, V.S. Leontiev, A.M. Plokhotnichenko, and S. Roth, Spectroscopy study of SWNT in aqueous solution with different surfactants, *AIP Conf. Proc.* **685**, No. 1, 202–206 (2003).

294. G.L. Hwang and K.C. Hwang, Carbon nanotube reinforced ceramics, *J. Mater. Chem.* **11**, 1722–1725 (2001).

295. K.D. Ausman, M.J. O'Connell, P. Boul, L.M. Ericson, M.J. Casavant, D.A. Walters, C. Huffman, R. Saini, Y. Wang, E. Haroz, E.W. Billups, and R.E. Smalley, Roping and wrapping carbon nanotubes [http://smalley.rice.edu/rick's%20publications/ausman.pdf].

296. M.J. O'Connell, P. Boul, L.M. Ericson, C. Huffman, Y. Wang, E. Haroz, C. Kuper, J. Tour, K.D. Ausman, and R.E. Smalley, Reversible water-solubilization of single-walled carbon nanotubes by polymer wrapping *Chem. Phys. Lett.* **342**, 265–271 (2001).

297. D.W. Schaefer, J.M. Brown, D.P. Anderson, J. Zhao, K. Chokalingam, D. Tomlin, and J. Ilavsky, Structure and dispersion of carbon nanotubes, *J. Appl. Cryst.* **36**, 553–557 (2003).

298. D.W. Schaefer, J. Zhao, J.M. Brown, D.P. Anderson, and D.W. Tomlin, Morphology of dispersed carbon single-walled nanotubes, *Chem. Phys. Lett.* **375**, 369–375 (2003).

299. G.Z. Chen, M.S.P. Shaffer, D. Coleby, G. Dixon, W. Zhou, D.J. Fray, and A.H. Windle, Carbon nanotube and polypyrrole composites: coating and doping, *Adv. Mater.* **12**, 522–526 (2000).

300. B.Z. Tang and H. Xu, Preparation, alignment, and optical properties of soluble poly(phenylacetylene)-wrapped carbon nanotubes, *Macromolecules* **32**, 2569–2576 (1999).

301. D. Li, H. Wang, J. Zhu, X. Wang, L. Lu, and X. Yang, Dispersion of carbon nanotubes in aqueous solutions containing poly(diallyldimethylammonium chloride), *J. Mater. Sci. Lett.* **22**, 253–255 (2003).

302. B. Kim, H. Park, and W.M. Sigmund, Electrostatic interactions between shortened multiwall carbon nanotubes and polyelectrolytes, *Langmuir* **19**, 2525–2527 (2003).

303. J. Chen, M.J. Dyer, and M.-F. Yu, Cyclodextrin-mediated soft cutting of single-walled carbon nanotubes, *J. Am. Chem. Soc.* **123**, 6201–6202 (2001).

304. H. Dodziuk, A. Ejchart, W. Anczewski, H. Ueda, E. Krinichnaya, G. Dolgonos, and W. Kutner, Determination of the number of different types of SWNTs by complexation with η-cyclodextrin, *Chem. Commun.* 986 (2003).

305. Z.H. Wang, G.A. Luo, and S.F. Xiao, Functionalization of cyclodextrins-incorporated carbon nanotube electrodes for neutral nitrophenol recognition, *Sensors, 2003. Proceedings of IEEE.* Vol. 2, 941–945 (2003).

306. T. Takahashi, K. Tsunoda, H. Yajima, and T. Ishii, Isolation of single-wall carbon nanotube bundles through gelatin wrapping and unwrapping processes, *Chem. Lett.* **31**, 690–691 (2002).

307. H. Li, D.Q. Wang, H.L. Chen, B.L. Liu, and L.Z. Gao, A novel gelatin-carbon nanotubes hybrid hydrogel, *Macromol. Biosci.* **3**, 720–724 (2003).

308. A. Star, D.W. Steuerman, J.R. Heath, and J.F. Stoddart, Starched carbon nanotubes, *Angew. Chem. Int. Edit.* **41**, 2508–2512 (2003).

309. O.-K. Kim, J. Je, J.W. Baldwin, S. Kooi, P.E. Pehrsson, and L.J. Buckley, Solubilization of single-wall carbon nanotubes by supramolecular encapsulation of helical amylose, *J. Am. Chem. Soc.* **125**, 4426–4427 (2003).

310. M. Numata, M. Asai, K. Kaneko, T. Hasegawa, N. Fujita, Y. Kitada, K. Sakurai, and S. Shinkai, Curdlan and schizophylan (β-1,3-glucans) can entrap single-wall carbon nanotubes in their helical superstructure, *Chem. Lett.* **33**, 232–233 (2004).

311. G.R. Dieckmann, A.B. Dalton, P.A. Johnson, J. Razal, J. Chen, G.M. Giordano, E. Muñoz, I.H. Musselman, R.H. Baughman, and R.K. Draper, Controlled assembly of carbon nanotubes by designed amphiphilic peptide helices, *J. Am. Chem. Soc.* **125**, 1770–1777 (2003).

312. F. Torrens, Calculation on solvent dispersions of carbon nanotubes, *The Electrochem. Soc. 205th Meet.*, Abs. 457 (2004).

313. Z. Shi, Y. Lian, X. Zhou, Z. Gu, Y. Zhang, S.Iijima, Q. Gong, H. Li, and S.-L. Zhang, Single-wall carbon nanotube colloids in polar solvents, *Chem. Commun.* 461–462 (2000).

314. X. Gao, T. Hu, L. Liu, and Z. Guo, Self-assembly of modified carbon nanotubes in toluene, *Chem. Phys. Lett.* **370**, 661–664 (2003).

315. M.E. Itkis, D.E. Perea, S. Niyogi, S.M. Rickard, M. Hamon, H. Hu, B. Zhao, and R.C. Haddon, Purity evaluation of as-prepared single-walled carbon nanotube soot by use of solution-phase near-IR spectroscopy, *Nano Lett.* **3**, 309–314 (2003).

316. A.B. Dalton, W.J. Blau, G. Chambers, J.N. Coleman, K. Henderson, S. Lefrant, B. McCarthy, C. Stephan, and H.J. Byrne, A functional conjugated polymer to process, purify and selectively interact with single wall carbon nanotubes, *Synth. Met.* **121**, 1217–1218 (2001).

317. D.W. Steuerman, A. Star, R. Narizzano, H. Choi, R.S. Ries, C. Nicolini, J.F. Stoddart, and J.R. Heath, Interactions between conjugated polymers and single-walled carbon nanotubes, *J. Phys. Chem.* **B 106**, 3124–3130 (2002).

318. A. Star, Y. Liu, K. Grant, L. Ridvan, J.F. Stoddart, D.W. Steuerman, M.R. Diehl, A. Boukai, and J.R. Heath, Noncovalent sidewall functionalization of single-walled carbon nanotubes, *Macromolecules* **36**, 553–560 (2003).

319. J.N. Coleman, D.F. O'Brien, M. in het Panhius, A.B. Dalton, B. McCarthy, R.C. Barklie, and W.J. Blau, Solubility and purity of nanotubes in arc discharge carbon powder, *Synth. Met.* **121**, 1229–1230 (2001).

320. J. Chen, H. Liu, W.A. Weimer, M.D. Halls, D.H. Waldeck, and G.C. Walker, Noncovalent engineering of carbon nanotube surface by rigid functional conjugated polymers, *J. Am. Chem. Soc.* **124**, 9034–9035 (2002).

321. J.-M. Nam, M.A. Ratner, X. Liu, and C.A. Mirkin, Single-walled carbon nanotubes and C_{60} encapsulated by a molecular macrocycles, *J. Phys. Chem.* **B 107**, 4705–4710 (2003).

322. Z. Jin, L. Huang, S.H. Goh, G. Xu, and W. Ji, Size-dependent optical limiting behavior of multi-walled carbon nanotubes, *Chem. Phys. Lett.* **352**, 328–333 (2002).

323. V.A. Davis, L.M. Ericson, R. Saini, R. Sivarajan, R.H. Hauge, R.E. Smalley, and M. Pasquali, Rheology, phase behavior, and fiber spinning of carbon nanotube dispersion. Paper prepared for presentation at the 2001 AIChE annual meeting, November 9, 2001, T7019 session.

324. V.A. Davis, L.M. Ericson, A.N.G. Parra-Vasquez, H. Fan, Y. Wang, V. Prieto, J.A. Longoria, S. Ramesh, R.K. Saini, C. Kittrell, W.E. Billups, W.W. Adams, R.H. Hauge, R.E. Smalley, and M. Pasquali, Phase behavior and rheology of SWNTs in superacids, *Macromolecules* **37**, 154–160 (2004).

325. W. Zhou, J. Vavro, C. Guthy, K.I. Winey, J.E. Fisher, L.M. Ericson, S. Ramesh, R. Saini, V.A. Davis, C. Kittrell, M. Pasquali, R.H. Hauge, and R.E. Smalley, Single wall carbon nanotube fibers extruded from super-acid suspensions: preferred orientation, electrical, and thermal transport, *J. Appl. Phys.* **95**, 649–655 (2004).

326. T.V. Sreekumar, T. Liu, S. Kumar, L.M. Ericson, R.H. Hauge, and R.E. Smalley, Single-wall carbon nanotube films, *Chem. Mater.* **15**, 175–178 (2003).

327. H. Dai, J. Kong, C. Zhou, N. Franklin, T. Tombler, A. Cassell, S. Fan, and M. Chapline, Controlled chemical routes to nanotube architectures, physics, and devices, *J. Phys. Chem.* **B 103**, 11246–11255 (1999).

328. X. Yu, T. Mu, H. Huang, Z. Liu, and N. Wu, The study of the attachment of a single-walled carbon nanotube to a self-assembled monolayer using X-ray photoelectron spectroscopy, *Surf. Sci.* **461**, 199–207 (2000).

329. J. Liu, M.J. Casavant, M. Cox, D.A. Walters, P. Boul, W. Lu, A.J. Rimberg, K.A. Smith, D.T. Colbert, and R.E. Smalley, Controlled deposition of individual single-walled carbon nanotubes on chemically functionalized templates, *Chem. Phys. Lett.* **303**, 125–129 (1999).

330. K.H. Choi, J.P. Bourgoin, S. Auvray, D. Esteve, G.S. Duesberg, S. Roth, and M. Burghard, Controlled deposition of carbon nanotubes on a patterned substrate, *Surf. Sci.* **462**, 195–202 (1999).

331. S.G. Rao, L. Huang, W. Setyawan, and S. Hong, Large-scale assembly of carbon nanotubes, *Nature* **425**, 36 (2003).

332. B. Wincheski, J. Smits, M. Namkung, J. Ingram, N. Watkins, J.D. Jordan, and R. Louie, Nanomanipulation and lithography for carbon nanotube based nondestructive evaluation sensor development [http://techreports.larc.nasa.gov/ltrs/PDF/2002/mtg/NASA-2002-sem-bw.pdf].

333. B. Kim and W.M. Sigmund, Self-alignment of shortened multiwall carbon nanotubes on polyelectrolyte layers, *Langmuir* **19**, 4848–4851 (2003).

334. M. Sano, A. Kamino, J. Okamura, and S. Shinkai, Noncovalent self-assembly of carbon nanotubes for construction of "cages," *Nano Lett.* **2**, 531–533 (2002).

335. J. Chen and W.A. Weimer, Room-temperature assembly of directional carbon nanotube strings, *J. Am. Chem. Soc.* **124**, 758–759 (2002).

336. Y. Yang, H. Zou, B. Wu, J. Zhang, Z. Liu, X. Guo, and Z. Du, Directional carbon nanotube assembly via solvent evaporation induced mechanical stretching, *Advanced Nanomaterials Nanodevices (IUMRS-ICEM 2002, Xian, China, June 10–14, 2002).* 12–22 (2002).

337. Y.-H. Li, C. Xu, B. Wei, X. Zhang, M. Zheng, D. Wu, and P.M. Ajayan, Self-organized ribbons of aligned carbon nanotubes, *Chem. Mater.* **14**, 483–485 (2002).

338. H. Shimoda, S.J. Oh, H.Z. Geng, R.J. Walker, X.B. Zhang, L.E. McNeil, and O. Zhou, Self-assembly of carbon nanotubes, *Adv. Mater.* **14**, 899–901 (2002).

339. H. Shimoda, L. Fleming, K. Horton, and O. Zhou, Formation of macroscopically ordered carbon nanotube membranes by self-assembly, *Physica* **B 323**, 135–136 (2002).

340. N. Chakrapani, B. Wei, A. Carrillo, P.M. Ajayan, and R.S. Kane, Capillary-driven assembly of two-dimensional cellular carbon nanotube foams, *Proc. Nat. Acad. Sci.* **101**, 4009–4012 (2004).

341. I. Dierking, G. Scalia, P. Morales, and D. LeClere, Aligning and re-orienting carbon nanotubes by nematic liquid crystals [http://www.mat.casaccia.enea.it/link6.pdf].

342. M.D. Lynch and D.L. Patrick, Organizing carbon nanotubes with liquid crystals, *Nano Lett.* **2**, 1197–1201 (2002).

343. M. Senthil Kumar, T.H. Kim, S.H. Lee, S.M. Song, J.W. Yang, K.S. Nahm, and E.-K. Suh, Influence of electric field type on the assembly of single walled carbon nanotubes, *Chem. Phys. Lett.* **383**, 235–239 (2004).

344. M.R. Diehl, S.N. Yaliraki, R.A. Beckman, M. Barahona, and J.R. Heath, Self-assembled, deterministic carbon nanotube wiring networks, *Angew. Chem. Int. Ed.* **41**, 353–356 (2002).

345. B. Gao, G.Z. Yue, Q. Qiu, Y. Cheng, H. Shimoda, L. Fleming, and O. Zhou, Fabrication and electron field emission properties of carbon nanotube films by electrophoretic deposition, *Adv. Mater.* **13**, 1770–1774 (2001).

346. J. Tang, B. Gao, H. Geng, O.D. Velev, L.-C. Qin, and O. Zhou, Assembly of 1D nanostructures into sub-micrometer diameter fibrils with controlled and variable length by electrophoresis, *Adv. Mater.* **15**, 1352–1355 (2003).

347. M. Burghard, G. Duesberg, G. Philipp, J. Muster, and S. Roth, Controlled adsorption of carbon nanotubes on chemically modified electrode arrays, *Adv. Mater.* **10**, 584–588 (1998).

348. Y. Guo, J. Wu, and Y. Zhang, Manipulation of single-wall carbon nanotubes into aligned molecular layers, *Chem. Phys. Lett.* **362**, 314–318 (2002).

349. L. Feng, H. Li, F. Li, Z. Shi, and Z. Gu, Functionalization of carbon nanotubes with amphiphilic molecules and their Langmuir-Blodgett films, *Carbon* **41**, 2385–2391 (2003).

350. N.P. Armitage, J.-C.P. Gabriel, and G. Grüner, Quasi-Langmuir-Blodgett thin film deposition of carbon nanotubes, *J. Appl. Phys.* **95**, 3328–3330 (2004).

351. V. Bavastrello, S. Carrara, M.K. Ram, and C. Nicolini, Optical and electrochemical properties of poly(o-toluidine) multiwalled carbon nanotubes composite Langmuir-Schaefer films, *Langmuir* **20**, 969–973 (2004).

352. T. Fukushima, A. Kosaka, Y. Ishimura, T. Yamamoto, T. Takigawa, N. Ishii, and T. Aida, Molecular ordering of organic molten salts triggered by single-walled carbon nanotubes, *Science* **300**, 2972–2974 (2003).

353. B. Vigolo, P. Poulin, M. Lukas, P. Launois, and P. Bernier, Improved structure and properties of single-wall carbon nanotube spun fibers, *Appl. Phys. Lett.* **81**, 1210–1212 (2002).

354. F. Ko, Y. Gogotsi, A. Ali, N. Naguib, H. Ye, G. Yang, C. Li, and P. Willis, Electrospinning of continuous carbon nanotube filled nanofiber yarns, *Adv. Mater.* **15**, 1164–1165 (2003).

355. R. Sen, B. Zhao, D. Perea, M.E. Itkis, H. Hu, J. Love, E. Bekyarova, and R.C. Haddon, Preparation of single-walled carbon nanotube reinforced polystyrene and polyurethane nanofibers and membranes, *Nano Lett.* **4**, 459–464 (2004).

356. T.E. Müller, D.G. Reid, W.K. Hsu, J.P. Hare, H.W. Kroto, and D.R.M. Walton, Synthesis of nanotubes via catalytic pyrolysis of acetylene: a SEM study, *Carbon* **35**, 951–966 (1997).

357. Y.K. Chen, A. Chu, J. Cook, M.L.H. Green, P.J.F. Harris, R. Heesom, M. Humphries, J. Sloan, S.C. Tsang, and J.F.C. Turner, Synthesis of carbon nanotubes containing metal oxides and metals of the d-block and f-block transition metals and related studies, *J. Mater. Sci.* **7**, 545–549 (1997).

358. C.T. Kuo, C.H. Lin, and A.Y. Lo, Feasibility studies of magnetic particle-embedded carbon nanotubes for perpendicular recording media, *Diamond Rel. Mater.* **12**, 799–805 (2003).

359. H. Liu, J.-M. Dong, M.-C. Qian, and X.-G. Wan, Dependence of in-tube doping on the radius and helicity of single-wall carbon nanotubes, *Chinese Phys.* **12**, 542–547 (2003).

360. D. Bera, S.C. Kuiry, M. McCutchen, A. Kruize, H. Heinrich, M. Meyyappan, and S. Seal, In-situ synthesis of palladium nanoparticles-filled carbon nanotubes using arc-discharge in solution, *Chem. Phys. Lett.* **386**, 364–368 (2004).

361. E. Dujardin, C. Meny, P. Panissod, J.-P. Kintzinger, N. Yao, and T.W. Ebbesen, Interstitial metallic residues in purified single shell carbon nanotubes, *Solid State Commun.* **114**, 543–546 (2000).

362. X.P. Gao, Y. Zhang, X. Chen, G.L. Pan, J. Yan, F. Wu, H.T. Yuan, and D.Y. Song, Carbon nanotubes filled with metallic nanowires, *Carbon* **42**, 47–52 (2004).

363. C.H. Liang, G.W. Meng, L.D. Zhang, N.F. Shen, and X.Y. Zhang, Carbon nanotubes filled partially or completely with nickel, *J. Cryst. Growth* **218**, 136–139 (2000).

364. B.C. Satishkumar, A. Govindaraj, P.V. Vanitha, A.K. Raychaudhuri, and C.N.R. Rao, Barkhausen jumps and related magnetic properties of iron nanowires encapsulated in aligned carbon nanotube bundles, *Chem. Phys. Lett.* **362**, 301–306 (2002).

365. A. Leonhardt, M. Ritschel, R. Kozhuharova, A. Graff, T. Mühl, R. Huhle, I. Mönch, D. Elefant, and C.M. Schneider, Synthesis and properties of filled carbon nanotubes, *Diamond Rel. Mater.* **12**, 790–793 (2003).

366. N. Grobert, M. Mayne, M. Terrones, J. Sloan, R.E. Dunin-Borkowski, R. Kamalakaran, T. Seeger, H. Terrones, M. Rühle, D.R.M. Walton, H.W. Kroto, and J.L. Hutchison, Alloy nanowires: invar inside carbon nanotubes, *Chem. Commun.* 471–472 (2001).

367. X. Li, Z. Lei, R. Ren, J. Liu, X. Zuo, Z. Dong, H. Wang, and J. Wang, Characterization of carbon nanohorn encapsulated Fe particles, *Carbon* **41**, 3068–3072 (2003).

368. R. Sen, A. Govindaraj, and C.N.R. Rao, Metal-filled and hollow carbon nanotubes obtained by the decomposition of metal-containing free precursor molecules, *Chem. Mater.* **9**, 2078–2081 (1997).

369. B.H. Liu, J. Ding, Z.Y. Zhong, Z.L. Dong, T. White, and J.Y. Lin, Large-scale preparation of carbon-encapsulated cobalt nanoparticles by the catalytic method, *Chem. Phys. Lett.* **358**, 96–102 (2002).

370. G.Y. Zhang and E.G. Wang, Cu-filled carbon nanotubes by simultaneous plasma-assisted copper incorporation, *Appl. Phys. Lett.* **82**, 1926–1928 (2003).

371. C.Y. Zhi, D.Y. Zhong, and E.G. Wang, GaN-filled carbon nanotubes: synthesis and photoluminescence, *Chem. Phys. Lett.* **381**, 715–719 (2003).

372. A.K. Sinha, D.W. Hwang, and L.-P. Hwang, A novel approach to bulk synthesis of carbon nanotubes filled with metal by a catalytic chemical vapor deposition method, *Chem. Phys. Lett.* **332**, 455–460 (2000).

373. R. Ma, Y. Bando, and T. Sato, Coaxial nanocables: Fe nanowires encapsulated in BN nanotubes with intermediate C layers, *Chem. Phys. Lett.* **350**, 1–5 (2001).

374. S. Liu and J. Zhu, Carbon nanotubes filled with long continuous cobalt nanowires, *Appl. Phys.* **A 70**, 673–675 (2000).

375. S. Liu, S. Zhu, Y. Mastai, I. Felner, and A. Gedanken, Preparation and characteristics of carbon nanotubes filled with cobalt, *Chem. Mater.* **12**, 2205–2211 (2000).

376. Y. Gogotsi, N. Naguib, and J.A. Libera, In situ chemical experiments in carbon nanotubes, *Chem. Phys. Lett.* **356**, 354–360 (2002).

377. J.L. Rivera, C. McCabe, and P.T. Cummings, Layering behavior and axial phase equilibria of pure water and water + carbon dioxide inside single wall carbon nanotubes, *Nano Lett.* **2**, 1427–1431 (2002).

378. E. Dujardin, T.W. Ebbesen, H. Hiura, and K. Tanigaki, Capillarity and wetting of carbon nanotubes, *Science* **265**, 1850–1852 (1994).

379. E. Dujardin, T.W. Ebbesen, A. Krishnan, and M.M. Treacy, Wetting of single shell carbon nanotubes, *Adv. Mater.* **10**, 1472–1475 (1998).

380. G. Hummer, J.C. Rasaiah, and J.P. Noworyta, Water conduction through the hydrophobic channel of a carbon nanotube, *Nature* **414**, 188–190 (2001).

381. M.S.P. Sansom and P.C. Biggin, Biophysics: water at the nanoscale, *Nature* **414**, 156 (2001).

382. J. Martí and M.C. Gordillo, Temperature effects on the static and dynamic properties of liquid water inside nanotubes. *Phys. Rev.* **E 64**, 21504.1–6 (2001).

383. K. Koga, G.T. Gao, H. Tanaka, and X.C. Zeng, Formation of ordered ice nanotubes inside carbon nanotubes, *Nature* **412**, 802–805 (2001).

384. W.H. Noon, K.D. Ausman, R.E. Smalley, and J. Ma, Helical ice-sheets inside carbon nanotubes in the physiological condition, *Chem. Phys. Lett.* **355**, 445–448 (2002).

385. A. Waghe, J.C. Rasaiah, and G. Hummer, Filling and emptying kinetics of carbon nanotubes in water, *J. Chem. Phys.* **117**, 10789–10795 (2002).

386. A. Kalra, S. Garde, and G. Hummer, Osmotic water transport through carbon nanotube membranes, *Proc. Nat. Acad. Sci. USA* **100**, 10175–10180 (2003).

387. M.P. Rossi, H. Ye, Y. Gogotsi, S. Babu, P. Ndungu, and J.-C. Bradley, Environmental scanning electron microscopy study of water in carbon nanopipes, *Nano Lett.* **4**, 989–993 (2004).

388. S. Joseph, R.J. Mashl, E. Jacobsson, and N.R. Aluru, Ion channel based biosensors: ionic transport in carbon nanotubes, *Nanotech 2003*, **1**, 158–161 (2003) [http://www.nsti.org/procs/Nanotech2003v1/8/T23.05].

389. S.C. Tsang, Y.K. Chen, P.J.F. Harris, and M.L. Green, A simple chemical method of opening and filling carbon nanotubes, *Nature* **372**, 159–162 (1994).

390. J. Chancolon, F. Archaimbault, S. Delpeux, A. Pineau, M.L. Saboungi, and S. Bonnamy, Filling of multiwalled carbon nanotubes with oxides and metals, *AIP Conf. Proc.* **633**, No. 1, 131–134 (2002).

391. P. Corio, A.P. Santos, P.S. Santos, M.L.A. Temperini, V.W. Brar, M.A. Pimenta, and M.S. Dresselhaus, Characterization of single-wall carbon nanotubes filled with silver and with chromium compounds, *Chem. Phys. Lett.* **383**, 475–480 (2004).

392. C. Pham-Huu, N. Keller, C. Estournès, G. Ehret, and M.J. Ledoux, Synthesis of $CoFe_2O_4$ nanowire in carbon nanotubes. A new use of confinement effect, *Chem. Commun.* 1882–1883 (2002).

393. C. Pham-Huu, N. Keller, C. Estournès, G. Ehret, J.M. Grenèche, and M.J. Ledoux, Microstructural investigation and magnetic properties of $CoFe_2O_4$ nanowires synthesized inside carbon nanotubes, *Phys. Chem. Chem. Phys.* **5**, 3716–3723 (2003).

394. N. Keller, C. Pham-Huu, C. Estournès, J.-M. Grenèche, G. Ehret, and M.J. Ledoux, Carbon nanotubes as a template for mild synthesis of magnetic $CoFe_2O_4$ nanowires, *Carbon* **42**, 1395–1399 (2004).

395. J. Sloan, J. Hammer, M. Zwiefka-Sibley, and M.L.H. Green, The opening and filling of single walled carbon nanotubes (SWTs), *Chem. Commun.* 347–348 (1998).

396. B. Rajesh, K.R. Thampi, J.-M. Bonard, and B. Viswanathan, Preparation of a Pt–Ru bimetallic system supported on carbon nanotubes, *J. Mater. Chem.* **10**, 1757–1759 (2000).

397. B. Rajesh, K.R. Thampi, J.-M. Bonard, and B. Visvanathan, Preparation of Pt-Ru bimetallic catalyst supported on carbon nanotubes, *Bull. Mater. Sci.* **23**, 341–344 (2000).

398. B. Rajesh, V. Karthik, S. Karthikeyan, K.R. Thampi, J.-M. Bonard, and B. Viswanathan, Pt–WO_3 supported on carbon nanotubes as possible anodes for direct methanol fuel cells, *Fuel* **81**, 2177–2190 (2002).

399. B. Rajesh, K.R. Thampi, J.-M. Bonard, N. Xanthopoulos, H.J. Mathieu, and B. Viswanathan, Carbon nanotubes generated from template carbonization of polyphenyl acetylene as the support for electrooxidation of methanol, *J. Phys. Chem.* **B 107**, 2701–2708 (2003).

400. J. Bao, Q. Zhou, J. Hong, and Z. Xu, Synthesis and magnetic behavior of an array of nickel-filled carbon nanotubes, *Appl. Phys. Lett.* **81**, 4592–4594 (2002).

401. K. Matsui, T. Kyotani, and A. Tomita, Hydrothermal synthesis of nano-sized iron oxide crystals in the cavity of carbon nanotubes, *Mol. Cryst. Liq. Cryst.* **387**, 1–5 (2002).

402. J. Mittal, M. Monthioux, H. Allouche, and O. Stephan, Room temperature filling of single-wall carbon nanotubes with chromium oxide in open air, *Chem. Phys. Lett.* **339**, 311–318 (2001).

403. H.-Q. Wu, X.-W. Wei, M.-W. Shao, J.-S. Gu, and M.-Z. Qu, Synthesis of copper oxide nanoparticles using carbon nanotubes as templates, *Chem. Phys. Lett.* **364**, 152–156 (2002).

404. L. Jiang and L. Gao, Modified carbon nanotubes: an effective way to selective attachment of gold nanoparticles, *Carbon* **41**, 2923–2929 (2003).

405. X.-R. Ye, Y. Lin, C. Wang, and C.M. Wai, Supercritical fluid fabrication of metal nanowires and nanorods templated by multiwalled carbon nanotubes, *Adv. Mater.* **15**, 316–319 (2003).

406. D.A. Britz, A.N. Khlobystov, J. Wang, A.S. O'Neil, M. Poliakoff, A. Ardavan, and G.A.D. Briggs, Selective host-guest interaction of single-walled carbon nanotubes with functionalized fullerenes, *Chem. Commun.* 176–177 (2004).

407. F. Simon, H. Kuzmany, H. Rauf, T. Pichler, J. Bernardi, H. Peterlik, L. Korecz, F. Fülöp, and A. Jánossy, Low temperature fullerene encapsulation in single-wall carbon nanotubes: synthesis of N@C$_{60}$@SWCNT, *Chem. Phys. Lett.* **383**, 362–367 (2004).

408. M. Yudasaka, K. Ajima, K. Suenaga, T. Ichihashi, A. Hashimoto, and S. Iijima, Nano-extraction and nano-condensation for C$_{60}$ incorporation into a single-wall carbon nanotubes in liquid phases, *Chem. Phys. Lett.* **380**, 42–46 (2003).

409. T. Takenobu, T. Takano, M. Shiraishi, Y. Murakami, M. Ata, H. Kataura, Y. Achiba, and Y. Iwasa, Stable and controlled amphoteric doping by encapsulation of organic molecules inside carbon nanotubes, *Nat. Mater.* **2**, 683–688 (2003).

410. H. Gao, Y. Kong, D. Cui, and C.S. Ozkan, Spontaneous insertion of DNA oligonucleotides into carbon nanotubes, *Nano Lett.* **3**, 471–473 (2003).

411. S. Friedrichs, R.R. Meyer, J. Sloan, A.I. Kirkland, J.L. Hutchison, and M.L.H. Green, Complete characterization of a Sb$_2$O$_3$/(21,–8)SWNT inclusion composite, *Chem. Commun.* 929–930 (2001).

412. S. Friedrichs, J. Sloan, M.L.H. Green, J.L. Hutchison, R.R. Meyer, and A.I. Kirkland, Simultaneous determination of inclusion crystallography and nanotube conformation for a Sb$_2$O$_3$/single-walled nanotube composite, *Phys. Rev.* **B 64**, 045406.1–8 (2001).

413. J. Sloan, D.M. Wright, H.-G. Woo, S. Bailey, G. Brown, A.P.E. York, K.S. Coleman, J.L. Hutchison, and M.L.H. Green, Capillarity and silver nanowire formation observed in single-walled carbon nanotubes, *Chem. Commun.* 699–700 (1999).

414. J. Sloan, A.I. Kirkland, J.L. Hutchison, and M.L.H. Green, Integral atomic layer architectures of 1D crystals inserted into single-walled carbon nanotubes, *Chem. Commun.* 1319–1332 (2002).

415. C. Xu, J. Sloan, G. Brown, S. Bailey, V.C. Williams, S. Friedrichs, K.S. Coleman, E. Flahaut, J.L. Hutchison, R.E. Dunin-Borkowski, and M.L.H. Green, 1D lanthanide halide crystals inserted into single-walled carbon nanotubes, *Chem. Commun.* 2427–2428 (2000).

416. C.-H. Kiang, J.-S. Choi, T.T. Tran, and A.D. Bacher, Molecular nanowires of 1 nm diameter from capillary filling of single-walled nanotubes, *J. Phys. Chem.* **B 103**, 7449–7451 (1999).

417. C.-H. Kiang, Electron irradiation induced dimensional change in bismuth-filled carbon nanotubes, *Carbon* **38**, 1699–1701 (2000).

418. X. Fan, E.C. Dickey, P.C. Eklund, K.A. Williams, L. Grigorian, R. Buczko, S.T. Pantelides, and S.J. Pennycook, Atomic arrangement of iodine atoms inside single-walled carbon nanotubes, *Phys. Rev. Lett.* **84**, 4621–4624 (2000).

419. W. Zhou, S. Xie, L. Sun, D. Tang, Y. Li, Z. Liu, L. Ci, X. Zou, G. Wang, P. Tan, X. Dong, B. Xu, and B. Zhao, Raman scattering and thermogravimetric analysis of iodine-doped multiwall carbon nanotubes, *Appl. Phys. Lett.* **80**, 2553–2555 (2002).

420. Y. Gao and Y. Bando, Carbon nanothermometer containing gallium, *Nature* **415**, 599 (2002).

421. A. Fujiwara, K. Ishii, H. Suematsu, H. Kataura, Y. Maniwa, S. Suzuki, and Y. Achiba, Gas adsorption in the inside and outside of single-walled carbon nanotubes, *Chem. Phys. Lett.* **336**, 205–211 (2001).

422. Z. Mao, A. Garg, and S.B. Sinnott, Molecular dynamics simulations of the filling and decorating of carbon nanotubules, *Nanotechnology* **10**, 273–277 (1999).

423. Z. Mao and S.B. Sinnott, A computational study of molecular diffusion and dynamic flow through carbon nanotubes, *J. Phys. Chem.* **B 104**, 4618–4624 (2000).

424. M. Li, M. Lu, C. Wang, and H. Li, Preparation of well-aligned carbon nanotubes/silicon nanowires core-sheath composite structure arrays in porous anodic aluminium oxide templates, *Sci. China* **45 B**, No. 4, 435–444 (2002).

425. G. Brown, S.R. Bailey, J. Sloan, C. Xu, S. Friedrichs, E. Flahaut, K.S. Coleman, J.L. Hutchison, R.E. Dunin-Borkowski, and M.L.H. Green, Electron beam induced *in situ* characterization of 1D ZrCl$_4$ chains within single-walled carbon nanotubes, *Chem. Commun.* 845–846 (2001).

426. H. Kataura, Y. Maniwa, T. Kodama, K. Kikuchi, K. Hirahara, K. Suenaga, S. Iijima, S. Suzuki, Y. Achiba, and W. Krätschmer, High-yield fullerene encapsulation in single-wall carbon nanotubes, *Synth. Met.* **121**, 1195–1196 (2001).

427. K. Hirahara, K. Suenaga, S. Bandow, H. Kato, T. Okazaki, H. Shinohara, and S. Iijima, One-dimensional metallofullerene crystal generated inside single-walled carbon nanotubes, *Phys. Rev. Lett.* **85**, 5384–5387 (2000).

428. K. Hirahara, S. Bandow, K. Suenaga, H. Kato, T. Okazaki, H. Shinohara, and S. Iijima, Electron diffraction study of one-dimensional crystals of fullerenes, *Phys. Rev.* **B 64**, 115420.1–5 (2001).

429. A.R. Harutyunyan, B.K. Pradhan, G.U. Sumanasekera, E.Yu. Korobko, A.A. Kuznetsov, Carbon nanotubes for medical applications, *Eur. Cells Mater.* **3**, Suppl. 2, 84–87 (2002).

430. B.K. Pradhan, T. Toba, T. Kyotani, and A. Tomita. Inclusion of crystalline iron oxide nanoparticle in uniform carbon nanotubes prepared by a template carbonization method, *Chem. Mater.* **10**, 2510–2515 (1998).

431. B.K. Pradhan, T. Kyotani, and A. Tomita, Nickel nanowires of 4 nm diameter in the cavity of carbon nanotubes, *Chem. Commun.* 1317–1318 (1999).

432. K. Matsui, B.K. Pradhan, T. Kyotani, and A. Tomita, Formation of nickel oxide nanoribbons in the cavity of carbon nanotubes, *J. Phys. Chem.* **B 105**, 5682–5688 (2001).

433. X. Li, L. Liu, Y. Qin, W. Wu, Z.-X. Guo, L. Dai, and D. Zhu, C_{60} modified single-walled carbon nanotubes, *Chem. Phys. Lett.* **377**, 32–36 (2003).

434. G.-H. Jeong, A.A. Farajian, T. Hirata, R. Hatakeyama, K. Tohji, T.M. Briere, H. Mizuseki, and Y. Kawazoe, Encapsulation of cesium inside single-walled carbon nanotubes by plasma-ion irradiation method, *Thin Solid Films* **435**, 307–311 (2003).

435. F. Banhart, The formation of a connection between carbon nanotubes in an electron beam, *Nano Lett.* **1**, 329–332 (2001).

436. Z. Wang, L. Yu, W. Zhang, Y. Ding, Y. Li, Z. Zhu, J. Han, H. Xu, G. He, Y. Chen, and G. Hu, Amorphous molecular junctions by ion irradiation on carbon nanotubes, *Phys. Lett.* **A 324**, 321–325 (2004).

437. D. Ding and J. Wang, Electron irradiation of multiwalled carbon nanotubes with encapsulated Ni particles, *Carbon* **40**, 787–803 (2002).

438. R. Che, L.-M. Peng, Q. Chen, X.F. Duan, B.S. Zou, and Z.N. Gu, Controlled synthesis and phase transformation of ferrous nanowires inside carbon nanotubes, *Chem. Phys. Lett.* **375**, 59–64 (2003).

439. J.S. Lee, B. Min, K. Cho, S. Kim, J. Park, Y.T. Lee, N.S. Kim, M.S. Lee, S.O. Park, and J.T. Moon, Al_2O_3 nanotubes and nanorods fabricated by coating and filling of carbon nanotubes with atomic-layer deposition, *J. Cryst. Growth* **254**, 443–448 (2003).

440. Z. Liu, Y. Gao, and Y. Bando, Highly effective metal vapor absorbents based on carbon nanotubes, *Appl. Phys. Lett.* **81**, 4844–4846 (2002).

441. B.W. Smith, M. Monthioux, and D.E. Luzzi, Encapsulated C_{60} in carbon nanotubes, *Nature* **396**, 323–324 (1998).

442. J. Sloan, R.E. Dunin-Borkowski, J.L. Hutchison, K.S. Coleman, V.C. Williams, J.B. Claridge, A.P.E. York, C. Xu, S.R. Bailey, G. Brown, S. Friedrichs, and M.L.H. Green, The size distribution, imaging and obstructing properties of C_{60} and higher fullerenes formed within arc-grown single-walled carbon nanotubes, *Chem. Phys. Lett.* **316**, 191–198 (2000).

443. R. Pfeiffer, Ch. Kramberger, Ch. Schaman, A. Sen, M. Holzweber, H. Kuzmany, T. Pichler, H. Kataura, and Y. Achiba, Defect free inner tubes in DWCNTs, *AIP Conf. Proc.* **685**, No. 1, 297–301 (2003).

444. H. Kataura, Y. Maniwa, M. Abe, A. Fujiwara, T. Kodama, K. Kikuchi, M. Imahori, Y. Misaki, S. Suzuki, and Y. Achiba, Optical properties of fullerene and non-fullerene peapods, *Appl. Phys.* **A 74**, 349–354 (2002).

445. S. Bandow, M. Takizawa, K. Hirahata, M. Yudasaka, and S. Iijima, Raman scattering study of double-wall carbon nanotubes derived from the chains of fullerenes in single-wall carbon nanotubes, *Chem. Phys. Lett.* **337**, 48–54 (2001).

446. S. Bandow, T. Hiraoka, T. Yumura, K. Hirahara, H. Shinohara, and S. Iijima, Raman scattering study on fullerene derived intermediates formed within single-wall carbon nanotube: from peapod to double-wall carbon nanotube, *Chem. Phys. Lett.* **384**, 320–325 (2004).

447. T. Okazaki, K. Suenaga, K. Hirahara, S. Bandow, S. Iijima, and H. Shinohara, Real time reaction dynamics in carbon nanotubes, *J. Am. Chem. Soc.* **123**, 9673–9674 (2001).

448. C.M. Megaridis, A. Güvenç Yaziciroglu, J.A. Libera, and Y. Gogotsi, Attoliter fluid experiments in individual closed-end carbon nanotubes: liquid film interface dynamics, *Phys. Fluids* **14**, L5–8 (2002).

449. H. Ye, N. Naguib, Y. Gogotsi, A.G. Yaziciroglу, and C.M. Megaridis, Wall structure and surface chemistry of hydrothermal carbon nanofibres, *Nanotechnology* **15**, 232–236 (2004).

450. M. Green, *Carbon nanotubes. Chemistry in a carbon cage*. The Institute of Applied Catalysis, 2000. [http://www.iac.org.uk/pages/vis3.html].

451. Z. Liu, X. Dai, J. Xu, B. Han, J. Zhang, Y. Wang, Y. Huang, and G. Yang, Encapsulation of polystyrene within carbon nanotubes with the aid of supercritical CO_2, *Carbon* **42**, 423–460 (2004).

452. Q. Yang, L. Li, H. Cheng, M. Wang, and J. Bai, Inner-tubular physicochemical processes of carbon nanotubes, *Chin. Sci. Bull.* **48**, 2395–2403 (2003).

453. Q. Jiang, N. Aya, and F.G. Shi, Nanotube size-dependent melting of single crystals in carbon nanotubes, *Appl. Phys.* **A 64**, 627–629 (1997).

454. J. Sloan, S. Friedrichs, R.R. Meyer, A.I. Kirkland, J.L. Hutchison, and M.L.H. Green, Structural changes induced in nanocrystals of binary compounds confined within single-wall carbon nanotubes: a brief review, *Inorg. Chim. Acta* **330**, 1–12 (2002).

455. J. Sloan, A.I. Kirkland, J.L. Hutchison, and M.L.H. Green, Structural characterization of atomically regulated nanocrystals formed within single-walled carbon nanotubes using electron microscopy, *Acc. Chem. Res.* **35**, 1054–1062 (2002).

456. J. Sloan, M.C. Novotny, S.R. Bailey, G. Brown, C. Xu, V.C. Williams, S. Friedrichs, E. Flahaut, R.R. Callender, A.P.E. York, K.S. Coleman, M.L.H. Green, R.E. Dunin-Borkowski, and J.L. Hutchison, Two layer 4:4 co-ordinated KI crystals grown within single-walled carbon nanotubes, *Chem. Phys. Lett.* **329**, 61–65 (2000).

457. M. Wilson, Structure and phase stability of novel "twisted" crystal structures in carbon nanotubes, *Chem. Phys. Lett.* **366**, 504–509 (2002).

458. M. Wilson, The formation of low-dimensional ionic crystallites in carbon nanotubes, *J. Chem. Phys.* **116**, 3027–3041 (2002).

459. Y. Maniwa, H. Kataura, M. Abe, A. Fujiwara, R. Fujiwara, H. Kira, H. Tou, S. Suzuki, Y. Achiba, E. Nishibori, M. Takata, M. Sakata, and H. Suematsu, C_{70} molecular stumbling inside single-walled carbon nanotubes, *J. Phys. Soc. Jpn.* **72**, 45–48 (2002).

460. A.C. Dillon, K.M. Jones, T.A. Bekkedahl, C.H. Kiang, D.S. Bethune, and M.J. Heben, Storage of hydrogen in single-walled carbon nanotubes, *Nature* **386**, 377–379 (1997).

461. A. Chambers, C. Park, R.T.K. Baker, and N.M. Rodriguez, Hydrogen storage in graphite nanofibers, *J. Phys. Chem.* **B 102**, 4253–4256 (1998).

462. P. Chen, X. Wu, J. Lin, and K.L. Tan, High H_2 uptake by alkali-doped carbon nanotubes under ambient pressure and moderate temperatures, *Science* **285**, 91–93 (1999).

463. Y. Chen, D.T. Shaw, X.D. Bai, E.G. Wang, C. Lund, W.M. Lu, and D.D.L. Chung, Hydrogen storage in aligned carbon nanotubes, *Appl. Phys. Lett.* **78**, 2128–2130 (2001).

464. C. Nützenadel, A. Züttel, D. Chartouni, and L. Schlapbach, Electrochemical storage of hydrogen in nanotube materials, *Electrochem. Solid-State Lett.* **2**, 30–32 (1999).

465. N. Rajalakshmi, K.S. Dhathathreyan, A. Govindaraj, and B.C. Satishkumar, Electrochemical investigation of single-walled carbon nanotubes for hydrogen storage, *Electrochim. Acta* **45**, 4511–4515 (2000).

466. M. Hirscher, M. Becher, M. Haluska, U. Dettlaff-Weglikowska, A. Quintel, G.S. Duesberg, Y.-M. Choi, P. Downes, M. Hulman, S. Roth, I. Stepanek, and P. Bernier, Hydrogen storage in sonicated carbon materials, *Appl. Phys.* **A 72**, 129–132 (2001).

467. G.-P. Dai, C. Liu, M. Liu, M.-Z. Wang, and H.-M. Cheng, Electrochemical hydrogen storage behavior of ropes of aligned single-walled carbon nanotubes, *Nano Lett.* **2**, 503–506 (2002).

468. F.J. Owens and Z. Iqbal, Electrochemical functionalization of carbon nanotubes with hydrogen [www.asc2002.com/summaries/1/LP-11.pdf].

469. H. Takagi, H. Hatori, Y. Soneda, N. Yoshizawa, and Y. Yamada, Adsorptive hydrogen storage in carbon and porous materials, *Mater. Sci. Eng.* **B 108**, 143–147 (2004).

470. A. Ansón, M.A. Callejas, A.M. Benito, W.K. Maser, M.T. Izquierdo, B. Rubio, J. Jagiello, M. Thommes, J.B. Parra, and M.T. Martínez, Hydrogen adsorption studies on single-wall carbon nanotubes, *Carbon* **42**, 1243–1248 (2004).

471. N. Nishimiya, K. Ishigaki, H. Takikawa, M. Ikeda, Y. Hibi, T. Sakakibara, A. Matsumoto, and K. Tsutsumi, Hydrogen sorption by single-walled carbon nanotubes prepared by a torch arc method, *J. Alloys Comp.* **339**, 275–282 (2002).

472. M. Shiraishi, T. Takenobu, and M. Ata, Gas-solid interactions in the hydrogen/single-walled carbon nanotube system, *Chem. Phys. Lett.* **367**, 633–636 (2003).

473. M. Shiraishi, T. Takenobu, H. Kataura, and M. Ata, Hydrogen adsorption and desorption in carbon nanotube systems and its mechanisms, *Appl. Phys.* **A 78**, 947–953 (2004).

474. B.P. Tarasov, J.P. Maehlen, M.V. Lototsky, V.E. Muradyan, and V.A. Yartys, Hydrogen sorption proper-
      ties of arc generated single-wall carbon nanotubes, *J. Alloys Comp.* **356–357**, 510–514 (2003).
475. P. David, T. Piquero, K. Metenier, Y. Pierre, J. Demoment, and A. Lecas-Hardit, Hydrogen adsorption
      in carbon materials [http://www.waterstof.org/2003072EHECP2-200.pdf].
476. C. Liu, Y.Y. Fan, M. Liu, H.T. Cong, H.M. Cheng, and M.S. Dresselhaus, Hydrogen storage in single-
      walled carbon nanotubes at room temperature, *Science* **286**, 1127–1129 (1999).
477. D. Luxembourg, G. Flamant, A. Guillot, and D. Laplaze, Hydrogen storage in solar produced single-
      walled carbon nanotubes, *Mater. Sci. Eng.* **B 108**, 114–119 (2004).
478. M.A. Callejas, A. Ansón, A.M. Benito, W. Maser, J.L.G. Fierro, M.L. Sanjuán, and M.T. Martínez,
      Enhanced hydrogen adsorption on single-wall carbon nanotubes by sample reduction, *Mater. Sci. Eng.*
      B 108, 120–123 (2004).
479. B.K. Pradhan, A.R. Harutyunyan, D. Stojkovic, J.C. Grossman, P. Zhang, M.W. Cole, V. Crespi,
      H. Goto, J. Fujiwara, and P.C. Eklund, Large cryogenic storage of hydrogen in carbon nanotubes at low
      pressures, *J. Mater. Res.* **17**, 2209–2216 (2002).
480. P. Sudan, A. Züttel, Ph. Mauron, Ch. Emmenegger, P. Wenger, and L. Schlapbach, Physisorption of
      hydrogen in single-walled carbon nanotubes, *Carbon* **41**, 2377–2383 (2003).
481. Y. Ye, C.C. Ahn, C. Witham, B. Fultz, J. Liu, A.G. Rinzler, D. Colbert, K.A. Smith, and R.E. Smalley,
      Hydrogen absorption and cohesive energy of single-walled carbon nanotubes, *Appl. Phys. Lett.* **74**,
      2307–2309 (1999).
482. X.B. Wu, P. Chen, J. Lin, and K.L. Tan, Hydrogen uptake by carbon nanotubes, *Int. J. Hydrogen*
      Energy **25**, 261–265 (2000).
483. X. Qin, X.P. Gao, H. Liu, H.T. Yuan, D.Y. Yan, W.L. Gong, and D.Y. Song, Electrochemical hydrogen
      storage of multiwalled carbon nanotubes, *Electrochem. Solid-State Lett.* **3**, 532–535 (2000).
484. A.K.M. Fazle Kibria, Y.H. Mo, K.S. Park, K.S. Nahm, and M.H. Yun, Electrochemical hydrogen storage
      behaviors of CVD, AD and LA grown carbon nanotubes in KOH medium, *Int. J. Hydrogen Energy*, **26**,
      823–829 (2001).
485. R.T. Yang, Hydrogen storage by alkali-doped carbon nanotubes — revisited, *Carbon* **38**, 623–641
      (2000).
486. G.Q. Ning, F. Wei, G.H. Luo, Q.X. Wang, Y.L. Wu, and H. Yu, Hydrogen storage in multiwall carbon
      nanotubes using samples up to 85 g, *Appl. Phys.* **A 78**, 955–959 (2004).
487. A. Cao, H. Zhu, X. Zhang, X. Li, D. Ruan, C. Xu, B. Wei, J. Liang, and D. Wu, Hydrogen storage of
      dense-aligned carbon nanotubes, *Chem. Phys. Lett.* **342**, 510–514 (2001).
488. X. Li, H. Zhu, L. Ci, C. Xu, Z. Mao, B. Wei, J. Liang, and D. Wu, Hydrogen uptake by graphitized
      multiwalled carbon nanotubes under moderate pressure and at room temperature, *Carbon* **39**,
      2077–2079 (2001).
489. H.W. Zhu, A. Chen, Z.Q. Mao, C.L. Xu, X. Xiao, B.Q. Wei, J. Liang, and D.H. Wu, The effect of sur-
      face treatments on hydrogen storage of carbon nanotubes, *J. Mater. Sci. Lett.* **19**, 1237–1239 (2000).
490. H. Zhu, A. Cao, X. Li, C. Xu, Z. Mao, D. Ruan, J. Liang, and D. Wu, Hydrogen adsorption in bundles
      of well-aligned carbon nanotubes at room temperature, *Appl. Surf. Sci.* **178**, 50–55 (2001).
491. W.Z. Huang, X.B. Zhang, J.P. Tu, F.Z. Kong, J.X. Ma, F. Liu, H.M. Lu, and C.P. Chen, The effect of
      pretreatments on hydrogen adsorption of multiwalled carbon nanotubes, *Mater. Chem. Phys.* **78**,
      144–148 (2002).
492. P.-X. Hou, S.-T. Xu, Z. Ying, Q.-H. Yang, C. Liu, and H.-M. Cheng, Hydrogen adsorption/desorption
      behavior of multiwalled carbon nanotubes with different diameters, *Carbon* **41**, 2471–2476 (2003).
493. K. Hanada, H. Shiono, and K. Matsuzaki, Hydrogen uptake of carbon nanofiber under moderate tem-
      perature and low pressure, *Diamond Rel. Mater.* **12**, 874–877 (2003).
494. D.J. Browning, M.L. Gerrard, J.B. Lakeman, I.M. Mellor, R.J. Mortimer, and M.C. Turpin, Studies into
      the storage of hydrogen in carbon nanofibers: proposal of a possible reaction mechanism, *Nano Lett.*
      2, 201–205 (2002).
495. K. Ichimura, K. Imaeda, and H. Inokuchi, Characteristic bonding of rare gases in solid carbon nan-
      otubes, *Synth. Met.* **121**, 1191–1192 (2001).
496. S. Challet, P. Azaïs, R.J.-M. Pellenq, L. Duclaux, D_2 adsorption in potassium-doped single-wall
      carbon nanotubes: a neutron diffraction and isotherms study, *Chem. Phys. Lett.* **377**, 544–550
      (2003).
497. M. Rzepka, P. Lamp, and M.A. de la Casa-Lillo, Physisorption of hydrogen on microporous carbon
      and carbon nanotubes, *J. Phys. Chem.* **B 102**, 10894–10898 (1998).

498. M.K. Kostov, M.W. Cole, J.C. Lewis, P. Diep, and J.K. Johnson, Many-body interactions among adsorbed atoms and molecules within carbon nanotubes and in free space, *Chem. Phys. Lett.* **332**, 26–34 (2000).

499. C.W. Bauschlicher Jr., Hydrogen and fluorine binding to the sidewalls of a (10,0) carbon nanotube, *Chem. Phys. Lett.* **322**, 237–241 (2000).

500. K.A. Williams and P.C. Eklund, Monte Carlo simulations on H_2 physisorption in finite-diameter carbon nanotube ropes, *Chem. Phys. Lett.* **320**, 352–358 (2000).

501. S.M. Lee, K.S. Park, Y.C. Choi, Y.S. Park, J.M. Bok, D.J. Bae, K.S. Nahm, Y.G. Choi, S.C. Yu, N.-G. Kim, T.Frauenheim, and Y.H. Lee, Hydrogen adsorption and storage in carbon nanotubes, *Synth. Met.* **113**, 209–216 (2000).

502. S.M. Lee and Y.H. Lee, Hydrogen storage in single-walled carbon nanotubes, *Appl. Phys. Lett.* **76**, 2877–2879 (2000).

503. S.-P. Chan, G. Chen, X.G. Gong, and Z.-F. Liu, Chemisorption of hydrogen molecules on carbon nanotubes under high pressure, *Phys. Rev. Lett.* **87**, 205502 (2001).

504. C. Gu, G.-H. Gao, Y.-X. Yu, and Z.-Q. Mao, Simulation study of hydrogen storage in single-walled carbon nanotubes, *Int. J. Hydrogen Energy* **26**, 691–696 (2001).

505. S.M. Lee, K.H. An, Y.H. Lee, G. Seifert, and T. Frauenheim, A hydrogen storage mechanism in single-walled carbon nanotubes, *J. Am. Chem. Soc.* **123**, 5059–5063 (2001).

506. S.M. Lee, K.H. An, W.S. Kim, Y.H. Lee, Y.S. Park, G. Seifert, and T. Frauenheim, Hydrogen storage in carbon nanotubes, *Synt. Met.* **121**, 1189–1190 (2001).

507. Y. Ren and D.L. Price, Neutron scattering study of H_2 adsorption in single-walled carbon nanotubes, *Appl. Phys. Lett.* **79**, 3684–3686 (2001).

508. H. Cheng, G.P. Pez, and A.C. Cooper, Mechanism of hydrogen sorption on single-walled carbon nanotubes, *J. Am. Chem. Soc.* **123**, 5845–5846 (2001).

509. C.W. Bauschlicher Jr., High coverages of hydrogen on (10,0) carbon nanotube, *Nano Lett.* **1**, 223–226 (2001).

510. M.K. Kostov, H. Cheng, A.C. Cooper, and G.P. Pez, Influence of carbon curvature in carbon-based materials: a force field approach, *Phys. Rev. Lett.* **89**, 146105 (2002).

511. M.K. Kostov, H. Cheng, R.M. Herman, M.W. Cole, and J.C. Lewis, Hindered rotation of H_2 adsorbed interstitially in nanotube bundles, *J. Chem. Phys.* **116**, 1720–1724 (2002).

512. Y. Ma, Y. Xia, M. Zhao, and M. Ying, Hydrogen storage capacity in single-walled carbon nanotubes, *Phys. Rev.* **B 65**, 155430.1–6 (2002).

513. R.A. Trasca, M.K. Kostov, and M.W. Cole, Isotopic and spin selectivity of H_2 adsorbed in bundles of carbon nanotubes, *Phys. Rev.* **B 67**, 035410.1–8 (2003).

514. J. Li, T. Furuta, H. Goto, T. Ohashi, Y. Fujiwara, and S. Yip, Theoretical evaluation of hydrogen storage capacity in pure carbon nanostructures, *J. Chem. Phys.* **119**, 2376–2385 (2003).

515. M. Haluska, M. Hirscher, M. Becher, U. Dettlaff-Weglikowska, X. Chen, and S. Roth, Interaction of hydrogen isotopes with carbon nanostructures, *Mater. Sci. Eng.* **B 108**, 130–133 (2004).

516. H.G. Schimmel, G. Nijkamp, G.J. Kearley, A. Rivera, K.P. de Jong, and F.M. Mulder, Hydrogen adsorption in carbon nanostructures compared, *Mater. Sci. Eng.* **B 108**, 124–129 (2004).

517. D. Chen, L. Chen, S. Liu, C.X. Ma, D.M. Chen, and L.B. Wang, Microstructure and hydrogen storage property of Mg/MWNTs composites, *J. Alloys Comp.* **372**, 231–237 (2004).

518. B.P. Tarasov, N.G. Goldshleger, and A.P. Moravsky, Hydrogen-containing carbon nanostructures: synthesis and properties, *Russ. Chem. Rev.* **70**, 131–146 (2001).

519. H.-M. Cheng, Q.-H. Yang, and C. Liu, Hydrogen storage in carbon nanotubes, *Carbon* **39**, 1447–1454 (2001).

520. F.L. Darkrim, P. Malbrunot, and G.P. Tartaglia, Review of hydrogen storage by adsorption in carbon nanotubes, *Int. J. Hydrogen Energy* **27**, 193–202 (2002).

521. V.V. Simonyan and J.K. Johnson, Hydrogen storage in carbon nanotubes and graphitic nanofibers, *J. Alloys Comp.* **330–332**, 659–665 (2002).

522. E. Frackowiak and F. Béguin, Electrochemical storage of energy in carbon nanotubes and nanostructured carbons, *Carbon* **40**, 1775–1787 (2002).

523. A. Züttel, P. Sudan, Ph. Mauron, T. Kiyobayashi, Ch. Emmenegger, and L. Schlapbach, Hydrogen storage in carbon nanostructures, *Int. J. Hydrogen Energy* **27**, 203–212 (2002).

524. M. Becher, M. Haluska, M. Hirscher, A. Quintel, V. Skakalova, U. Dettlaff-Weglikovska, X. Chen, M. Hulman, Y. Choi, S. Roth, V. Meregalli, M. Parrinello, R. Ströbel, L. Jörissen, M.M. Kappes, J. Fink, A. Züttel, I. Stepanek, and P. Bernier, Hydrogen storage in carbon nanotubes, *C. R. Physique* **4**, 1055–1062 (2003).

525. B. Viswanathan, M. Sankaran, and M.A. Scibioh, Carbon nanomaterials — are they appropriate candidates for hydrogen storage? *Bull. Catal. Soc. India* **2**, 12–32 (2003).

526. A. Züttel, P. Wenger, P. Sudan, P. Mauron, and S.-I. Orimo, Hydrogen density in nanostructured carbon, metals and complex materials, *Mater. Sci. Eng.* **B 108**, 9–18 (2004).

527. M. Conte, P.P. Prosini, and S. Passerini, Overview of energy/hydrogen storage: state-of-the-art technologies and prospects for nanomaterials, *Mater. Sci. Eng.* **B 108**, 2–8 (2004).

528. S. Talapatra, A.Z. Zambano, S.E. Weber, and A.D. Migone, Gases do not adsorb on the interstitial channels of closed-ended single-walled carbon nanotube bundles, *Phys. Rev. Lett.* **85**, 138–141 (2000).

529. M.M. Calbi, S.M. Gatica, M.J. Bojan, and M.W. Cole, Phases of neon, xenon, and methane adsorbed on nanotube bundles, *J. Chem. Phys.* **115**, 9975–9981 (2001).

530. M. Muris, N. Dupont-Pavlovsky, M. Bienfait, and P. Zeppenfeld, Where are the molecules adsorbed on single-walled nanotubes? *Surf. Sci.* **492**, 67–74 (2001).

531. J. Zhao, A. Buldum, J. Han, and J.P. Lu, Gas molecule adsorption in carbon nanotubes and nanotube bundles, *Nanotechnology* **13**, 195–200 (2002).

532. B.-Y. Wei, M.-C. Hsu, Y.-S. Yang, S.-H. Chien, and H.-M. Lin, Gases adsorption on single-walled carbon nanotubes measured by piezoelectric quartz crystal microbalance, *Mater. Chem. Phys.* **81**, 126–133 (2003).

533. A. Kleinhammes, S.-H. Mao, X.-J. Yang, X.-P. Tang, H. Shimoda, J.P. Lu, O. Zhou, and Y. Wu. Gas adsorption in single-walled carbon nanotubes studied by NMR, *Phys. Rev.* **B 68**, 075418.1–6 (2003).

534. A. Goldoni, R. Larciprete, L. Petaccia, and S. Lizzit, Single-wall carbon nanotube interaction with gases: sample contaminants and environmental monitoring, *J. Am. Chem. Soc.* **125**, 11329–11333 (2003).

535. S.-H. Jhi, S.G. Loui, and M.L. Cohen, Electronic properties of oxidized carbon nanotubes, *Phys. Rev. Lett.* **85**, 1710–1713 (2000).

536. S. Dag, O. Gülseren, and S. Ciraci, A comparative study of O_2 adsorbed carbon nanotubes, *Chem. Phys. Lett.* **380**, 1–5 (2003).

537. M. Cinke, J. Li, C.W. Bauschlicher Jr., A. Ricca, and M. Meyyappan, CO_2 adsorption in single-walled carbon nanotubes, *Chem. Phys. Lett.* **276**, 761–766 (2003).

538. C. Matranga, L. Chen, M. Smith, E. Bittner, J.K. Johnson, and B. Bockrath, Trapped CO_2 in carbon nanotube bundles, *J. Phys. Chem.* **B 107**, 12930–12941 (2003).

539. W.-L. Yim, O. Byl, J.T. Yates Jr., and J.K. Johnson, Vibrational behavior of adsorbed CO_2 on single-walled carbon nanotubes, *J. Chem. Phys.* **120**, 5377–5386 (2004).

540. H. Chang, J.D. Lee, S.M. Lee, Y.H. Lee, Adsorption of NH_3 and NO_2 molecules on carbon nanotubes, *Appl. Phys. Lett.* **79**, 3863–3865 (2001).

541. S.E. Weber, S. Talapatra, C. Journet, A. Zambano, and A.D. Migone, Determination of the binding energy of methane on single-walled carbon nanotube bundles, *Phys. Rev.* **B 61**, 13150–13154 (2000).

542. H. Tanaka, M. El-Merraoui, W.A. Steele, and K. Kaneko, Methane adsorption on single-walled carbon nanotube: a density functional theory model, *Chem. Phys. Lett.* **352**, 334–341 (2002).

543. S. Talapatra and A.D. Migone, Adsorption of methane on bundles of closed-ended single-wall carbon nanotubes, *Phys. Rev.* **B 65**, 454161–454166 (2002).

544. L. Valentini, I. Armentano, L. Lozzi, S. Santucci, and J.M. Kenny, Interaction of methane with carbon nanotube thin films: role of defects and oxygen adsorption, *Mater. Sci. Eng.* **C 24**, 527–533 (2004).

545. Gy. Onyestyák, J. Valyon, K. Hernádi, I. Kiricsi, and L.V.C. Rees, Equilibrium and dynamics of acetylene sorption in multiwalled carbon nanotubes, *Carbon* **41**, 1241–1248 (2003).

546. N. Chakrapani, Y.M. Zhang, S.K. Nayak, J.A. Moore, D.L. Carroll, Y.Y. Choi, and P.M. Ajayan, Chemisorption of acetone on carbon nanotubes, *J. Phys. Chem.* **B 107**, 9308–9311 (2003).

547. M.R. Babaa, N. Dupont-Pavlovsky, E. McRae, and K. Masenelli-Varlot, Physical adsorption of carbon tetrachloride on as-produced and on mechanically opened single-walled carbon nanotubes, *Carbon* **42**, 1549–1554 (2004).

548. C.-M. Yang, H. Kanoh, K. Kaneko, M. Yudasaka, and S. Iijima, Adsorption behaviors of HiPco single-walled carbon nanotube aggregates for alcohol vapors, *J. Phys. Chem.* **B 106**, 8994–8999 (2002).

549. J.A. Nisha, M. Yudasaka, S. Bandow, F. Kokai, K. Takahashi, and S. Iijima, Adsorption and catalytic properties of single-wall carbon nanohorns, *Chem. Phys. Lett.* **328**, 381–386 (2000).

550. H. Ago, R. Azumi, S. Ohshima, Y. Zhang, H. Kataura, and M. Yumura, STM study of molecular adsorption on single-wall carbon nanotube surface, *Chem. Phys. Lett.* **383**, 469–474 (2004).

551. M. Bienfait, B. Asmussen, M. Johnson, and P. Zeppenfeld, Methane mobility in carbon nanotubes, *Surf. Sci.* **460**, 243–248 (2000).

552. S.M. Cooper, B.A. Cruden, and M. Meyyappan, Gas transport characteristics through a carbon nanotubule, *Nano Lett.* **4**, 377–381 (2004).

553. P.G. Collins, K. Bradley, M. Ishigami, and A. Zettl, Extreme oxygen sensitivity of electronic properties of carbon nanotubes, *Science* **287**, 1801–1804 (2000).

554. R. Pati, Y. Zhang, S.K. Nayak, and P.M. Ajayan, Effect of H_2O adsorption on electron transport in a carbon nanotube, *Appl. Phys. Lett.* **81**, 2638–2640 (2002).

555. G.U. Sumanasekera, C.K.W. Adu, S. Fang, and P.C. Eklund, Effect of gas adsorption and collisions on electrical transport in single-walled carbon nanotubes, *Phys. Rev. Lett.* **85**, 1096–1099 (2000).

556. K. Bradley, S.-H. Jhi, P.G. Collins, J. Hone, M.L. Cohen, S.G. Louie, and A. Zettl, Is the intrinsic thermoelectric power of carbon nanotubes positive? *Phys. Rev. Lett.* **85**, 4361–4364 (2000).

557. G.U. Sumanasekera, B.K. Pradhan, H.E. Romero, K.W. Adu, and P.C. Eklund, Giant thermopower effects from molecular physisorption on carbon nanotubes, *Phys. Rev. Lett.* **89**, 166801 (2002).

558. J. Kong, N.R. Franklin, C. Zhou, M.C. Chapline, S. Peng, K. Cho, and H. Dai, Nanotube molecular wires as chemical sensors, *Science* **287**, 622–625 (2000).

559. S. Peng, K. Cho, P. Qi, and H. Dai, Ab initio study of CNT NO_2 gas sensor, *Chem. Phys. Lett.* **387**, 271–276 (2004).

560. C. Cantalini, L. Valentini, I. Armentano, J.M. Kenny, L. Lozzi, and S. Santucci, Carbon nanotubes as new materials for gas sensing applications, *J. Eur. Ceram. Soc.* **24**, 1405–1408 (2004).

561. L. Valentini, C. Cantalini, I. Armentano, J.M. Kenny, L. Lozzio, and S. Santucci, Highly sensitive and selective sensors based on carbon nanotubes thin films for molecular detection, *Diamond Rel. Mater.* **13**, 1301–1305 (2004).

562. L. Valentini, F. Mercuri, I. Armentano, C. Cantalini, S. Picozzi, L. Lozzi, S. Santucci, A. Sgamellotti, and J.M. Kenny, Role of the defects on the gas sensing properties of carbon nanotubes thin films: experiment and theory, *Chem. Phys. Lett.* **387**, 356–361 (2004).

563. O.K. Varghese, P.D. Kichambre, D. Gong, K.G. Ong, E.C. Dickey, and C.A. Grimes, Gas sensing characteristics of multiwall carbon nanotubes, *Sensors Actuators* **B 81**, 32–41 (2001).

564. K.G. Ong, K. Zeng, and C.A. Grimes, A wireless, passive carbon nanotube-based gas sensor, *IEEE Sensor J.* **2**, 82–88 (2002).

565. S.G. Wang, Q. Zhang, D.J. Yang, P.J. Sellin, and G.F. Zhong, Multiwalled carbon nanotube-based gas sensors for NH_3 detection, *Diamond Rel. Mater.* **13**, 1327–1332 (2004).

566. C.K.W. Adu, G.U. Sumanasekera, B.K. Pradhan, H.E. Romero, and P.C. Eklund, Carbon nanotubes: a thermoelectric nano-nose, *Chem. Phys. Lett.* **337**, 31–35 (2001).

567. S. Peng and K. Cho, Ab initio study of doped carbon nanotube sensors, *Nano Lett.* **3**, 513–517 (2003).

568. F. Villalpando-Páez, A.H. Romero, E. Muñoz-Sandoval, L.M. Martínez, H. Terrones, and M. Terrones, Fabrication of vapor and gas sensors using films of aligned CN_x nanotubes, *Chem. Phys. Lett.* **386**, 137–143 (2004).

569. J. Li, Y. Lu, Q. Ye, M. Cinke, J. Han, and M. Meyyappan, Carbon nanotube sensors for gas and organic vapor detection, *Nano Lett.* **3**, 929–933 (2003).

570. J. Kong, M.G. Chapline, and H. Dai. Functionalized carbon nanotubes for molecular hydrogen sensors, *Adv. Mater.* **13**, 1384–1386 (2001).

571. L. Dai, P. Soundarrajan, and T. Kim, Sensors and sensor arrays based on conjugated polymers and carbon nanotubes, *Pure Appl. Chem.* **74**, 1753–1772 (2002).

572. A. Huczko, H. Lange, E. Caiko, H. Grubek-Jaworska, and P. Droszcz, Physiological testing of carbon nanotubes: are they asbestos-like? *Fullerene Sci. Technol. (Fullerenes Nanotubes Carbon Nanostruct.)* **9**, 251–254 (2001).

573. M. Shim, N.W.S. Kam, R.J. Chen, Y. Li, and H. Dai, Functionalization of carbon nanotubes for biocompatibility and biomolecular recognition, *Nano Lett.* **2**, 285–288 (2002).

574. K. Besteman, J.-O. Lee, F.G.M. Wiertz, H.A. Heering, and C. Dekker, Enzyme-coated carbon nanotubes as single-molecule biosensors, *Nano Lett.* **3**, 727–730 (2003).

575. A. Bianco and M. Prato, Can carbon nanotubes be considered useful tools for biological applications? *Adv. Mater.* **15**, 1765–1768 (2003).

576. J.J. Davis, K.S. Coleman, B.R. Azamian, C.B. Bagshaw, and M.L.H. Green, Chemical and biochemical sensing with modified single walled carbon nanotubes, *Chem. Eur. J.* **9**, 3733–3739 (2003).

577. Q. Zhao, Z. Gan, and Q. Zhuang, Electrochemical sensors based on carbon nanotubes, *Electroanalysis* **14**, 1609–1613 (2002).

578. P.J. Britto, K.S.V. Santhanam, and P.M. Ajayan, Carbon nanotube electrode for oxidation of dopamine, *Bioelectrochem. Bioenergetics* **41**, 121–125 (1996).

579. J. Wang, M. Li, Z. Shi, N. Li, and Z. Gu, Electrocatalytic oxidation of norepinephrine at a glassy carbon electrode modified with single wall carbon nanotubes, *Electroanalysis* **14**, 225–230 (2002).

580. K. Wu, X. Ji, J. Fei, and S. Hu, The fabrication of a carbon nanotube film on a glassy carbon electrode and its application to determining thyroxine, *Nanotechnology* **15**, 287–291 (2004).

581. S.C. Tsang, Z. Guo, Y.K. Chen, M.L.H. Green, H.A.O. Hill, T.W. Hambley, and P.J. Sadler, Immobilization of platinated and iodinated oligonucleotides on carbon nanotubes, *Angew. Chem. Int. Edit.* **36**, 2200–2220 (1997).

582. Z. Guo, P.J. Sadler, and S.C. Tsang, Immobilization and vizualization of DNA and proteins on carbon nanotubes, *Adv. Mater.* **10**, 701–703 (1998).

583. R.J. Chen, Y. Zhang, D. Wang, and H. Dai, Noncovalent sidewall functionalization of single-walled carbon nanotubes for protein immobilization, *J. Am. Chem. Soc.* **123**, 3838–3839 (2001).

584. R.J. Chen, S. Bangsaruntip, K.A. Drouvalakis, N.W.S. Kam, M. Shim, Y. Li, W. Kim, P.J. Utz, and H. Dai, Noncovalent functionalization of carbon nanotubes for highly specific electronic biosensors, *Proc. Nat. Acad. Sci. USA* **100**, 4984–4989 (2003).

585. S.E. Baker, W. Cai, T.L. Lasseter, K.P. Weidkamp, and R.J. Hamers, Covalently bonded adducts of deoxyribonucleic acid (DNA) oligonucleotides with single-wall carbon nanotubes: synthesis and hybridization, *Nano Lett.* **2**, 1413–1417 (2002).

586. K.A. Williams, P.T.M. Veenhuizen, B.G. de la Torre, R. Eritja, and C. Dekker, Nanotechnology: carbon nanotube with DNA recognition, *Nature*, **420**, 761 (2002).

587. C.V. Nguyen, L. Delzeit, A.M. Cassell, J. Li, J. Han, and M. Meyyappan, Preparation of nucleic acid functionalized carbon nanotube arrays, *Nano Lett.* **2**, 1079–1081 (2002).

588. J. Li, H.T. Ng, A.M. Cassell, W. Fan, H. Chen, Q. Ye, J. Koehne, J. Han, and M. Meyappan, Carbon nanotube nanoelectrode array for ultrasensitive DNA detection, *Nano Lett.* **3**, 597–602 (2003).

589. J. Koehne, H. Chen, J. Li, A.M. Cassell, Q. Ye, H.T. Ng, J. Han, and M. Meyyappan, Ultrasensitive label-free DNA analysis using an electronic chip based on carbon nanotube nanoelectrode arrays, *Nanotechnology* **14**, 1239–1245 (2003).

590. H. Cai, X. Cao, Y. Jiang, P. He, and Y. Fang, Carbon nanotube-enhanced electrochemical DNA biosensor for DNA hybridization detection, *Anal. Bioanal. Chem.* **375**, 287–293 (2003).

591. M.L. Pedano and G.A. Rivas, Immobilization of DNA on glassy carbon electrodes for the development of affinity biosensors, *Biosens. Bioelectron.* **18**, 269–277 (2003).

592. A. Guiseppi-Elie, C. Lei, and R.H. Baughman, Direct electron transfer of glucose oxidase on carbon nanotubes, *Nanotechnology* **13**, 559–564 (2002).

593. S. Sotiropoulou and N.A. Chaniotakis, Carbon nanotube array-based biosensor, *Anal. Bioanal. Chem.* **375**, 103–105 (2003).

594. L. Wang and Z. Yuan, Direct electrochemistry of glucose oxidase at a gold electrode modified with single-wall carbon nanotubes, *Sensors* **3**, 544–554 (2003).

595. J.H.T. Luong, S. Hrapovic, D. Wang, F. Bensebaa, and B. Simard, Solubilization of multiwall carbon nanotubes by 3-aminopropyltriethoxysilane towards the fabrication of electrochemical biosensors with promoted electron transfer, *Electroanalysis* **16**, 132–139 (2004).

596. K.P. Loh, S.L. Zhao, and W.D. Zhang, Diamond and carbon nanotube glucose sensors based on electropolymerization, *Diamond Rel. Mater.* **13**, 1075–1079 (2004).

597. R.J. Chen, H.C. Choi, S. Bangsaruntip, E. Yenilmez, X. Tang, Q. Wang, Y.-L. Chang, and H. Dai, An investigation of the mechanisms of electronic sensing of protein adsorption on carbon nanotube devices, *J. Am. Chem. Soc.* **126**, 1563–1568 (2004).

598. A. Star, J.-C.P. Gabriel, K. Bradley, and G. Grüner, Electronic detection of specific protein binding using nanotube FET devices, *Nano Lett.* **3**, 459–463 (2003).

599. S.C. Tsang, J.J. Davis, M.L.H. Green, H.A.O. Hill, Y.C. Leung, and P.J. Sadler, Immobilization of small proteins in carbon nanotubes: high-resolution transmission electron microscopy study and catalytic activity, *Chem. Commun.* 1803–1804 (1995).

600. J.J. Davis, M.L.H. Green, H.A.O. Hill, Y.C. Leung, P.J. Sadler, J. Sloan, A.V. Xavier, and S.C. Tsang, The immobilization of proteins in carbon nanotubes, *Inorg. Chim. Acta* **272**, 261–266 (1998).

601. F. Balavoine, P. Schultz, C. Richard, V. Mallouh, T.W. Ebbesen, and C. Mioskowski, Helical crystallization of proteins on carbon nanotubes: a first step towards development of new biosensors, *Angew. Chem. Int. Ed.* **38**, 1912–1915 (1999).

602. B.R. Azamian, J.J. Davis, K.S. Coleman, C.B. Bagshow, and M.L.H. Green, Bioelectrochemical single-walled carbon nanotubes, *J. Am. Chem. Soc.* **124**, 12664–12665 (2002).

603. W. Huang, S. Taylor, K. Fu, Y. Lin, D. Zhang, T.W. Hanks, A.M. Rao, and Y.-P. Sun, Attaching proteins to carbon nanotubes via diimide-activated amidation, *Nano Lett.* **2**, 311–314 (2002).

604. X. Yu, D. Chattopadhyay, I. Galeska, F. Papadimitrakopoulos, and J.F. Rusling, Peroxidase activity of enzymes bound to the ends of single-wall nanotube forest electrodes, *Electrochem. Commun.* **5**, 408–411 (2003).

605. M. in het Panhuis, C. Salvador-Morales, E. Franklin, G. Chambers, A. Fonseca, J.B. Nagy, W.J. Blau, and A.I. Minetta, Characterization of an interaction between functionalized carbon nanotubes and an enzyme, *J. Nanosci. Nanotech.* **3**, 209–213 (2003).

606. K. Jiang, L.S. Schadler, R.W. Siegel, X. Zhang, H. Zhang, and M. Terrones, Protein immobilization on carbon nanotubes *via* a two-step process of diamide-activated amidation, *J. Mater. Chem.* **14**, 37–39 (2004).

607. D. Pantarotto, C.D. Partidos, R. Graff, J. Hoebeke, J.-P. Briand, M. Prato, and A. Bianco, Synthesis, structural characterization, and immunological properties of carbon nanotubes functionalized with peptides, *J. Am. Chem. Soc.* **125**, 6160–6164 (2003).

608. R.D. Klausner, Challenges and vision for nanoscience and nanotechnology in medicine: cancer as a model, *BECON Nanoscience and Nanotechnology Symp. Report*, June 2000, 8–9.

609. M. Meyyappan, D.J. Loftus, J. Han, A.M. Cassell, J. Kaysen, C.V. Nguyen, S. Liang, R.M.D. Stevens, R. Jaffe, A. Hermone, and S. Verma, Applications of nanotube nanotechnology to biosensors and cancer research, *Principal Investigators Meeting Unconventional Innovations Program*, June 28–29, 2000, Hyatt Dulles Hotel, Herdnon, Virginia, US [http://otir.nci.nih.gov/cgi-bin/uip_search.cgi?ABSTRACTID=UIP-00-015].

610. Y. Lin, F. Lu, and J. Wang, Disposable carbon nanotube modified screen-printed biosensor for amperometric detection of organophosphorus pesticides and nerve agents, *Electroanalysis* **16**, 145–149 (2004).

611. T.S. Huang, Y. Tzeng, Y.K. Liu, Y.C. Chen, K.R. Walker, R. Guntupalli, and C. Liu, Immobilization of antibodies and bacterial binding on nanodiamond and carbon nanotubes for biosensor applications, *Diamond Rel. Mater.* **13**, 1098–1102 (2004).

612. Y.Z. Guo and A.R. Guadalupe, Direct electrochemistry of horseradish peroxidase adsorbed on glassy carbon electrode from organic solutions, *Chem. Commun.* 1437–1438 (1997).

613. P. Král, Control of catalytic activity of proteins in vivo by nanotube ropes excited with infrared light, *Chem. Phys. Lett.* **382**, 399–403 (2003).

614. K. Rege, N.R. Raravikar, D.-Y. Kim, L.S. Schadler, P.M. Ajayan, and J.S. Dordick, Enzyme–polymer–single-walled carbon nanotube composites as biocatalytic films, *Nano Lett.* **3**, 829–832 (2003).

615. D. Pantarotto, J.-P. Briand, M. Prato, and A. Bianco, Translocation of bioactive peptides across cell membranes by carbon nanotubes, *Chem. Commun.* 16–17 (2004).

616. L. Zhang, G.-C. Zhao, X.-W. Wei, and Z.-S. Yang, Electroreduction of oxygen by myoglobin on multi-walled carbon nanotube-modified glassy carbon electrode, *Chem. Lett.* **33**, 86–87 (2004).

617. J.-S. Ye, Y. Wen, W.D. Zhang, H.-F. Cui, L.M. Gan, G.Q. Xu, and F.-S. Sheu, Application of multi-walled carbon nanotubes functionalized with hemin for oxygen detection in neutral solution, *J. Electroanal. Chem.* **562**, 241–246 (2004).

618. M.P. Mattson, R.C. Haddon, and A.M. Rao, Molecular functionalization of carbon nanotubes and use as substrates for neuronal growth, *J. Mol. Neurosci.* **14**, 175–182 (2000).

619. H. Hu, Y. Ni, V. Montana, R.C. Haddon, and V. Parpura, Chemically functionalized carbon nanotubes as substrates for neuronal growth, *Nano Lett.* **4**, 507–511 (2004).

620. C. Richard, F. Balavoine, P. Schultz, T.W. Ebbesen, and C. Mioskowski, Supramolecular self-assembly of lipid derivatives on carbon nanotubes, *Science* **300**, 775–778 (2003).

621. N. Nakashima, S. Okuzono, H. Marukami, T. Nakai, and K. Yoshikawa, DNA dissolves single-walled carbon nanotubes in water, *Chem. Lett.* **32**, 456–457 (2003).

622. K. Matsuura, K. Hayashi, and N. Kimizuka, Lectin-mediated supramolecular junctions of galactose-derivatized single-walled carbon nanotubes, *Chem. Lett.* **32**, 212–213 (2003).

623. M. Zheng, A. Jagota, M.S. Strano, A.P. Santos, P. Barone, S.G. Chou, B.A. Diner, M.S. Dresselhaus, R.S. Mclean, G.B. Onoa, G.G. Samsonidze, E.D. Semke, M. Ursey, and D.J. Walls, Structure-based carbon nanotube sorting by sequence-dependent DNA assembly, *Science* **302**, 1545–1548 (2003).

624. M. Zheng, A. Jagota, E.D. Semke, B.A. Diner, R.S. McLean, S.R. Lustig, R.E. Richardson, and N.G. Tassi, DNA-assisted dispersion and separation of carbon nanotubes, *Nat. Mater.* **2**, 338–342 (2003).

625. J.N. Barisci, M. Tahhan, G.G. Wallace, S. Badaire, T. Vaugien, M. Maugey, and P. Poulin, Properties of carbon nanotube fibers spun from DNA-stabilized dispersions, *Adv. Funct. Mater.* **14**, 133–138 (2004).

626. H. Xin and A.T. Woolley, DNA-templated nanotube localization, *J. Am. Chem. Soc.* **125**, 8710–8711 (2003).

627. K. Keren, R.S. Berman, E. Buchstab, U. Sivan, and E. Braun, DNA-templated carbon nanotube field-effect transistor, *Science* **302**, 1380–1382 (2003).

628. C. Dwyer, M. Guthold, M. Falvo, S. Washburn, R. Superfine, and D. Erie, DNA-functionalized single-walled carbon nanotubes, *Nanotechnology* **13**, 601–604 (2002).

629. R. Czerw, M. Terrones, J.-C. Charlier, X. Blase, B. Foley, R. Kamalakaran, N. Grobert, H. Terrones, D. Tekleab, P.M. Ajayan, W. Blau, M. Rühle, and D.L. Carroll, Identification of electron donor states in N-doped carbon nanotubes, *Nano Lett.* **1**, 457–460 (2001).

630. M. Terrones, N. Grobert, M. Terrones, H. Terrones, P.M. Ajayan, F. Banhart, X. Blase, D.L. Carroll, R. Czerw, B. Foley, J.C. Charlier, B. Foley, R. Kamalakaran, P.H. Kohler-Redlich, M. Rühle, and T. Seeger, Doping and connecting carbon nanotubes, *Mol. Cryst. Liq. Cryst.* **387**, 51–62 (2002).

631. A.H. Nevidomskyy, G. Csányi, and M.C. Payne, Chemically active substitutional nitrogen impurity in carbon nanotubes, *Phys. Rev. Lett.* **91**, 105502-1–4 (2003).

632. Y. Huang, J. Gao, and R. Liu, Structure and electronic properties of nitrogen-containing carbon nanotubes, *Synth. Met.* **113**, 251–255 (2000).

633. R. Droppa Jr., P. Hammer, A.C.M. Carvalho, M.C. dos Santos, and F. Alvarez, Incorporation of nitrogen in carbon nanotubes, *J. Non-Cryst. Solids* **299–302**, 874–879 (2002).

634. M. Glerup, J. Steinmetz, D. Samaille, O. Stéphan, S. Enouz, A. Loiseau, S. Roth, and P. Bernier, Synthesis of N-doped SWNT using the arc-discharge procedure, *Chem. Phys. Lett.* **387**, 193–197 (2004).

635. R. Sen, B.C. Satishkumar, A. Govindaraj, K.R. Harikumar, G. Raina, J.-P. Zhang, A.K. Cheetham, and C.N.R. Rao, B–C–N, C–N and B–N nanotubes produced by the pyrolysis of precursor molecules over Co catalysts, *Chem. Phys. Lett.* **287**, 671–676 (1998).

636. K. Suenaga, M. Yudasaka, C. Colliex, and S. Iijima, Radially modulated nitrogen distribution in CN$_x$ nanotubular structures prepared by CVD using Ni phtalocyanine, *Chem. Phys. Lett.* **316**, 365–372 (2000).

637. R. Kurt and A. Karimi, Influence of nitrogen on the growth mechanism of decorated C:N nanotubes, *Chem. Phys. Chem.* **2**, 388–392 (2001).

638. R. Kurt, J.-M. Bonard, and A. Karimi, Structure and field emission properties of decorated C/N nanotubes tuned by diameter variations, *Thin Solid Films* **398–399**, 193–198 (2001).

639. R. Kurt, C. Klinke, J.-M. Bonard, K. Kern, and A. Karimi, Tailoring the diameter of decorated C–N nanotubes by temperature variations using HF-CVD, *Carbon* **39**, 2163–2172 (2001).

640. M. Terrones, P.M. Ajayan, F. Banhart, X. Blase, D.L. Carroll, J.C. Charlier, R. Czerw, B. Foley, N. Grot, R. Kamalakaran, P. Kohler-Redlich, M. Rühle, T. Seeger, and H. Terrones, N-doping and coalescence of carbon nanotubes: synthesis and electronic properties, *Appl. Phys.* **A 74**, 355–361 (2002).

641. X. Wang, Y. Liu, D. Zhu, L. Zhang, H. Ma, N. Yao, and B. Zhang, Controllable growth, structure, and low field emission of well-aligned CN$_x$ nanotubes, *J. Phys. Chem.* **B 106**, 2186–2190 (2002).

642. C.J. Lee, S.C. Lyu, H.-W. Kim, J.H. Lee, and K.I. Cho, Synthesis of bamboo-shaped carbon-nitrogen nanotubes using C$_2$H$_2$–NH$_3$–Fe(CO)$_5$ system, *Chem. Phys. Lett.* **359**, 115–120 (2002).

643. T.-Y. Kim, K.-R. Lee, K.Y. Eun, and K.-H. Oh, Carbon nanotube growth enhanced by nitrogen incorporation, *Appl. Phys. Lett.* **372**, 603–607 (2003).

644. H. Yan, Q. Li, J. Zhang, and Z. Liu, The effect of hydrogen on the formation of nitrogen-doped carbon nanotubes via catalytic pyrolysis of acetonitrile, *Adv. Nanomat. Nanodevices (8th International Conference on Electronic Materials*, IUMRS-ICEM 2002, Xi'an, China, June 10–14, 2002), preprint [http://nanotechweb.org/dl/nanomaterials/Xian_article_01_was153468.pdf].

645. H. Yan, Q. Li, J. Zhang, and Z. Liu, The effect of hydrogen on the formation of nitrogen-doped carbon nanotubes via catalytic pyrolysis of acetonitrile, *Chem. Phys. Lett.* **380**, 347–351 (2003).

646. C.H. Lin, H.L. Chang, C.M. Hsu, A.Y. Lo, and C.T. Kuo, The role of nitrogen in carbon nanotube formation, *Diamond Rel. Mater.* **12**, 1851–1857 (2003).

647. V.D. Blank, E.V. Polyakov, D.V. Batov, B.A. Kulnitskiy, U. Bangert, A. Gutiérrez-Sosa, A.J. Harvey, and A. Seepujak, Formation of N-containing C-nanotubes and nanofibres by carbon resistive heating under high nitrogen pressure, *Diamond Rel. Mater.* **12**, 864–869 (2003).

648. J. Maultzsch, S. Reich, C. Thomsen, S. Webster, R. Czerw, D.L. Carroll, S.M.C. Vieira, P.R. Birkett, and C.A. Rego, Raman characterization of boron-doped multiwalled carbon nanotubes, *Appl. Phys. Lett.* **81**, 2647–2649 (2002).

649. J. Xu, M. Xiao, R. Czerw, and D.L. Carroll, Optical limiting and enhanced optical nonlinearity in boron-doped carbon nanotubes, *Chem. Phys. Lett.* **389**, 247–250 (2004).

650. B.C. Satishkumar and C.N.R. Rao, Boron-carbon nanotubes from the pyrolysis of C_2H_2–B_2H_6 mixtures, *Chem. Phys. Lett.* **300**, 473–477 (1999).

651. C.F. Chen, C.L. Tsai, and C.L. Lin, The characterization of boron-doped carbon nanotube arrays, *Diamond Rel. Mater.* **12**, 1500–1504 (2003).

652. J.Y. Lao, W.Z. Li, J.G. Wen, and Z.F. Ren, Boron carbide nanolumps on carbon nanotubes, *Appl. Phys. Lett.* **80**, 500–502 (2002).

653. E. Borowiak-Palen, T. Pichler, G.G. Fuentes, A. Graff, R.J. Kalenczuk, M. Knupfer, and J. Fink, Efficient production of B-substituted single-wall carbon nanotubes, *Chem. Phys. Lett.* **378**, 516–520 (2003).

654. E. Borowiak-Palen, T. Pichler, A. Graff, R.J. Kalenczuk, M. Knupfer, and J. Fink, Synthesis and electronic properties of B-doped single wall carbon nanotubes, *Carbon* **42**, 1123–1126 (2004).

655. Y.-J. Lee, H.-H. Kim, and H. Hatori, Effects of substitutional B on oxidation of carbon nanotubes in air and oxygen plasma, *Carbon* **42**, 1053–1056 (2004).

656. W. Han, Y. Bando, K. Kurashima, and T. Sato, Synthesis of boron nitride nanotubes from carbon nanotubes by a substitution reaction, *Appl. Phys. Lett.* **73**, 3085–3087 (1998).

657. W. Han, Y. Bando, K. Kurashima, and T. Sato, Boron-doped carbon nanotubes prepared through a substitution reaction, *Chem. Phys. Lett.* **299**, 368–373 (1999).

658. D. Golberg, Y. Bando, K. Kurashima, and T. Sato, Synthesis, HRTEM and electron diffraction studies of B/N-doped C and BN nanotubes, *Diamond Rel. Mater.* **10**, 63–67 (2001).

659. X.D. Bai, J.D. Guo, J. Yu, E.G. Wang, J. Yuan, and W. Zhou, Synthesis and field-emission behavior of highly oriented boron carbonitride nanofibers, *Appl. Phys. Lett.* **76**, 2624–2626 (2000).

660. D. Golberg, Y. Bando, L. Bourgeois, K. Kurashima, and T. Sato, Large-scale synthesis and HRTEM analysis of single-walled B- and N-doped carbon nanotube bundles, *Carbon* **38**, 2017–2027 (2000).

661. D. Golberg, P. Dorozhkin, Y. Bando, M. Hasegawa, and Z.-C. Dong, Semiconducting B–C–N nanotubes with few layers, *Chem. Phys. Lett.* **359**, 220–228 (2002).

662. R.J. Baierle, S.B. Fagan, R. Mota, A.J.R. da Silva, and A. Fazzio, Electronic and structural properties of silicon-doped carbon nanotubes, *Phys. Rev.* **B 64**, 085413.1–4 (2001).

663. S.B. Fagan, R. Mota, A.J.R. da Silva, and A. Fazzio, Substitutional Si doping in deformed carbon nanotubes, *Nano Lett.* **4**, 975–977 (2004).

664. Y. Zhang, T. Ichihashi, E. Landree, F. Nihey, and S. Iijima, Heterostructures of single-walled carbon nanotubes and carbide nanorods, *Science* **285**, 1719–1722 (1999).

665. Q. Li, S. Fan, W. Han, C. Sun, and W. Liang, Coating of carbon nanotube with nickel by electroless plating method, *Jpn. J. Appl. Phys.* **36 B**, L501–L503 (1997).

666. Z. Shi, X. Wang, and Z. Ding, The study of electroless deposition of nickel on graphite fibers, *Appl. Surf. Sci.* **140**, 106–110 (1999).

667. L.M. Ang, T.S.A. Hor, G.Q. Xu, C.H. Tung, S.P. Zhao, and J.L.S. Wang, Decoration of activated carbon nanotubes with copper and nickel, *Carbon* **38**, 363–372 (2000).

668. F.Z. Kong, X.B. Zhang, W.Q. Xiong, F. Liu, W.Z. Huang, Y.P. Sun, J.P. Tu, and X.W. Chen, Continuous Ni-layer on multiwall carbon nanotubes by an electroless plating method, *Surf. Coat. Technol.* **155**, 33–36 (2002).

669. Q. Xu, L. Zhang, J. Zhu, Controlled growth of composite nanowires based on coating Ni on carbon nanotubes by electrochemical deposition method, *J. Phys. Chem.* **B 107**, 8294–8296 (2003).

670. S. Arai, M. Endo, and N. Kaneko, Ni-deposited multi-walled carbon nanotubes by electrodeposition, *Carbon* **42**, 641–644 (2004).

671. T.V. Reshetenko, L.B. Avdeeva, Z.R. Ismagilov, and A.L. Chuvilin, Catalytic filamentous carbon as supports for nickel catalysts, *Carbon* **42**, 143–148 (2004).

672. Y. Zhang, W. Franklin, R.J. Chen, and H. Dai, Metal coating on suspended carbon nanotubes and its implication to metal-tube interaction, *Chem. Phys. Lett.* **331**, 35–41 (2000).

673. Y. Zhang and H.J. Dai, Formation of metal nanowires on suspended single-walled carbon nanotubes, *Appl. Phys. Lett.* **77**, 3015–3017 (2000).

674. P. Chen, X. Wu, J. Lin, and K.L. Tan, Synthesis of Cu nanoparticles and microsized fibers by using carbon nanotubes as a template, *J. Phys. Chem.* **B 103**, 4559–4561 (1999).

675. X. Chen, J. Xia, J. Peng, W. Li, and S. Xie, Carbon-nanotube metal-matrix composites prepared by electroless plating, *Composites Sci. Technol.* **60**, 301–306 (2000).

676. Z.-J. Liu, Z. Xu, Z.-Y. Yuan, W. Chen, W. Zhou, and L.-M. Peng, A simple method for coating carbon nanotubes with Co–B amorphous alloy, *Mater. Lett.* **57**, 1339–1344 (2003).

677. B.C. Satishkumar, E.M. Vogl, A. Govindaraj, and C.N.R. Rao, The decoration of carbon nanotubes by metal nanoparticles, *J. Phys.* **D 29**, 3173–3176 (1996).

678. W. Li, C. Liang, J. Qiu, W. Zhou, H. Han, Z. Wei, G. Sun, and Q. Xin, Carbon nanotubes as support for cathode catalyst of a direct methanol fuel cell, *Carbon* **40**, 791–794 (2002).

679. W. Li, C. Liang, W. Zhou, J. Qiu, H. Li, G. Sun, and Q. Xin, Homogeneous and controllable Pt particles deposited on multiwall carbon nanotubes as cathode catalyst for direct methanol fuel cells, *Carbon* **42**, 436–439 (2004).

680. R. Yu, L. Chen, Q. Liu, J. Lin, K.-L. Tan, S.C. Ng, H.S.O. Chan, G.-Q. Xu, and T.S.A. Hor, Platinum deposition on carbon nanotubes via chemical modification, *Chem. Mater.* **10**, 718–722 (1998).

681. V. Lordi, N. Yao, and J. Wei, Method for supporting platinum on single-walled carbon nanotubes for a selective hydrogenation catalyst, *Chem. Mater.* **13**, 733–737 (2001).

682. S.H. Joo, S.J. Choi, I. Oh, J. Kwak, Z. Liu, O. Terasaki, and R. Ryoo, Ordered nanoporous arrays of carbon supporting high dispersions of platinum nanoparticles, *Nature* **412**, 169–172 (2001).

683. B. Xue, P. Chen, Q. Hong, J. Lin, and K. Tan. Growth of Pd, Pt, Ag and Au nanoparticles on carbon nanotubes, *J. Mater. Chem.* **11**, 2378–2381 (2001).

684. H.C. Choi, M. Shim, S. Bangsaruntip, and H. Dai, Spontaneous reduction of metal ions on the side-walls of carbon nanotubes, *J. Am. Chem. Soc.* **124**, 9058–9059 (2002).

685. X.R. Ye, Y. Lin, and C.M. Wai, Decorating catalytic palladium nanoparticles on carbon nanotubes in supercritical carbon dioxide, *Chem. Commun.* 642–643 (2003).

686. B.R. Azamian, K.S. Coleman, J.J. Davis, N. Hanson, and M.L.H. Green, Directly observed covalent coupling of quantum dots to single-wall carbon nanotubes, *Chem. Commun.* 366–367 (2002).

687. A.V. Ellis, K. Vijayamohanan, R. Goswami, N. Chakrapani, L.S. Ramanathan, P.M. Ajayan, and G. Ramanath, Hydrophobic anchoring of monolayer-protected gold nanoclusters to carbon nanotubes, *Nano Lett.* **3**, 279–282 (2003).

688. L. Liu, T. Wang, J. Li, Z.-X. Guo, L. Dai, D. Zhang, and D. Zhu, Self-assembly of gold nanoparticles to carbon nanotubes using a thiol-terminated pyrene as interlinker, *Chem. Phys. Lett.* **367**, 747–752 (2003).

689. A. Fási, I. Pálinkó, J.W. Seo, Z. Kónya, K. Hernadi, and I. Kiricsi, Sonication assisted gold deposition on multiwall carbon nanotubes, *Chem. Phys. Lett.* **372**, 848–852 (2003).

690. S. Fullam, D. Cottell, H. Rensmo, and D. Fitzmaurice, Carbon nanotube templated self-assembly and thermal processing on gold nanowires, *Adv. Mater.* **12**, 1430–1432 (2000).

691. W.S. Yun, J. Kim, K.H. Park, J.S. Ha, Y.-J. Ko, K. Park, S.K. Kim, Y.-J. Doh, H.-J. Lee, J.P. Salvetat, and L. Forró, Fabrication of metal nanowire using carbon nanotubes as a mask, *J. Vac. Sci. Technol.* **A 18**, 1329 (2000).

692. B.C. Satishkumar, A. Govindaraj, E.M. Vogl, L. Basumallick, and C.N.R. Rao, Oxide nanotubes prepared using carbon nanotubes as templates, *J. Mater. Res.* **12**, 604–606 (1997).

693. T. Seeger, Ph. Redlich, N. Grobert, M. Terrones, D.R.M. Walton, H.W. Kroto, and M. Rühle, SiO_x-coating of carbon nanotubes at room temperature, *Chem. Phys. Lett.* **339**, 41–46 (2001).

694. M. Rühle, T. Seeger, Ph. Redlich, N. Grobert, M. Terrones, D.R.M. Walton, and H.W. Kroto, Novel SiO_x-coated carbon nanotubes, *J. Ceram. Process. Res.* **3**, 1–5 (2002).

695. T. Seeger, Th. Kühler, Th. Frauenheim, N. Grobert, M. Rühle, M. Terrones, and G. Seifert, Nanotube composites: novel SiO_2 coated nanotubes, *Chem. Commun.* 34–35 (2002).

696. K. Hernadi, E. Ljubović, J.W. Seo, and L. Forró, Synthesis of MWNT-based composite materials with inorganic coating, *Acta Mater.* **51**, 1447–1452 (2003).

697. E.A. Whitsitt and A.R. Barron, Silica coated single walled carbon nanotubes, *Nano Lett.* **3**, 775–778 (2003).

698. Q. Fu, C. Lu, and J. Han, Selective coating of single-wall carbon nanotubes with thin SiO_2 layer, *Nano Lett.* **2**, 329–332 (2002).

699. S.W. Lee and W.M. Sigmund, Formation of anatase TiO_2 nanoparticles on carbon nanotubes, *Chem. Commun.* 780–781 (2003).

700. A. Jitianu, T. Cacciaguerra, R. Benoit, S. Delpeux, F. Béguin, and S. Bonnamy, Synthesis and characterization of carbon nanotubes — TiO_2 nanocomposites, *Carbon* **42**, 1147–1151 (2004).

701. J. Sun, M. Iwasa, L. Gao, and Q. Zhang, Single-walled carbon nanotubes coated with titania nanoparticles, *Carbon* **42**, 895–899 (2004).

702. A. Brioude, P. Vincent, C. Journet, J.C. Plenet, and S.T. Purcell, Synthesis of sheated carbon nanotube tips by the sol-gel technique, *Appl. Surf. Sci.* **221**, 4–9 (2004).
703. Y.-H. Li, S. Wang, A. Cao, D. Zhao, X. Zhang, C. Xu, Z. Luan, D. Ruan, J. Liang, D. Wu, and B. Wei, Adsorption of fluoride from water by amorphous alumina supported on carbon nanotubes, *Chem. Phys. Lett.* **350**, 412–416 (2001).
704. K. Hernadi, E. Couteau, J.W. Seo, and L. Forró, Al(OH)₃/multiwalled carbon nanotubes composite: homogeneous coverage of Al(OH)₃ on carbon nanotube surfaces, *Langmuir* **19**, 7026–7029 (2003).
705. Y. Zhang, J. Liu, R. He, Q. Zhang, X. Zhang, and J. Zhu, Synthesis of alumina nanotubes using carbon nanotubes as template, *Chem. Phys. Lett.* **360**, 579–584 (2002).
706. M.H. Chen, Z.C. Huang, G.T. Wu, G.M. Zhu, J.K. You, and Z.G. Lin, Synthesis and characterization of SnO-carbon nanotube composite as anode material for lithium-ion batteries, *Mater. Res. Bull.* **38**, 831–836 (2003).
707. W.-Q. Han and A. Zettl, Coating single-walled carbon nanotubes with tin oxide, *Nano Lett.* **3**, 681–683 (2003).
708. L. Zhao and L. Gao, Coating of multiwalled carbon nanotubes with thick layers of tin(IV) oxide, *Carbon* **42**, 1858–1861 (2004).
709. H. Kim and W. Sigmund, Zinc oxide nanowires on carbon nanotubes, *Appl. Phys. Lett.* **81**, 2085–2087 (2002).
710. J. Ding, Y. Li, C. Xu, and D. Wu, Depositing CeO₂ nano-particles on surface of carbon nanotubes, *J. Chin. Rare Earth Soc.* **21**, 441–444 (2003) (in Chinese).
711. B.C. Satishkumar, A. Govindaraj, M. Nath, and C.N.R. Rao, Synthesis of metal oxide nanorods using carbon nanotubes as templates, *J. Mater. Chem.* **10**, 2115–2119 (2000).
712. C.N.R. Rao and A. Govindaraj, Nanotubes and nanowires, *Proc. Indian Acad. Sci.* **113**, 375–392 (2001).
713. J. Shi, Y. Qin, W. Wu, X. Li, Z.-X. Guo, and D. Zhu, In situ synthesis of CdS nanoparticles on multiwalled carbon nanotubes, *Carbon* **42**, 455–458 (2004).
714. S. Ravindran, S. Chaudhary, B. Colburn, M. Ozkan, and C.S. Ozkan, Covalent coupling of quantum dots to multiwalled carbon nanotubes for electronic device applications, *Nano Lett.* **3**, 447–453 (2003).
715. H. Kim and W. Sigmund, Zinc sulfide nanocrystals on carbon nanotubes, *J. Cryst. Growth* **255**, 114–118 (2003).
716. J.W. Liu, D.Y. Zhong, F.Q. Xie, M. Sun, E.G. Wang, and W.X. Liu, Synthesis of SiC nanofibers by annealing carbon nanotubes covered with SiC, *Chem. Phys. Lett.* **348**, 357–360 (2001).
717. E. Muñoz, A.B. Dalton, S. Collins, A.A. Zakhidov, R.H. Baughman, W.L. Zhou, J. He, C.J. O'Connor, B. McCarthy, and W.J. Blau, Synthesis of SiC nanorods from sheets of single-walled carbon nanotubes, *Chem. Phys. Lett.* **359**, 397–402 (2002).
718. Y-J. Lee, Formation of silicon carbide on carbon fibers by carbothermal reduction of silica, *Diamond Rel. Mater.* **13**, 383–388 (2004).
719. K.C. Chin, A. Gohel, H.I. Elim, W. Ji, G.L. Chong, K.Y. Lim, C.H. Sow, and A.T.S. Wee, Optical limiting properties of amorphous SiₓNᵧ and SiC coated carbon nanotubes, *Chem. Phys. Lett.* **383**, 72–75 (2004).
720. B.Q. Wei, J.W. Ward, R. Vajtai, P.M. Ajayan, R. Ma, and G. Ramanath, Simultaneous growth of silicon carbide nanorods and carbon nanotubes by chemical vapor deposition, *Chem. Phys. Lett.* **354**, 264–268 (2002).
721. M. Gao, S. Huang, L. Dai, G. Wallace, R. Gao, and Z. Wang, Aligned coaxial nanowires of carbon nanotubes sheathed with conducting polymers, *Angew. Chem. Int. Ed.* **39**, 3664–3667 (2000).
722. D. Shi, J. Lian, P. He, L.M. Wang, W.J. van Ooij, M. Schulz, Y. Liu, and D.B. Mast, Plasma deposition of ultrathin polymer films on carbon nanotubes, *Appl. Phys. Lett.* **81**, 5216–5218 (2002).
723. A.B. Artyukhin, O. Bakajin, P. Stroeve, and A. Noy, Layer-by-layer electrostatic self-assembly of polyelectrolyte nanoshells on individual carbon nanotube templates, *Langmuir* **20**, 1442–1448 (2004).
724. L. Cao, H.-Z. Chen, H.-B. Zhou, L. Zhu, J.-Z. Sun, X.-B. Zhang, J.-M. Xu, and M. Wang, Carbon-nanotube-templated assembly of rare-earth phthalocyanine nanowires, *Adv. Mater.* **15**, 909–913 (2003).
725. Z.-X. Jin, G.Q. Xu, and S.H. Goh, A preferentially ordered accumulation of bromine on multi-wall carbon nanotubes, *Carbon* **38**, 1135–1139 (2000).
726. L. Grigorian, K.A. Williams, S. Fang, G.U. Sumanasekera, A.L. Loper, E.C. Dickey, S.J. Pennycook, and P.C. Eklund, Reversible intercalation of charged iodine chains into carbon nanotube ropes, *Phys. Rev. Lett.* **80**, 5560–5563 (1998).

727. M.T. Martínez, M.A. Callejas, A.M. Benito, M. Cochet, T. Seeger, A. Ansón, J. Schreiber, C. Gordon, C. Marhic, O. Chauvet, J.L.G. Fierro, and W.K. Maser, Sensitivity of single wall carbon nanotubes to oxidative processing: structural modification, intercalation and functionalisation, *Carbon* **41**, 2247–2256 (2003).

728. R. Graupner, J. Abraham, A. Vencelová, T. Seyller, F. Hennrich, M.M. Kappes, A. Hirsch, and L. Ley, Doping of single-walled carbon nanotube bundles by Brønsted acids, *Phys. Chem. Chem. Phys.* **5**, 5472–5476 (2003).

729. A. Kukovecz, T. Pichler, R. Pfeiffer, C. Kramberger, and H. Kuzmany, Diameter selective doping of single-wall carbon nanotubes, *Phys. Chem. Chem. Phys.* **5**, 582–587 (2003).

730. Y. Liu, H. Yukawa, and M. Morinaga, First-principles study on lithium absorption in carbon nanotubes, *Computat. Mater. Sci.* **30**, 50–56 (2004).

731. S. Suzuki, Y. Watanabe, T. Ogino, S. Heun, L. Gregoratti, A. Barinov, B. Kaulich, M. Kiskinova, W. Zhu, C. Bower, and O. Zhou, Extremely small diffusion constant of Cs in multiwalled carbon nanotubes, *J. Appl. Phys.* **92**, 7527–7531 (2002).

732. A.S. Claye, J.E. Fischer, C.B. Huffman, A.G. Rinzler, and R.E. Smalley, Solid-state electrochemistry of the Li-single-wall carbon nanotube system, *J. Electrochem. Soc.* **147**, 2845–2852 (2000).

733. B. Gao, A. Kleinhammes, X.P. Tang, C. Bower, L. Fleming, Y. Wu, and O. Zhou, Electrochemical intercalation of single-walled carbon nanotubes with lithium, *Chem. Phys. Lett.* **307**, 153–157 (1999).

734. H. Shimoda, B. Gao, X.P. Tang, A. Kleinhammes, L. Fleming, Y. Wu, and O. Zhou, Lithium intercalation into etched single-wall carbon nanotubes, *Physica* **B 323**, 133–134 (2002).

735. H. Shimoda, B. Gao, X.P. Tang, A. Kleinhammes, L. Fleming, Y. Wu, and O. Zhou, Lithium intercalation into opened single-wall carbon nanotubes: storage capacity and electronic properties, *Phys. Rev. Lett.* **88**, 015502.1–4 (2002).

736. B. Gao, C. Bower, J.D. Lorentzen, L. Fleming, A. Kleinhammes, X.P. Tang, L.E. McNeil, Y. Wu, and O. Zhou, Enhanced saturation lithium composition in ball-milled single-walled carbon nanotubes, *Chem. Phys. Lett.* **327**, 69–75 (2000).

737. C.H. Mi, G.S. Cao, and X.B. Zhao, A non-GIC mechanism of lithium storage in chemical etched MWNTs, *J. Electroanal. Chem.* **526**, 217–221 (2004).

738. Z. Yang, S. Sang, K. Huang, and H.-Q. Wu, Lithium insertion into the raw multiwalled carbon nanotubes pre-doped with lithium — an electrochemical impedance study, *Diamond Rel. Mater.* **13**, 99–105 (2004).

3 Graphite Whiskers, Cones, and Polyhedral Crystals

Svetlana Dimovski and Yury Gogotsi
Department of Materials Science and Engineering,
Drexel University, Philadelphia, Pennsylvania

CONTENTS

Abstract 109
3.1   Preface 110
3.2   Graphite Whiskers and Cones 110
        3.2.1   Synthetic Whiskers and Cones 111
            3.2.1.1   Whiskers 111
            3.2.1.2   Cones 113
        3.2.2   Occurrence of Graphite Whiskers and Cones in Nature 116
        3.2.3   Structure: Geometrical Considerations 117
        3.2.4   Properties and Applications 121
            3.2.4.1   Electronic Properties of Synthetic Whiskers and Cones 121
            3.2.4.2   Raman Spectra 122
3.3   Graphite Polyhedral Crystals — Polygonal Multiwall Tubes 123
        3.3.1   Synthesis 123
        3.3.2   Structure of Polygonal Tubes 125
        3.3.3   Properties and Applications 128
            3.3.3.1   Electronic Band Structure 128
            3.3.3.2   Raman Spectra 129
            3.3.3.3   Chemical, Thermal, and Mechanical Stability 130
3.4   Conclusions 131
Acknowledgment 132
References 132

ABSTRACT

Carbon nanotubes and fullerenes have been extensively studied and are described in separate chapters in this book. However, the world of carbon nanostructures is not limited to these two groups of materials. Carbon nanocones, whiskers, and larger polygonized nanotubes, called graphite polyhedral crystals, are a whole new class of carbon nanomaterials; which, along with vapor-grown carbon fibers, can be placed between graphite and fullerene families of carbon. They are elongated and, typically, axially symmetric structures. This chapter provides a comparative study of several such graphitic nanomaterials. The discussion covers the gamut of such materials — from the first graphite cones and whiskers discovered long before the beginning of "nano" age, to the latest additions to this family of nano- and micromaterials. Some of these particles are as large as several micrometers, and they provide a bridge between carbon nanomaterials such as nanotubes, and conventional materials such as carbon fibers or planar graphite. Their structure, properties, Raman spectra, and potential applications are discussed in detail in this chapter.

3.1 PREFACE

Both planar graphite[1] and carbon nanotubes[2] have been extensively studied, and their structure and properties are well documented in the literature. This section is a review of the current understanding of some less common nonplanar graphitic materials, such as graphite whiskers, cones, and polygonized carbon nanotubes (graphite polyhedral crystals). Although nonplanar graphitic microstructures in the shape of cones were reported as early as 1957,[3,4] it is only recently[5–8] that attention has been paid to these exotic classes of graphitic materials. There is no doubt that this growing interest has been triggered by the discovery of fullerenes[9] and nanotubes,[2,10] which has stimulated very intensive research on carbonaceous nanomaterials over the past 20 years. While fullerenes and nanotubes have been discussed in several books during the past decade, carbon cones, whiskers, and other similar structures have received much less attention. Our intention here is to give an overview of the current understanding of their structure, synthesis methods, properties, and potential applications.

Some of the engineering disciplines that could benefit from the emergence of these forms of carbon are:

- *Materials engineering*: graphitic cones and polyhedral crystals will enable the development of new functional nanomaterials and fillers for nanocomposites.
- *Chemistry and biomedicine*: in the development of new chemical sensors, cellular probes, and micro-/nanoelectrodes.
- *Analytical tools and instrumentation development*: cones and polyhedral crystals can act as probes for atomic force and scanning tunneling microscopes.
- *Energy, transportation, and electronic devices*: as materials for energy storage, field emitters, and components for nanoelectromechanical systems.

The common features of carbon whiskers, cones, scrolls, and graphite polyhedral crystals, besides their chemistry and the graphitic nature of their bonds, are their morphology and the high length-to-diameter aspect ratio, which places them between graphite and carbon nanotube materials. In the following sections, we will examine various types of these materials, and we will show the effect of structural conformation as well as describe their properties and potential applications.

3.2 GRAPHITE WHISKERS AND CONES

Graphite whiskers, cones, and polyhedral crystals are all needle-like structures — meaning that their length is considerably larger than their width or diameter. The major difference between the graphitic cones and polyhedral crystals, besides their shape, is in their texture, i.e., in the orientation of the atomic planes within the structure. While graphite polyhedral crystals (GPCs) comprise of (0 0 0 1) planes parallel to their main axis, carbon cones and whiskers may have various textures. Therefore, we will consider GPCs separately from cones and whiskers. The orientation and the stacking arrangement of the planes is closely related to the nucleation mechanism and the growth conditions of the cones. The following section explains currently available methods for synthesis of carbon whiskers and cones. Also, two different types of cones have also been observed in natural deposits of carbon. These are described briefly in a separate section. A detailed explanation of their structure and properties in relation to their potential applications follows.

3.2.1 SYNTHETIC WHISKERS AND CONES

3.2.1.1 Whiskers

Graphite whiskers are the first known nonplanar graphitic structures that were obtained through a controlled preparation. Bacon[11] succeeded in growing high-strength graphite whiskers on carbon electrodes using a DC arc under an argon pressure of 92 atm. The temperatures developed in the arc were sufficiently close to the sublimation point of graphite (above 3600°C), which enabled carbon to vaporize from the tip of the positively charged electrode, and form cylindrical deposits embedded with whiskers of up to 3 cm in length and a few microns in diameter.[12] Carbon deposition under extreme conditions, such as a "flash CVD" process, also resulted in the growth of very peculiar micron-sized tree-like carbon structures.[13]

Whiskers and filaments of graphite have also been observed to form during pyrolytic deposition of various hydrocarbon materials. Hillert and Lange[14] studied the thermal decomposition of n-heptane and reported the formation of filamentous graphite on iron surfaces at elevated temperatures. Pyrolysis of methane,[15] and carbon monoxide[16] on iron surfaces or heated carbon filaments[17] also resulted in the formation of similar structures as well as the thermal decomposition of acetylene on Nichrome wires below 700°C.[18]

Whisker growth during pyrolytic deposition process is generally considered as being catalyzed by metals.[19,20] Haanstra et al.[21] however, observed noncatalytic columnar growth of carbon on β-SiC crystals by pyrolysis of carbon monoxide at 1 atm pressure above 1800°C (Figure 3.1). The experiments showed that the growth of carbon whiskers in that case was defined by rotation twinning and stacking faults on {1 1 1} habit plane of the β-SiC substrate. The cylindrical carbon columns formed by this mechanism were observed to consist of parallel conical graphitic layers stacked along the column axis. Most specimens of a run had the diameter between 3 and 6 μm, and they were several tens of microns long. The apex angle of the conical mantle was measured to be about 141°. The conical nuclei of these columns were produced on defects in twinned β-SiC, such as the dislocation with a screw component perpendicular to the surface.

Very similar "needle"- or "spine"-like graphitic materials were also reported by Knox et al.[22] In an attempt to synthesize porous graphitic carbon material that would be capable of withstanding considerable shear forces, such as those seen in high-performance liquid chromatography, they produced porous glassy carbon spheres, which in most cases contained graphitic needles. Knox et al. impregnated the high-porosity silica gel spheres with a melt of phenol and hexamethylenetetramine (hexamine) in a 6:1 weight ratio. Impregnated material was first heated gradually to 150°C to form

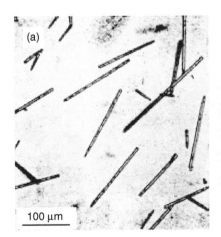
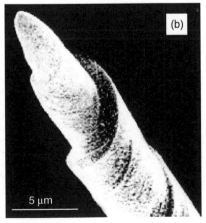

FIGURE 3.1 Electron micrographs of (a) pencil-like carbon columns; (b) columnar carbon specimen with screw-like markings on side.[21]

phenol-formaldehyde resin within the pores of the silica gel, and then carbonized slowly at 900°C in a stream of oxygen-free nitrogen. The silica was then dissolved with hot aqueous potassium hydroxide (at least 99% complete), and the remaining porous glassy carbon was consequently heated up to 2500°C in oxygen-free argon.

Besides the expected glassy carbon structure, the resulting product often contained considerable amounts of needle-like material that was determined to have a three-dimensional graphitic structure. The graphitic whiskers resulting from the experiments were usually a few microns long and about 1 μm thick. Electron diffraction and transmission electron microscopy (TEM) revealed the twinned structure of the whiskers. The angle between the layers was measured to be about 135°.

Since this material was a side product of their experiment, Knox et al. neither provided any further details of its structure nor explained the nucleation mechanism. It is, however, highly probable that their graphitic needles were nucleated and grown in a pretty much the same way as were Haanstra's whiskers shown in Figure 3.1. Incomplete dissolution of the silica matrix could have caused formation of twinned β-SiC phase during the glassy carbon pyrolysis between 1000 and 2500°C, which further induced growth of columnar graphite from disproportionated CO within the porous glassy carbon spheres. The 135° whiskers have also been synthesized recently at 2100°C from gaseous CO and ball-milled natural graphite[23,24] contaminated with zirconia particles during milling. When heated above 1900°C, the zirconia particles react with the carbon to form ZrC.[25] The growth of the graphitic whiskers was probably initiated by screw dislocations on the surfaces of ZrC particles.

Similarly, Gillot et al.[26] studied the heat treatment of products of martensite electrolytic dissolution, and observed the formation of "cigar"-shaped crystals of graphite at 2800°C. The model of the texture they obtained is shown in Figure 3.2a. The length of the crystals ranged from a few microns to 250 μm with a length-to-diameter ratio of about 10. It was suggested that the growth mechanism of the "cigars" involved mass transfer through the gas phase. The graphite layers in the whisker had the shape of an obtuse cone, the axis of which was coincident with the axis of the whisker, and had basically the same structure noted but not fully described by Knox et al. Such whiskers were assumed to be formed by a single graphene sheet coiled around the axis in a helix, each turn of the helix having the shape of a cone (Figure 3.2b). The angular shift θ of the $(h\,k\,0)$ crystallographic directions from one whorl to the next one in the helix was measured to be $\theta \approx 60°$. All graphitic layers were found to have the same stacking arrangement as of a perfect graphite crystal. Later, Double and Hellawell[27] proposed the cone-helix growth mechanism of such

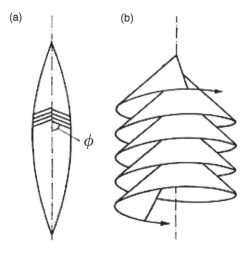

FIGURE 3.2 Model illustrating the formation of cigar-like graphite: (a) longitudinal cross section of the cigars showing their texture; (b) cone-helix structure of graphitic filaments.[26]

structures, which relies on the formation of a negative wedge disclination within a graphene sheet. This model will be explained in detail in the following section.

3.2.1.2 Cones

Ge and Sattler,[6] Sattler,[7] and Krishnan et al.[8] were among the first to observe and study fullerene nanocones, i.e., seamless conical structures formed when one or more pentagonal rings are incorporated into a graphene network. Incorporation of pentagonal and heptagonal defects into graphene sheets and nanotubes had at that time been already discussed by Iijima et al.,[28] Ajayan,[29,30] Ebbesen,[31] Ebbesen and Takada,[32] and others,[33–35] to explain conical morphologies of carbon nanotube tips observed by high-resolution transmission electron microscopy (HRTEM). The importance of the pentagonal defects in the formation of three-dimensional conical graphitic structures, however, was not fully recognized till the thorough investigation of their electron diffraction patterns by Amelinckx et al.,[5,36,37] who studied helically wound conical graphite whiskers (Figure 3.3a) by electron microscopy and electron diffraction. Whiskers gave rise to unusual diffraction effects consisting of periodically interrupted circular ring patterns (Figure 3.3b). Very similar diffraction patterns had been previously obtained from whiskers described in Ref. 21. Amelinckx et al. proposed a growth mechanism whereby the initial graphite layer adopts a slitted dome-shaped configuration (Figures 3.4a and b) by removing a sector β and introducing a five-fold carbon ring in the sixfold carbon network (Figure 3.4c). Successive graphene sheets were then rotated with respect to the previous one over a constant angle, thus realizing a helical cone around a "disclination," with a five-fold carbon ring core. The model explains the morphological features and the particular diffraction effects observed on these reproducibly prepared columnar graphite crystals and it also builds on the other cone models.[21,27]

The first true multishell fullerene graphitic cones consisting of seamless axially stacked conical surfaces (Figure 3.5) were observed in the products of chlorination of silicon carbide at temperatures above 1000°C in 1972[38] and then reported by Millward and Jefferson[39] in 1978. Since these structures were rather singular observations in the products of the reaction, they were not recognized as a new material until much later.[38] Similar structures in large quantities were for the first time successfully produced by Ge and Sattler.[6] Up to 24 nm in length and 8 nm in base diameter, these nanometer-sized structures were generated by vapor condensation of carbon atoms on a highly oriented pyrolytic graphite substrate. All of the cones had the same apex angle ~19°, which is the smallest among five possible opening angles for perfect graphitic cones (Figure 3.6a). The growth of these nanostructures is thought to be initiated exclusively by fullerene-type nucleation seeds with

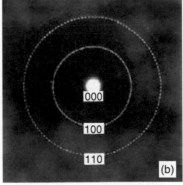

FIGURE 3.3 Conical graphite whiskers. (a) SEM micrograph of a cleavage fragment of conically wound graphite whisker. Note the 140° apex angle of the conical cleavage plane. (b) Electron diffraction pattern with the incident electron beam along the normal to the cleavage 'plane' of the conically wound whiskers. Note the 126-fold rotation symmetry of the pattern.[36]

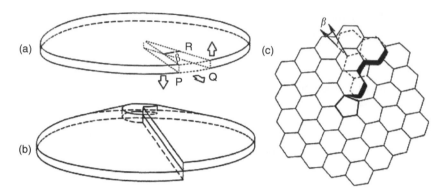

FIGURE 3.4 Model illustrating the formation of a conical helix. (a) Sector β is removed from a disc. (b) The angular gap is closed and a cone is formed. (c) Twisted nucleus of the conical helix containing one pentagonal ring in the graphene network. Conical helix is formed through rotation of successive graphene sheets over a constant angle.[36]

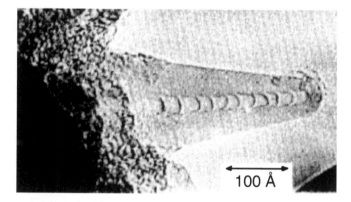

FIGURE 3.5 Carbon cone showing separation of layers in fullerenic end cups.[38]

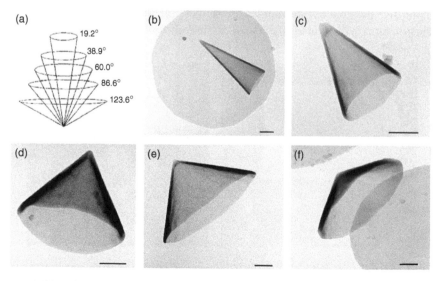

FIGURE 3.6 Fullerene cones. (a) The five possible seamless graphitic cones, with cone angles of 19.2, 38.9, 60, 86.6, and 123.6°.[6,7] Electron micrographs of the corresponding five types of cones (scale bars in b–f, 200 nm). Apex angles: (b) 19.2°, (c) 38.9°, (d) 60.0°, (e) 84.6°, and (f) 112.9°.[8]

different number of pentagons. Fullerene cones of other apex angles corresponding to 1 to 4 pentagons were produced and reported 3 years later by Krishnan et al.[8] (Figures 3.6c–f). They also reproduced the ~19° cone (Figure 3.6b).

Graphite conical crystals of very small apex angles (from ~3 to ~20°) and perfectly smooth surfaces (Figure 3.7) have been reported to form in the pores of glassy carbon at high temperatures,[40,41] in addition to other various axial graphitic nano- and microcrystals.[42] Graphitic structures from the glassy carbon pores were produced from carbon-containing gas formed during decomposition of phenol formaldehyde. The size of these graphite conical crystals ranged from about 100 to 300 nm in the cone base diameter, and their lengths ranged from about 500 nm to several micrometers. Similarly, a few other conical structures of graphite were produced by thermal decomposition of hydrocarbons[43] with or without the aid of a catalyst, or by employing various thermochemical routes.[44] The structure of the majority of catalyst-free cones observed is consistent with the cone-helix growth model; however, some of the small apex angle cones (~2.7°), as seen in Figure 3.7c,[40,41] do not conform to this rule. These are most likely carbon scroll structures.[45,46] Orientation of layers in catalytically produced cones is closely related to and resembles the shape of the catalyst particle.[47–49] Catalytically produced cones can adopt open,[49–51] helical,[49,51] or close-shell structures.[49]

Several other types of cones have been reported that are actually composed of cylindrical graphite sheets.[52,53] So-called tubular graphite cones (TGCs) (Figure 3.8) have been synthesized on an iron needle using a microwave-plasma-assisted chemical vapor deposition (MWCVD) method[52] in a CH_4/N_2 gaseous environment. Corn-shape carbon nanofibers with metal-free tips have also been synthesized by a MWCVD method using CH_4 and H_2 gases.[53] Graphitic coils wound around a tapered carbon nanotube core have also been produced by the same technique using different substrate material.[54] What

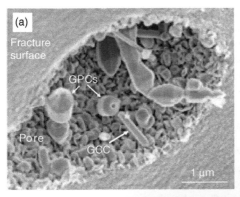
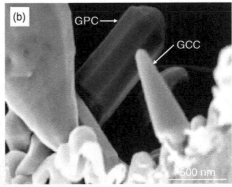
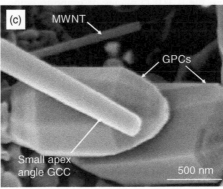

FIGURE 3.7 SEM micrographs of carbon nano- and microcrystals found in pores of glassy carbon. (a) Fracture surface, showing crystals in a pore. (b) Graphite polyhedral crystals (GPCs) and graphite conical crystals (GCCs). (c) A small apex angle GCC growing along with GPCs, and a stylus-like multiwall carbon nanotube (MWNT).[41]

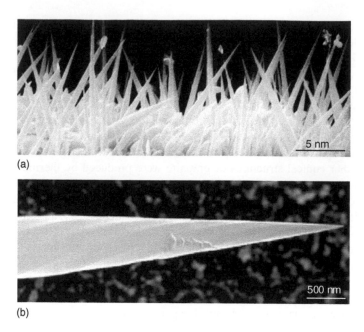

(a)

(b)

FIGURE 3.8 Tubular graphite cones. (a) Aligned TGCs grown on an iron needle surface. (b) A high-resolution view of one TGC shows the faceted and helical appearance.[52]

makes these and similar structures cone-shaped is not purely an inclination of their graphitic layers with respect to the cone axis, but rather the continuous shortening of graphitic wall layers from the interior to the exterior of the structure,[55] or a combination of both mechanisms, as in the case of carbon nanopipettes.[54] Although their morphology resembles a cone, intrinsically their microstructure is that of the multiwall carbon nanotubes. This implies different mechanical and electronic properties. Tailoring carbon nanotubes to cone shapes can now be done routinely.[56–59]

3.2.2 Occurrence of Graphite Whiskers and Cones in Nature

Graphite whiskers and cones have also been observed growing on natural Ticonderoga graphite crystals,[60] Gooderham carbon aggregates,[61] and friable, radially aligned fibers of Kola graphite.[62] In their brief communication, Patel and Deshapande[60] reported the growth of 65 to 125-μm-thick graphite whiskers in (0 0 0 1) direction, the (0 0 0 1) planes of graphite being perpendicular to the whisker axis. The growth of the whiskers was presumed to be a result of a screw dislocation mechanism during the growth of graphite, but no details indicating the relationship of the structure and the geological origin of the sample were given. Several other exotic forms of graphite have been observed recently from two different geological environments: arrays of graphite cones in calcite from highly sheared metamorphic rocks in eastern Ontario (Gooderham graphite, Figure 3.9a),[61] cones, and scrolls of tubular graphite in syenitic igneous rock from the Kola Peninsula of Russia (Figures 3.9b–d).[62]

In a few geological occurrences, graphite forms compact spherical aggregates with radial internal textures,[62–66] similar to those observed in graphite spheres in cast iron.[27] One prominent natural occurrence is in metasedimentary rocks exposed at a roadcut, south of Gooderham, Ontario, Canada.[67,68] In this region, graphite crystallizes in calcite in various forms of tabular flakes, spherical, spheroidal, and triskelial polycrystalline aggregates,[62,69] some of which were found to contain large arrays of graphitic cones dominating the surfaces of the samples.[61] Cone heights ranged from less than a micron to 40 μm, and unlike most laboratory-produced cones, they showed a wide distribution of apex angles. The apex angles were found to vary from 38 to ~140°, with 60° being the most common. The cone structure can be well described by the Double and Hellawell[27] disclination

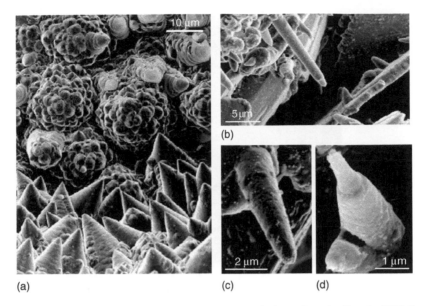

FIGURE 3.9 FESEM images of graphite cones from (a) Gooderham, Ontario, Canada. [61] (b–d) Graphite cones, scrolls, and tubes from Hackman Valley, Kola Peninsula, Russia. A scroll-type structure is suggested in (b). Some of the Kola cones appear to be hollow, as indicated by a fractured structure (d).[62]

model. Other than full and solid cones, some Gooderham samples also revealed partly conical hollow structures composed of curved graphite shells ("protocones").[62] These indicate a possible earlier growth stage for the cones reported in Ref. 61. Unlike large solid cones, many of these graphitic structures have partly faceted surfaces (Figure 3.10). The tips of the polygonal cones typically have six facets, and these facets only extend part way down the surfaces of the cones, which maintain a circular base.[62] The faceted cones are reminiscent of the polyhedral graphite crystals from glassy carbon pores.[42] The morphology and the surface topography of the cones and petrologic relations of the samples suggest that the cones formed from metamorphic fluids.

Numerous scroll-type graphite whiskers, up to 15 μm in length and up to 1 μm in diameter (Figures 3.9b–d), were discovered to cover inner and outer surfaces of channels comprised of tabular graphite crystals. They have been found in samples of alkaline syenitic pegmatite of Kola Peninsula, Russia.[70] The surfaces of cavities in the host rock were coated with fine-grained graphite layer comprised solely of such whiskers. Some of the Kola natural graphite whiskers are cigar-like (Figure 3.9b), while others exhibit true conical (Figure 3.9c) morphologies with dome-shaped tips. The conical whiskers appear to be significantly larger and more abundant than the tube-like whiskers. Many Kola cones show distinct spiral growth steps at the surfaces of their tips, suggesting that they have a scroll-type structure, as seen previously in other synthetic whiskers.[11,17,21,26] SEM images of some broken cones reveal that they are hollow (Figure 3.9d).

3.2.3 STRUCTURE: GEOMETRICAL CONSIDERATIONS

We have seen in the previous section that, on the basis of their structure, a distinction can be made between the two major classes of graphitic cones. One type has a "scroll-helix" structure, while the second type comprises seamless conical graphene layers stacked over each other along their axis (therefore called "fullerene cones"). This classification may be considered as an equivalent to differentiating between "scroll" and "Russian-doll" type of multiwall carbon nanotubes.[71]

Pure "scroll-helix" cones are made up of a single graphene sheet that coils around an axis, each layer having a cone shape. The nucleation of this kind of structure is generally controlled by a line defect (dislocation), although we will see later that in addition, it always involves a screw dislocation

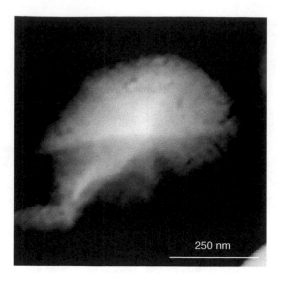

250 nm

FIGURE 3.10 FESEM micrograph of a graphite "protocone" having a faceted tip.[62]

and some kind of point defect at the terminated side of the dislocation line, as indicated in Figure 3.4. On the other hand, an ideal "fullerene" cone contains only point defects in the form of pentagonal, heptagonal, or lower/higher order carbon rings and their various combinations. It is also possible that some of the actual graphitic cones are neither purely helical nor purely fullerene structures, but rather a combination of the two.

Euler's theorem[72] has been found particularly useful in explaining geometrical aspects and generation of fullerenes and fullerene cones. Suppose that a polyhedral object is formed by enclosing a space with polygons. The number of polygons is therefore equal to the number of faces (F) of such object. If V is the number of vertexes, E the number of edges, and g a genus of the structure, then the four parameters correlate as follows:

$$V - E + F = 2(1 - g) \tag{3.1}$$

For bulk three-dimensional solids, g is equivalent to the number of cuts required to transform a solid structure into a structure topologically equivalent to a sphere (for instance, $g = 0$ for a polygonal sphere such as C_{60} or C_{70}, and $g = 1$ for a torus). Suppose, further, that the object is formed of polygons having different (i) number of sides. The total number of faces (F) is then

$$F = \sum N_i \tag{3.2}$$

where N_i is the number of polygons with i sides. Each edge, by definition, is shared between two adjacent faces, and each vertex between three adjacent faces, which is represented as

$$E = (1/2) \sum i N_i \tag{3.3}$$

and

$$V = (2/3)E \tag{3.4}$$

By substituting Equations (3.2) to (3.4) into (3.1), Euler's postulate for $i \geq 3$ is represented as

$$3N_3 + 2N_4 + N_5 - N_7 - 2N_8 - 3N_9 - \cdots - = 12(1 - g) \tag{3.5}$$

$$\sum (6 - i)N_i = 12(1 - g) \tag{3.6}$$

It can be observed from Equation (3.5) that the number of hexagons does not play a role, and a balance between the number of pentagons and higher order polygons ($i \geq 7$) is required in order to form an enclosed structure. If each vertex is considered an atomic site containing an sp^2-hybridized C atom, and each edge is assigned to one C–C bond, then according to Equations (3.5) and (3.6), only 12 pentagons are needed to form a fullerene or a nanotube. If one heptagon is present, then 13 pentagons will close the structure.

The total disclination in a completely closed structure, such as a sphere, is 720° (i.e., 4π). Each out of 12 pentagons contributes a positive disclination of 720°/12 = 60°, and a heptagon, similarly, creates a negative 60° disclination. Incorporation of a heptagon in a graphene sheet will, therefore, produce a saddle-like deformation,[73] while adding pentagons will result in conical structures. Exactly five different cones (Figure 3.6) are generated by having respectively 1 to 5 pentagonal rings in their structure, as experimentally observed[6,7] and mentioned in a previous section. Careful examination of such cones suggests that pentagons are isolated from each other by hexagonal rings, as in fullerene molecules and fullerene nanotube caps. The apex angles for these cones can be calculated from the following relation:[74]

$$\sin(\theta/2) = 1 - (N_5/6) \tag{3.7}$$

where N_5 is the number of pentagons in the cone structure.

Topo-combinatoric conformations of i-polygonal carbon rings (where i = 1, 2, 3, 4, 5, 7, 9, ...) within a hexagonal carbon network had been studied in detail even before the discovery of fullerenes, carbon nanotubes, and graphitic cones.[75,76] Growing interest in this topic resulted in the number of publications[73,77–80] that revealed the fine structure of the cone tip, such as the reconfiguration of carbon atoms and distribution of defects in the near vicinity of the tip. It had also been shown that the pentagons separated by hexagons (Figure 3.11)[74] make the most stable conformation of the cone tip structure, as observed experimentally. Establishing valid theoretical models of structure later helped in calculating the electronic properties of cones and curved carbon surfaces.[81–84]

Apart from seamless cones, there are conical structures that are formed by introducing a wedge disclination (Figure 3.12a) and a screw dislocation (Figure 3.12b) in a graphite sheet, as observed experimentally by various groups.[4,5,21,22,26,36] The cone-helix model[27] is based on growth around a positive disclination with a screw dislocation component (Figure 3.12c). As a graphene sheet wraps around the disclination, adjacent overlapping layers are rotated with respect to one another by an angle equal to the disclination angle. Among practically unlimited number of disclination angles, some of them should be energetically more favorable (Figure 3.12c and inset table in Figure 3.13). Their value can be calculated from the following equation:

$$\alpha = n \times 60°, \quad \text{or} \quad \alpha = n \times 60° \pm \omega \tag{3.8}$$

where n = 0, 1, 2, ... , 6, and ω = 13.2, 21.8, 27.8°, ... are expected low-energy (0 0 1) twist grain-boundary angles based on lattice coincides, which are a measure of "goodness of fit," but do not

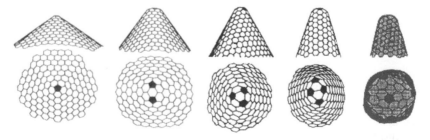

FIGURE 3.11 Distribution of pentagonal defects within the cone tip. The apex angle changes with the number of pentagons.[74]

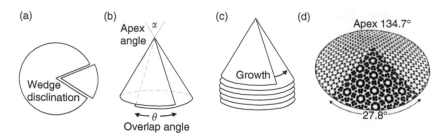

FIGURE 3.12 Formation of helical cones. (a) Positive wedge disclination is created after a sector is removed from a graphene sheet and (b) the cone is formed by an overlap through a screw dislocation. (c) Model illustrating growth of columnar carbon, and (d) one of several energetically preferred stacking arrangements.[27]

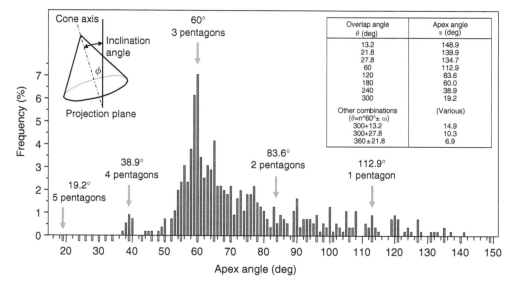

FIGURE 3.13 Frequency of occurrence of various apex angles for natural graphite cones. The maximum is observed at 60°. Apex angles that correspond to "goodness" of fit are listed in the table. When measuring apex angles of cones, the inclination of cones to the projection plane of the microscope has been taken into account (inset, left).[61]

account for atomistic interactions and the curvature of the sheets. Disclinations with overlap angles equal to integer multiples of 60° should be energetically the most favorable, because they preserve the graphite crystal structure without stacking faults, provided the screw component of the disclination has a Burgers vector corresponding to an even multiple of the graphite's c-axis interplanar spacing. Values of corresponding apex angles are calculated from the following relation:

$$\theta = 2 \sin^{-1}(1 - \alpha/360°) \qquad (3.9)$$

and they range from 6 to 149°. Graphitic cones having such apex angles should predominate over others.[61] The apex-angle distribution in a sample of natural cones is shown in Figure 3.13. Among all possible apex angles, the 60° angle is found to be the most frequent. Cones with smaller apex angles may be disfavored because of the higher elastic energy due to bending required to form the corresponding disclinations.

The dislocation line usually terminates with a point defect that includes bond recombination within the hexagonal network to form a pentagon or some other kind of polygon, as shown in Figure 3.4c.

3.2.4 Properties and Applications

3.2.4.1 Electronic Properties of Synthetic Whiskers and Cones

Carbon nanotubes are known to be either metallic or semiconducting, depending on their diameter and chirality.[85–89] The role of pentagon, heptagon, or pentagon–heptagon pair topological defects in structural and electronic properties of nanotubes has also been studied theoretically,[81,84,90] and experimentally by means of scanning tunneling microscopy (STM) and scanning tunneling spectroscopy (STS).[84] Special attention has been paid to curved surfaces of capped carbon nanotube tips, since these can be considered as regions of high density of defects. As the density of defect states increases at the tube ends, it can be expected that the electronic band structure of the end differs significantly from that elsewhere on the tube.[81–84] This has been successfully demonstrated by means of spatially resolved STM/STS carried out on a conically shaped tube end (Figures 3.14a and b),[84] which in fact is a fullerene-type carbon nanocone structure.

An STM image of one such conical tip is shown in Figure 3.14a. The apex of the cone has a diameter of 2.0 nm. The tunneling spectra were acquired at four different positions along the tube (marked with white letters in Figure 3.14a). Local densities of states (LDOS), derived from the scanning tunneling spectra, are represented in Figure 3.14b. In addition, the tight-binding calculations performed on two different tip morphologies are given in Figures 3.14c and d.

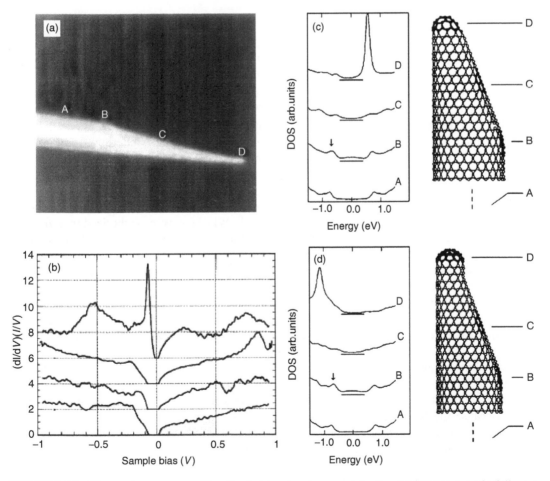

FIGURE 3.14 Electronic structure and localized states at carbon nanotube tips: (a) STM image of a fullerene carbon cone; (b) local densities of states for a cone tip derived from scanning tunneling spectra at four (A–D) points along the tip; and (c–d) tight-binding calculations for two different configurations of cone tips.[84]

As we move along the tube from position A to position D, the density of topological defects increases, since the topological defects are concentrated in a smaller volume. As a result, the effect of confinement on electronic structure becomes more and more pronounced. This does not seem to be very striking in the case of the conduction band, where only a slight and broad enhancement has been noted in the LDOS at the cone apex. The valence band, however, is found to alter considerably, exhibiting sharp resonant states at the cone tip (Figure 3.14b, curve D). The strength and position of these resonant states with respect to the Fermi level is, in addition, very sensitive to the distribution and position of defects within the cone. This is illustrated with two models of cones having different morphologies obtained by altering the position of pentagons within the tip structure (Figures 3.14c and d). In the two examples, the (A), (B), and (C) LDOS calculated by the tight-binding method are very similar. Strong and sharp peaks in (D) LDOS have different shape and position in the case of models I and II. The values calculated for model II show better fit to the experimental values given in Figure 3.14b. The distribution of the defects and their effect on electronic properties of the cones have been studied in detail elsewhere.[91] LDOS of helix-type carbon cones are obtained by establishing the tight-binding model of a screw dislocation in graphite.[81]

Localized resonant states are very important in predicting the electronic behavior of carbon cones. They can also strongly influence the field emission properties of cones.

3.2.4.2 Raman Spectra

Owing to its sensitivity to changes in the atomic structure of carbons, Raman spectroscopy has proven a useful tool in understanding the vibrational properties and the microstructure of graphitic crystals and various disordered carbon materials.[92–97] The relationship between the spectra and the structure has been extensively discussed in the literature, and the studies cover a wide range of carbon materials, such as pyrolytic graphite (PG)[94,95] and highly oriented pyrolytic graphite (HOPG),[95,98,99] microcrystalline graphite, amorphous carbon and glassy carbon, fullerenes, carbon onions, nanotubes, etc. Little work is carried out on the Raman scattering from graphite whiskers,[100–102] which usually consist of carbon layers oriented parallel to the growth axes. For such structures, it is expected that their Raman spectra will be similar to those of disordered graphite crystals and carbon fibers.

Figure 3.15 shows the Raman spectra of an individual graphite whisker and turbostratically stacked particles, using 632.8 nm excitation wavelength. Whiskers were synthesized in a graphitization furnace using a high-temperature heat-treatment method.[23] Carbon layers in these whiskers are almost perpendicular to their growth axes. Most of the first- and second-order Raman modes in whiskers, such as the D, G, and D′ modes at ~1333, 1582, and 1618 cm^{-1}, respectively, can be assigned to the corresponding modes in HOPG and PG.

In contrast to other carbon materials, the Raman spectra of whiskers exhibit several distinct characteristics. For example, the intensity of the 2D overtone is found to be 13 times stronger than that of the first-order G mode in whiskers. The strong enhancement of the D and 2D modes is also found in the Raman spectra of whiskers with 488.0 and 514.5 nm laser excitations.[100] Second, there are two additional low-frequency sharp peaks located around 228 and 355 cm^{-1}, and two additional strong modes (around 1833 and 1951 cm^{-1}) observed in the second-order frequency region. The line widths of the D, G, D′, 2D, and 2D′ modes in whiskers are 17, 18, 10, 20, and 14 cm^{-1}, respectively. Because the frequencies of the L_1 and L_2 modes are in the frequency region of acoustic modes, these two modes are supposed to be the resonantly excited acoustic modes in the transverse-acoustic and longitudinal-acoustic phonon branches. The two high-frequency modes at 1833 and 1951 cm^{-1} are designated as $L_1 + D′$ and $L_2 + D′$ modes, respectively. The observed excitation-energy dependence (140 cm^{-1} eV^{-1}) of the 1833 cm^{-1} mode is in excellent agreement with the theoretical value of 139 cm^{-1} eV^{-1} of the $L_1 + D′$ mode.[100]

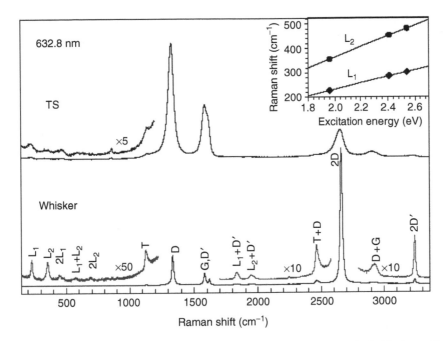

FIGURE 3.15 Raman spectra of turbostratically stacked (TS) particles and an individual graphite whisker excited with 632.8 nm laser excitation. The inset gives the energy dependence of the frequencies of the L_1 and L_2 modes.[102]

The intensity enhancement of the dispersive modes indicates that double-resonance Raman scattering may be responsible for this phenomenon.[103] Such enhancement of the 2D mode is also observed in GPCs (Figure 3.17) that have a similar loop-edge structure in brim regions.[42]

Raman spectra from tubular, helix-type, and naturally occurring carbon cones are available elsewhere in the literature.[61,102]

3.3 GRAPHITE POLYHEDRAL CRYSTALS — POLYGONAL MULTIWALL TUBES

3.3.1 SYNTHESIS

The structure of single- and multiwall carbon nanotubes, and single-wall carbon nanotube ropes have been widely studied over the last 10 years.[2,31,71,104–108]

While the ability of carbon to form multiwall tubular nanostructures is well known and these tubes have been studied extensively, very little information is available about carbon nanotube structures having polygonal cross sections. Although an occurrence of polygonal vapor-grown carbon fibers with a core carbon nanotube protrusion was noted by Speck et al.[109] as early as 1989, no details were given about core fiber structure and its polygonization.

Zhang et al.[110] have studied the structure of an arc-discharge-produced carbon soot by the HRTEM, and they were the first to indicate the possibility of polygonal multiwall carbon nanotubes, assuming that the tubes consisted of closed coaxial concentric layers. The first evidence for occurrence of polygonized carbon nanotubes came from Liu and Cowley,[104,105,108] who used nanodiffraction in conjunction with HRTEM and selected area electron diffraction to investigate the structures of carbon nanotubes having diameters of a few nanometers. Nanodiffraction is a form of convergent beam electron diffraction, which allows one to obtain a diffraction pattern from regions of the specimen about 1 nm or less in diameter. The tubes used in this study were produced by a variant of Kratschmer–Huffman arc-discharge method[111] in helium gas at a pressure of 550 Torr. The DC voltage applied to electrodes was 26–28 V and the corresponding current was 70 A. The carbon nanotubes obtained at the given experimental conditions consisted of 3 to 30 carbon sheets and had a length of up to 1 μm. The inner diameters of these tubes ranged from 2.2 to 6 nm, and the outer diameters ranged from 5 to 26 nm. In addition to nanotubes of circular cylindrical cross section, with zero, one or several helix angles, there were many tubes having polygonal cross sections, made up of flat regions joined by regions of high and uniform curvature.[105] An HRTEM image of one such structure is given in Figure 3.16a. Polygonization of the cross section is observed indirectly through formation of uneven patterns of lattice fringes on the two sides of the tube, with spacings varying from 0.34 nm from the circular cylinder tubes, to 0.45 nm from the regions of high curvature (Figure 3.16b).

In their study of the intershell spacing of multiwall carbon nanotubes prepared by the same Kratschmer–Huffman arc-discharge method, Kiang et al.,[112] similarly, found that the intershell spacing in carbon nanotubes ranged from 0.34 to 0.39 nm among different nanotubes, decreasing with the increase in the tube diameter (Figure 3.16c). Some other reports have also shown variation of the values from 0.344 nm (obtained by the electron and powder x-ray diffraction measurements)[113] to 0.375 nm (based on the HRTEM images).[114]

Faceted multiwall carbon nanotubes with larger diameters, called graphite polyhedral crystals (GPCs), have been reported to grow at high temperatures in the pores of a glassy carbon material (Figures 3.17a and b).[42] The glassy carbon containing polyhedral tubes was made from a thermoset phenolic resin by carbonization at 2000°C in N_2 atmosphere at ~10 Torr. The density of glassy carbon was 1.48 g/cm^3 with an open porosity of ~1%; its microstructure and properties are typical of other glassy carbons. After the structure of the matrix was set and some closed pores were formed, polyhedral nanotubes grew from C–H–O (N_2) gas trapped within these pores during the resin carbonization phase.

Graphite polyhedral crystals have a very complex morphology. Their size ranges from 100 to 1000 nm in diameter and up to few micrometers in length. The number of facets can vary from 5 to 14 and more, and they may possess a helical habit or be axially true. Many of the crystals terminate with a thin protruding needle that appears to be a multiwall nanotube (Figure 3.17b), typically with

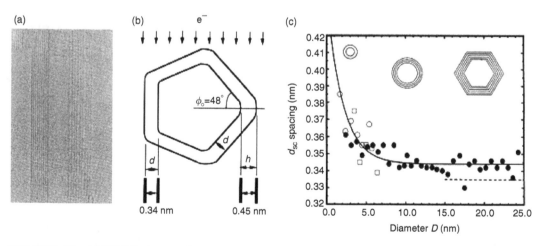

FIGURE 3.16 (a) HRTEM image of a nine-sheet nonsymmetric tube. A d spacing of 0.34 nm is found on the left side and a d spacing of 0.45 nm is seen on the right side.[108] (b) Model illustrating the formation of the non-symmetric fringes from a tube (a) with polygonal cross section.[108] (c) The graphitic interplanar spacing decreases as the tube diameter increases, and approaches 0.344 nm at roughly $D = 10$ nm. The data were measured from three different nanotubes indicated by different symbols. Hollow circles: from a seven-shell tube with innermost diameter $D_{min} = 1.7$ nm. For large D, graphitization may occur resulting in a polygonal cross section. The broken line indicates the expected decrease in interplanar spacing owing to local graphitic stacking.[112]

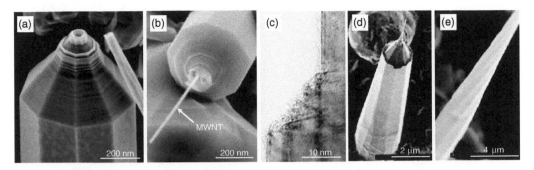

FIGURE 3.17 Graphite polyhedral crystals (GPCs). (a) SEM micrograph of a faceted GPC.[102] (b) A carbon nanotube stylus is connected to a microsize body. (c) TEM image of a GPC's lattice fringes indicates that GPCs are highly graphitic and that the basal planes are terminated by a closed-loop structure.[102] (d–e) GPCs produced by using the flame combustion method.[115]

a core diameter of about 5 to 20 nm and a conical, dome-capped, or semitoroidal tip. There is no evidence about catalytic nucleation of the graphite polyhedral crystals. Formation of highly ordered structures is promoted with the high temperature of treatment, the supersaturation of the environment with carbon atoms, slow reaction kinetics, and the presence of active species such as hydrogen and oxygen atoms that balance the crystal growth rate with the surface etching rate. This explains the surprisingly large number of ordered carbon layers (up to 1500) growing on the core nanotube, resulting in complex axis-symmetric structures. GPCs of somewhat less perfect structures (Figures 3.17c and d) have been successfully produced recently by using the flame combustion method.[115]

Annealing of carbon nanotubes with a circular cross section at high temperatures causes polygonization of their walls. An HRTEM image of a CVD carbon nanotube sample before and after annealing is shown in Figure 3.18. The tubes were annealed for 3 h in a 10^{-6} Torr vacuum at 2000°C. High-temperature annealing of carbon nanotubes in a vacuum or an inert environment allows for the transformation of circular tubes into polygonal ones. However, polygonization will not be uniform along the tube, nor will the cross section take the shape of a regular polygon.

To the best of our knowledge, natural counterparts of polygonal carbon multiwall nanotubes have not been observed so far, but it would not come as a surprise if they are discovered in the near future. Very short needle-like polygonal multiwall carbon nanotubes have also been synthesized from a supercritical C–H–O fluid by hydrothermal treatment of various carbon precursor materials[116,117] with and without the aid of a metal catalyst. Hollow carbon nanotubes, with multiwall structures, comprising of well-ordered concentric graphitic layers, have been produced by treating amorphous carbon in pure water at 800°C and 100 MPa.[117] HRTEM analysis of the reaction products indicates the presence of carbon nanotubes with polygonal cross sections (varying contrast and lattice spacing along the tube diameter) within these samples. The experimental conditions for hydrothermal synthesis of nanotubes resemble to a great extent the conditions of geological metamorphic fluids, and it is possible that some polygonal tubes are present in the Earth's crust along with the natural graphitic cones and tubules but have not been found yet.

3.3.2 STRUCTURE OF POLYGONAL TUBES

One of the earliest works dealing with the polygonization of the cross section of carbon nanotube, is a report by Zhang et al.[110] Based on experimental observations, it was suggested that the fine structure of carbon nanotubes is determined by two competing accommodation mechanisms (Figure 3.19a). As a result of one of the mechanisms, the successive tube can adopt different helical shapes to accommodate the change in circumference, therefore keeping an orientationally disordered (turbostratic) stacking. As a result of the second mechanism, the rows of hexagons in successive tubes

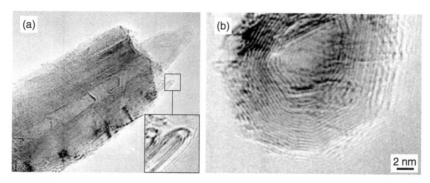

FIGURE 3.18 Transformation of multiwall carbon nanotubes into polygonal GPC-like structures by annealing in vacuum at 2000°C. (a) TEM image of a tube with a core nanotube in the form of stylus extension. Arched semitoroidal structures, similar to that of GPCs, have been also formed (inset) through elimination of dangling bonds at high temperature. (b) Polygonized cross section and graphitic walls of an annealed hollow tube. TEM micrographs courtesy of H. Ye.

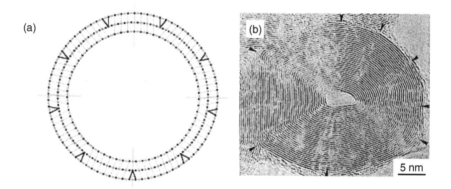

FIGURE 3.19 (a) Schematic model of a nanotube cross section. "Interfacial dislocations" (bold lines) are introduced to accommodate the strains on the tube surfaces. The graphite stacking is maintained in the tube walls (as indicated by lines). The full circles represent the atoms in the paper plane and the open circles are projected positions of the atoms of the paper plane. (b) Defect regions in the HRTEM image of a carbon "onion." The defect regions are characterized by their abnormal image contrast.[110]

remain parallel and adopt a graphitic stacking, thus inducing some regions of stacking faults due to the deformation of hexagons. The regions of stacking fault are assumed to be evenly distributed along the tube circumference,[110] and are separated from the graphitic structure by interfacial dislocations (bold lines in Figure 3.19a). Polygonization of carbon nanotubes therefore may appear as a result of the necessity to allow graphitic stacking of the layers, as often seen in carbon onions (Figure 3.19b).[110,118,119] It is easy to envision a trade-off between the energy associated with turbostratic stacking vs. the strain energy associated with shape changes and stacking faults. The mechanism prevailing depends strongly on the tube size. The relative strain required to maintain graphitic order is smaller with increasing tube size; therefore thicker tubes are expected to have graphitic ordering and polygonal cross sections with discrete regions of stacking faults, while thinner tubes are more likely to retain turbostratic concentric structures with cylindrical cross section and varying helicity between the individual shells.

In the near-planar regions of the polygonized tubes, an ordering of the stacking sequence of the carbon layers gives rise to hexagonal and possibly rhombohedral graphite structure.[108] The near-planar regions are connected into seamless shells through the regions of high curvature (Figure 3.16b).

A small value for the radius of curvature is preferred in regions of bending of the carbon sheets between the extended near-planar regions because of the nature of local perturbations of the carbon bonding arrangement.[104,105,108]

A tube structure model has been proposed to explain the variation of intershell spacing as a function of tube diameter (Figure 3.20a). Individual intershell spacings as a function of tube diameters were measured in real space from HRTEM images of various nanotubes.[112] The empirical equation for the best fit to these data is given as

$$\hat{d}_{002} = 0.344 + 0.1e^{-D/2} \tag{3.10}$$

The function is plotted in Figure 3.20b. The large full circles show experimental values.

The increase in the intershell spacing with decreased nanotube diameter is attributed to the high shell curvature of small diameter tubes and it has also been suggested that polygonization of the tube cross section will occur for inner tube diameters larger than ~12 nm (see Figure 3.16c). This observation is in agreement with the model suggested by Zhang et al.[110]

Furthermore, it has also been proposed that multiwall nanotubes most likely consist of circular core shells and polygonal outer layers.[120] In their pioneering work on nanodiffraction from carbon nanotubes, Liu[104] and Cowley[105] noted a possibility that there might be some nanotubes with neither entirely polygonal nor fully cylindrical cross sections. Such tubes could be considered as a mixture of the two possible morphologies, with structure varying along the tube length and the shell diameter. A schematic illustrating this model, taking into account variations of intershell spacing, is given in Figure 3.20.

In order to obtain direct evidence of tube microstructure, there have been several cross-sectional TEM studies.[121–124] These studies were conducted on both carbon nanotubes produced by arc-discharge method and the tubes produced by a chemical vapor deposition. A large number of defects in CVD tubes is very common, and it is an intrinsic property of the CVD process and, therefore, will not be discussed further here. HRTEM images of the cross sections of tubes reveal their nested structure, but they do not confirm models proposed by Liu,[105] Zhang,[110] and others.[112,120] Instead, rather random dislocation lines extending radially have been recorded.[121]

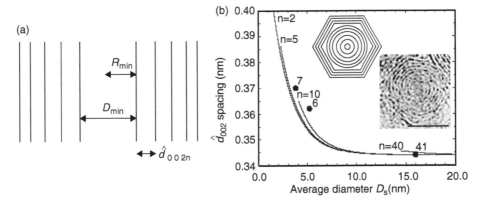

FIGURE 3.20 Effect of tube diameter on interplanar spacing. (a) Model for a nanotube crystal with a varying intershell spacing. (b) The d spacing is given an exponential function of tube diameter (Equation 3.10). The intershell spacing d is plotted as a function of the average tube diameter (D_a), where n is the number of shells in a nanotube. The curves are calculated for $n = 2$, 5, 10, and 40, using the above model and Equation 3.10. The three data points shown by the large full circles were obtained based on experimental observations.[112] Insets: Model illustrating change of interplanar spacing d and polygonization of tube cross section with increase of tube diameter, as observed in some TEM micrographs. (TEM image: courtesy of S. Welz; scale bar, 5 nm).

Aside from their polygonal cross section, GPCs[42,115] possess another important feature that may affect their electrical, chemical, and mechanical properties to a great extent. This is the several nanometer thick loop-like layer (Figure 3.17c and inset in Figure 3.18a) formed by zipping of the adjacent graphitic shells at their terminations.[125,126] This phenomenon is also observed in edge planes of some high-temperature planar graphites,[125,127] and on the surfaces of cup-like multiwalled carbon nanotubes annealed in argon atmosphere above 900°C.[128] In the case of planar graphite, zipping of graphitic layers (also known as "lip–lip" interactions) forms nanotube-like sleeves, while in the case of multiwall nanotubes, the resulting structure resembles concentric polygonal hemitoroidal structures.[126] "Lip–lip" interactions are especially pronounced when samples are annealed at temperatures above 1600°C.[42,51,125,128–130] The reactive edge sites transform into stable multiloops through the elimination of dangling bonds due to enhanced carbon mobility at higher temperatures. Multiloops are typically built by 2 to 6 adjacent graphitic layers. Typically, single-loop structures are formed between 900 and 1200°C, while 1500°C is considered as the threshold for the formation of multilayer loops.[128] The radius of curvature of the outer layer is similar to the average radius of double-walled nanotubes.[131]

3.3.3 Properties and Applications

3.3.3.1 Electronic Band Structure

Electronic properties of cylindrical single- and multiwall carbon nanotubes have been widely studied both theoretically and experimentally over the past 15 years, and findings have been summarized in several books about carbon nanotubes,[71,132] as well as in Chapter 1 of this book.

Electronic structure of polygonal single-wall carbon nanotubes has been investigated theoretically within a tight-binding and *ab initio* frameworks,[133,134] and it has been found that polygonization changes the electronic band structure qualitatively and quantitatively. An example of a zigzag nanotube is given. Considered is the (10, 0) tube with: a circular (a) and pentagonal (b) cross section (Figure 3.21). The (10, 0) carbon nanotube with a circular cross section is a semiconductor, with a band gap of 0.82 eV. In calculating the band structure of a polygonal tube, it is reasonable to assume that the zones of strong curvatures near the edges of the polygonal tube will introduce a $\sigma^*–\pi^*$ hybridization of carbon bonds. This local variation of bonding with strong sp^3 character in the folds creates a sort of defect line in the sp^2 carbon network.[135] In addition to the effect of bond hybridization, polygonization of the cross section lowers the symmetry from a ten- to a fivefold

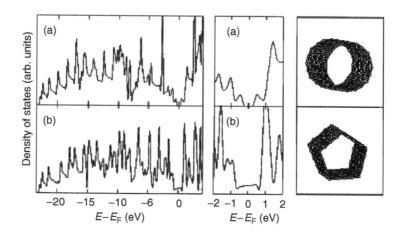

FIGURE 3.21 Tight-binding densities of states (states/eV/cell) for the (10, 0) cylindrical (a), and the (10, 0)⁵ pentagonal (b) cross-section nanotubes. The Fermi level is positioned at zero energy. Both nanotubes are also represented in the inset on the right of their respective DOS.[134]

axis. Furthermore, out-of-plane bending of the hexagonal carbon rings along the polygonal edges brings new pair of atoms closer than the second-neighbor distance in graphite. All this leads to the modification of the electronic band structure, and as a consequence, the semiconducting band gap of the (10, 0) polygonal tube is almost completely closed (Figure 3.21b). The *ab initio* calculations[134] confirm these tight-binding results and predict a gap of 0.08 eV for the pentagonal cross section. Electronic behavior of metallic armchair nanotubes is not so strongly altered with polygonization because the $\sigma^*-\pi^*$ hybridization is not possible in the case of armchair configurations.

Theoretical studies also suggest that the perturbation of electronic properties of carbon nanotubes will be different for various degrees of polygonization (i.e., various numbers of facets).[134] An example is given for a (12, 0) nanotube. Zigzag (12, 0) nanotube of circular cross section is metallic. When different polygonal cross sections (triangle, square, and hexagon) are considered, the results indicate that all kinds of electronic properties arise (Figure 3.22). The first two cases are metallic, while the third is a 0.5 eV band gap semiconductor. It is important to remember here that these calculations are given for a carbon nanotube comprised of a single shell.

3.3.3.2 Raman Spectra

Vibrational properties of GPCs have been studied by means of Raman spectroscopy,[42,102,115] and it has been confirmed that they are highly graphitized structures with the extinct disorder-induced (D) band and the strong graphitic (G) band of about the same full width at half maximum (FWHM = 14 cm^{-1}) as in crystals of natural graphite.[136] The selective micro-Raman spectra from the crystal's side face and tip are shown in Figure 3.23. Spectra from the crystal faces correspond to perfect graphite with a narrow G band at 1580 cm^{-1}. In addition to 1580 cm^{-1} peak, the spectra from the tips feature a weak D band at 1352 cm^{-1}, and an unusually strong 2-D (2706 cm^{-1}) overtone that exceeds the intensity of the G band. A similar effect was observed on graphite scrolls (Figure 3.15). Raman spectra of GPCs contain two additional bands in the second-order frequency range at ca. 1895 and 2045 cm^{-1} (Figure 3.23). A number of weak low-frequency bands including a doublet at 184/192 cm^{-1} has been observed in some samples.[42] These low-frequency bands, typical for single-wall nanotubes,[137] may come from the innermost carbon nanotube shells protruding

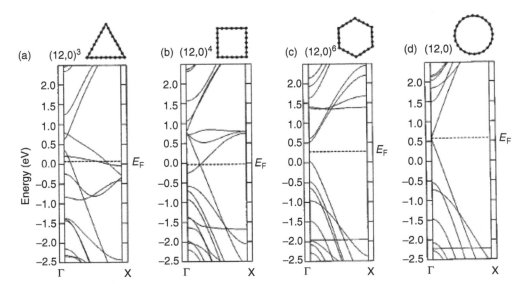

FIGURE 3.22 Tight-binding band structures of the metallic (12, 0) nanotube, illustrating the effect of the degree of polygonization of the cross section on electronic behavior. Given here are examples of (a) the triangle (12, 0)³, (b) square (12, 0)⁴, and (c) hexagonal (12, 0)⁶ geometries that are compared to the pure cylinder case tube (d).[134]

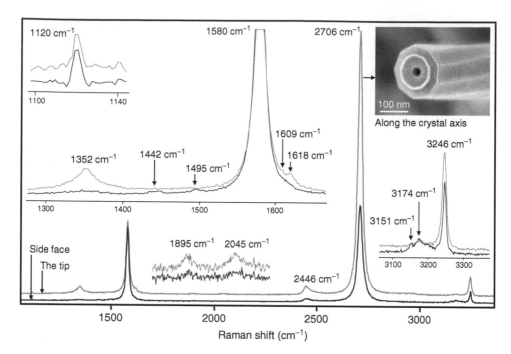

FIGURE 3.23 Fundamental modes, combination modes, and overtones in Raman spectra taken from the side face and the tip of an individual graphite polyhedral crystal (514.5 nm excitation). Inset: an SEM micrograph of a different crystal having structure similar to the one used to record Raman spectra.[102]

sometimes from the GPCs (see Figure 3.17b). Low-intensity bands are also present in the range 1440 to 1500 cm^{-1} (Figure 3.24), the origin of which has not been unambiguously determined yet.

3.3.3.3 Chemical, Thermal, and Mechanical Stability

Functionalization and chemical activity of carbon nanotubes is the subject of intensive study, and many breakthroughs have been made in this field over the past 5 years. Sidewalls of single-wall carbon nanotubes have been successfully functionalized with fluorine,[138] carboxylic acid groups,[139] isocyanate groups,[140] dichlorocarbene,[141] poly-methyl methacrylate,[142] and polystyrene.[143] Various functional groups have been attached to tube edge sites. This has made it possible to utilize carbon nanotubes and other tubular carbon materials in the fabrication of sorbents,[144] catalyst supports,[145,146] gas storage materials,[147] and polymer matrix composites.[148,149] The effect of polygonization on chemical behavior of carbon nanotubes has not been thoroughly investigated yet; however it is expected that polygonal tubes have chemical properties similar to those of circular multiwall carbon nanotubes and graphite, and that they may be more reactive (less stable) along the polygonal edges than in the extended near-planar regions.[120] Another interesting property of GPCs is polygonized hemitoroidal edge plane terminations.[150] Transformation of active sites into loops promotes the GPC into a more stable (or chemically inert) structure. Moreover, due to their specific spatial conformation, they allow for easier intercalation with foreign atoms such as lithium and others.[127,151]

Several methods have been utilized to probe the chemical activity of polygonal tubes. In one of them, GPCs were intercalated with 50:50 H$_2$SO$_4$/HNO$_3$ for 1 h, washed in deionized water, dried for 24 h and then exfoliated by rapid heating at about 980°C for about 15 sec or until maximum volume expansion is reached. GPCs survived very severe intercalation and exfoliation conditions, most of them retaining their original shape of faceted axial whiskers, although damage in the form of cracks along the axis and striations on the hemitoroidal surfaces was observed on most of the crystals.

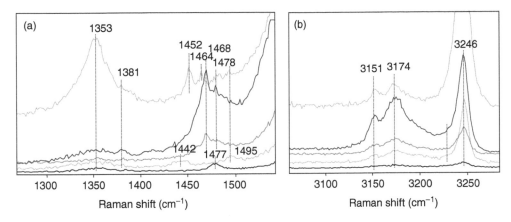

FIGURE 3.24 Raman spectra of graphite polyhedral crystals showing additional weak bands in the range: (a) 1400 to 1500 cm^{-1} and (b) 3100 to 3200 cm^{-1} (for 514.5 nm excitation).[102]

Corrosion studies of graphitic polyhedral crystals showed that exposure of GPCs to overheated steam at normal pressure for 1 h will cause their complete oxidation at temperatures between 600 and 700°C. Similarly, 700°C was determined as an onset temperature for oxidation in air. However, they were more stable than disordered glassy carbon in supercritical water.[42]

Polygonization of the cross section of a tube is not expected to significantly affect its tensile strength, nor should it drastically affect the bending modulus of the tube. However, for a helical polygonal multiwall nanotube, the pull-out strength is expected to be higher than in the case of its cylindrical counterpart, because of the more favorable stress distribution upon loading.

3.4 CONCLUSIONS

Extended growth of nanotubes or fullerene cones leads to formation of a myriad of nano- and microstructures, which are larger relatives of carbon nanotubes. Graphite cones and whiskers have much in common with carbon nanotubes and nanofibers. They inherit the conical structure of scrolled tubes (also known as "herring-bone" structure in carbon nanotubes). Seamless graphite nano- and microcones can have five different apex angles with "magic" numbers of 19.2, 38.9, 60.0, 83.6, and 112.9°, which correspond to 60, 120, 180, 240, and 300° disclinations in graphite. Scrolled conical structures may have virtually any apex angle, and very small (2 to 3°) angles have been observed. A few other specific values also occur more frequently as they are energetically preferred because they allow the registry between graphene sheets in the cone.

Polygonization of nanotubes accompanied by growth in the radial direction leads to the formation of GPCs. They have nanotube cores and graphite crystal faces. Unusual symmetries have been observed in GPCs. Most of them are built of co-axial graphite shells.

Cones and large nanotubes have been produced synthetically in the laboratory, and several methods for their synthesis are known. GPCs have only been discovered in synthetic carbon materials and can be grown by CVD or hydrothermally. However, since many natural graphites have been formed from hydrothermal deposits in nature, it would not be surprising to learn that GPCs exist in nature as well. Large carbon nanotubes have already been observed in natural deposits along with carbon cones.

Cones, whiskers, and GPC can bridge the nano- and microworlds and may have numerous applications, where sizes between nanotubes and carbon fibers are required. They may also have interesting electronic and mechanical properties determined by their geometry. However, while their structure has been well understood, very little is known about their properties. Properties need to be studied before their wide-scale applications can be explored.

ACKNOWLEDGMENT

This work was supported by the U.S. Department of Energy grant DE-FJ02-01ER45932.

REFERENCES

1. B. T. Kelly, *Physics of Graphite*, Applied Science Publishers, London, 1981.
2. S. Iijima, *Nature*, 354, 56, 1991.
3. T. Tsuzuku, *Proceedings of the 3rd Conferene on Carbon*, Pergamon Press, University of Buffalo, New York, 1957, p. 433.
4. T. Tsuzuku, *J. Phys. Soc. Jpn.*, 12, 778, 1957.
5. S. Amelinckx, W. Luyten, T. Krekels, et al., *J. Cryst. Growth*, 121, 543, 1992.
6. M. Ge and K. Sattler, *Chem. Phys. Lett.*, 220, 192, 1994.
7. K. Sattler, *Carbon*, 33, 915, 1995.
8. A. Krishnan, E. Dujardin, M. M. J. Treacy, et al., *Nature*, 388, 451, 1997.
9. H. W. Kroto, J.R.Heath, S.C.O'Brien, et al., *Nature*, 318, 162, 1985.
10. S. Iijima, *MRS Bull.*, 19, 43, 1994.
11. R. Bacon, *J. Appl. Phys.*, 31, 283, 1960.
12. D. W. McKee, *Annu. Rev. Mater. Sci.*, 03, 195, 1973.
13. P. M. Ajayan, J. M. Nugent, R. W. Siegel, et al., *Nature*, 404, 243, 2000.
14. M. Hillert and N. Lange, *Z. Kristallogr. Kristallgeometrie Krystallphys. Kristallchem.*, 111, 24, 1958.
15. S. D. Robertson, *Carbon*, 8, 365, 1970.
16. P. L. J. Walker, J. F. Rakszawski, and G. R. Imperial, *J. Phys. Chem.* 63, 133, 1959.
17. M. L. Lieberman, C. R. Hills, and C. J. Miglionico, *Carbon*, 9, 633, 1971.
18. P. A. Tesner, E. Y. Robinovich, I. S. Rafalkes, et al., *Carbon*, 8, 435, 1970.
19. W. R. Davis, R. J. Slawson, and G. R. Rigby, *Trans. Br. Ceram. Soc.*, 56, 67, 1957.
20. A. Fonseca, K. Hernadi, P. Piedigrosso, et al., *Appl. Phys. A: Mater. Sci. Process.*, 67, 11, 1998.
21. H. B. Haanstra, W. F. Knippenberg, and G. Verspui, *J. Cryst. Growth*, 16, 71, 1972.
22. J. H. Knox, B. Kaur, and G. R. Millward, *J. Chromatogr.*, 352, 3, 1986.
23. J. Dong, W. Shen, B. Zhang, et al., *Carbon*, 39, 2325, 2001.
24. J. Dong, W. Shen, F. Kang, et al., *J. Cryst. Growth*, 245, 77, 2002.
25. P. T. B. Shaffer, *Plenum Press Handbooks of High-Temperature Materials No. 1 Materials Index*, Plenum Press, New York, 1964.
26. J. Gillot, W. Bollman, and B. Lux, *Carbon*, 6, 381, 1968.
27. D. D. Double and A. Hellawell, *Acta Metall.*, 22, 481, 1974.
28. S. Iijima, T. Ichihashi, and Y. Ando, *Nature*, 356, 776, 1992.
29. P. M. Ajayan, T. Ichihashi, and S. Iijima, *Chem. Phys. Lett.*, 202, 384, 1993.
30. P. M. Ajayan and S. Iijima, *Nature*, 358, 23, 1992.
31. T. W. Ebbesen, *Annu. Rev. Mater. Sci.*, 24, 235, 1994.
32. T. W. Ebbesen and T. Takada, *Carbon*, 33, 973, 1995.
33. A. L. Mackay and H. Terrones, *Nature*, 352, 762, 1991.
34. H. Terrones and A. L. Mackay, *Carbon*, 30, 1251, 1992.
35. B. I. Dunlap, *Phys. Rev. B.*, 46, 1933, 1992.
36. W. Luyten, T. Krekels, S. Amelinckx, et al., *Ultramicroscopy*, 49, 123, 1993.
37. S. Amelinckx, A. Lucas, and P. Lambin, *Rep. Prog. Phys.*, 62, 1471, 1999.
38. H. P. Boehm, *Carbon*, 35, 581, 1997.
39. G. R. Millward and D. A. Jefferson, in *Chemistry and Physics of Carbon*, Vol. 14, P. L. J. Walker and P. A. Thrower, Eds., Dekker, New York, 1978, p. 1.
40. S. Dimovski, J. Libera, and Y. Gogotsi, *Mat. Res. Soc. Symp. Proc.*, 706, Z6.27.1, 2002.
41. Y. Gogotsi, S. Dimovski, and J. A. Libera, *Carbon*, 40, 2263, 2002.
42. Y. Gogotsi, J. A. Libera, N. Kalashinkov, et al., *Science*, 290, 2000.
43. N. Muradov and A. Schwitter, *Nano Lett.*, 2, 673, 2002.
44. J. Liu, W. Lin, X. Chen, et al., *Carbon*, 42, 669, 2004.
45. S. F. Braga, V. R. Coluci, S. B. Legoas, et al., *Nano Lett.*, 4, 881, 2004.
46. L. M. Viculis, J. J. Mack, and R. B. Kaner, *Science*, 299, 1361, 2003.

47. N. M. Rodriguez, A. Chambers, and R. T. K. Baker, *Langmuir*, 11, 3862, 1995.
48. V. I. Merkulov, A. V. Melechko, M. A. Guillorn, et al., *Chem. Phys. Lett.*, 350, 381, 2001.
49. V. I. Merkulov, M. A. Guillorn, D. H. Lowndes, et al., *Appl., Phys., Lett.*, 79, 1178, 2001.
50. H. Terrones, T. Hayashi, M. Muños-Navia, et al., *Chem. Phys. Lett.*, 343, 241, 2001.
51. M. Endo, Y. A. Kim, T. Hayashi, et al., *Appl. Phys. Lett.*, 80, 1267, 2002.
52. G. Y. Zhang, X. Jiang, and E. G. Wang, *Science*, 300, 472, 2003.
53. Y. Hayashi, T. Tokunaga, T. Soga, et al., *Appl. Phys. Lett.*, 84, 2886, 2004.
54. R. C. Mani, X. Li, M. K. Sunkara, et al., *Nano Lett.*, 3, 671, 2003.
55. P. Liu, Y. W. Zhang, and C. Lu, *Appl. Phys. Lett.*, 85, 1778, 2004.
56. Z. F. Ren, Z. P. Huang, D. Z. Wang, et al., *Appl. Phys. Lett.*, 75, 1086, 1999.
57. Q. Yang, C. Xiao, W. Chen, et al., *Diamond Relat. Mater.*, 13, 433, 2004.
58. H. Lim, H. Jung, and S.-K. Joo, *Microelectron. Eng.*, 69, 81, 2003.
59. Y. K. Yap, J. Menda, L. K. Vanga, et al., *Mat. Res. Soc. Symp. Proc.*, 821, P3.7.1, 2004.
60. A. R. Patel and S. V. Deshapande, *Carbon*, 8, 242, 1970.
61. J. A. Jaszczak, G. W. Robinson, S. Dimovski, et al., *Carbon*, 41, 2085, 2003.
62. S. Dimovski, J. A. Jaszczak, G. W. Robinson, et al. Extended Abstracts of Carbon 2004, Biennial Conference on Carbon, American Carbon Society, RI, 2004.
63. J. A. Jaszczak, in *Mesomolecules: From Molecules to Materials*, G. D. Mendenhall, A. Greenberg, and J. F. Liebman, Eds., Chapman & Hall, New York, 1995, p. 161.
64. V. N. Kvasnitsa and V. G. Yatsenko, *Mineralogicheskii Zh.*, 13, 95, 1991.
65. V. G. Kvasnitsa, V. N. Yatsenko, and V. M. Zagnitko, *Mineralogicheskii Zh.*, 20, 34, 1998.
66. C. Lemanski, *Picking Table*, 32, 1, 1991.
67. B. A. Van der Pluijm and K. A. Carlson, *Geology*, 17, 161, 1989.
68. Carlson, B. A. Van der Pluijm, and S. Hanmer, *Geol. Soc. Am. Bull.*, 102, 174, 1990.
69. J. A. Jaszczak and G. W. Robinson, *Rocks Miner.*, 75, 172, 2000.
70. S. N. Britvin, G. U. Ivanyuk, and V. N. Yakovenchuk, *World of Stones*, 5/6, 26, 1995.
71. P. J. F. Harris, *Carbon Nanotubes and Related Structures*, Cambridge University Press, Cambridge, 1999.
72. E. A. Lord and C. B. Wilson, *The Mathematical Description of Shape and Form*, Halsted Press, New York, 1984.
73. S. Ihara, S. Itoh, K. Akagi, et al., *Phys. Rev. B*, 54, 14713, 1996.
74. M. Endo, K. Takeuchi, K. Kobori, et al., *Carbon*, 33, 873, 1995.
75. J. R. Dias, *Carbon*, 22, 107, 1984.
76. A. T. Balaban, D. J. Klein, and X. Liu, *Carbon*, 32, 357, 1994.
77. D. J. Klein, *Phys. Chem. Chem. Phys.*, 4, 2099, 2002.
78. D. J. Klein, *Intl. J. Quantum Chem.*, 95, 600, 2003.
79. K. Kobayashi, *Phys. Rev. B*, 61, 8496, 2000.
80. Mizes and J. S. Foster, *Science*, 244, 559, 1989.
81. R. Tamura, K. Akagi, M. Tsukada, et al., *Phys. Rev. B*, 56, 1404, 1997.
82. R. Tamura and M. Tsukada, *Phys. Rev. B*, 52, 6015, 1995.
83. R. Tamura and M. Tsukada, *Phys. Rev. B*, 49, 7697, 1994.
84. D. L. Carroll, P. Redlich, P. M. Ajayan, et al., *Phys. Rev. Lett.*, 78, 2811, 1997.
85. B. I. Dunlap, *Phys. Rev. B*, 49, 5643, 1994.
86. J. W. Mintmire, B. I. Dunlap, and C. T. White, *Phys. Rev. Lett.*, 68, 631, 1992.
87. R. Saito, M. Fujita, G. Dresselhaus, et al., *Phys. Rev. B*, 46, 1804, 1992.
88. J.-C. Charlier and J.-P. Issi, *Appl. Phys. A: Mater. Sci. Process*, 67, 79, 1998.
89. J. W. G. Wilder, L. C. Venema, A. G. Rinzler, et al., *Nature*, 391, 1998.
90. J. C. Charlier, T. W. Ebbesen, and P. Lambin, *Phys. Rev. B*, 53, 11108, 1996.
91. S. Berber, Y.-K. Kwon, and D. Tománek, *Phys. Rev. B*, 62, 2291, 2000.
92. R. Vidano and D. B. Fischbach, *J. Am. Cer. Soc.*, 61, 13, 1978.
93. P. Lespade, R. Al-Jishi, and M. S. Dresselhaus, *Carbon*, 20, 427, 1982.
94. G. Katagiri, H. Ishida, and A. Ishitani, *Carbon*, 26, 565, 1988.
95. Y. Kawashima and G. Katagiri, *Phys. Rev. B*, 52, 10053, 1995.
96. G. G. Samsonidze, R. Saito, A. Jorio, et al., *Phys. Rev. Lett.*, 90, 027403, 2003.
97. F. Tuinstra and J. L. Koenig, *J. Chem. Phys.*, 53, 1126, 1970.
98. R. J. Nemanich and S. A. Solin, *Phys. Rev. B*, 20, 392, 1979.
99. P. H. Tan, Y. M. Deng, and Q. Zhao, *Phys. Rev. B*, 58, 5435, 1998.

100. P. H. Tan, C. Y. Hu, J. Dong, et al., *Phys. Rev. B*, 6421, 214301, 2001.
101. J. Dong, W. C. Shen, and B. Tatarchuk, *Appl. Phys. Lett.*, 80, 3733, 2002.
102. P. H. Tan, S. Dimovski, and Y. Gogotsi, *Phil. Trans. Royal Soc. Lond. A*, 362, 2289, 2004.
103. Thomsen, C. and Reich, S. *Phys. Rev. Lett.*, 85, 5214, 2000.
104. M. Q. Liu and J. M. Cowley, *Mater. Sci. Eng. A*, 185, 131, 1994.
105. M. Liu and J. M. Cowley, *Ultramicroscopy*, 53, 333, 1994.
106. S. Amelinckx, D. Bernaerts, X. B. Zhang, et al., *Science*, 267, 1334, 1995.
107. N. G. Chopra, L. X. Benedict, V. H. Crespi, et al., *Nature*, 377, 135, 1995.
108. M. Liu and J. M. Cowley, *Carbon*, 32, 393, 1994.
109. J. S. Speck, M. Endo, and M. S. Dresselhaus, *J. Cryst. Growth*, 94, 834, 1989.
110. X. F. Zhang, X. B. Zhang, G. Van Tendeloo, et al., *J. Cryst. Growth*, 130, 368, 1993.
111. W. Kratschmer, L. D. Lamb, K. Fostriopoulos, et al., *Nature*, 347, 354, 1990.
112. C.-H. Kiang, M. Endo, P. M. Ajayan, et al., *Phys. Rev. Lett.*, 81, 1869, 1998.
113. Y. Saito, T. Yoshikawa, S. Bandow, et al., *Phys. Rev. B*, 48, 1907, 1993.
114. M. Bretz, B. G. Demczyk, and L. Zhang, *J. Cryst. Growth*, 141, 304, 1994.
115. H. Okuno, A. Palnichenko, J.-F. Despres, et al., *Carbon*, 43, 692, 2005.
116. J. M. Calderon-Moreno and M. Yoshimura, *Mater. Trans.*, 42, 1681, 2001.
117. J. M. Calderon Moreno and M. Yoshimura, *J. Am. Chem. Soc.*, 123, 741, 2001.
118. S. Iijima, *J. Cryst. Growth*, 50, 675, 1980.
119. D. Ugarte, *Nature*, 359, 707, 1992.
120. Y. Maniwa, R. Fujiwara, H. Kira, et al., *Phys. Rev. B*, 64, 073105, 2001.
121. L. A. Bursill, P. Ju-Lin, and F. Xu-Dong, *Philos. Mag. A (Phys. Condens. Matter, Defects Mech. Prop.)*, 71, 1161, 1995.
122. S. Q. Feng, D. P. Yu, G. Hub, et al., *J. Phys. Chem. Solids*, 58, 1887, 1997.
123. G. Hu, X. F. Zhang, D. P. Yu, et al., *Solid State Commun.*, 98, 547, 1996.
124. J.-B. Park, Y.-S. Cho, S.-Y. Hong, et al., *Thin Solid Films*, 415, 78, 2002.
125. S. V. Rotkin and Y. Gogotsi, *Mater. Res. Innovations*, 5, 191, 2002.
126. A. Sarkar, H. W. Kroto, and M. Endo, *Carbon*, 33, 51, 1995.
127. K. Moriguchi, Y. Itoh, S. Munetoh, et al., *Phys. B: Condens. Matter*, 323, 127, 2002.
128. M. Endo, B. J. Lee, Y. A. Kim, et al., *New J. Phys.*, 5, 121.1, 2003.
129. H. Murayama and T. Maeda, *Nature*, 345, 791, 1990.
130. M. Endo, Y. A. Kim, T. Hayashi, et al., *Carbon*, 41, 1941, 2003.
131. Z. Zhou, L. Ci, X. Chen, et al., *Carbon*, 41, 337, 2003.
132. M. S. Dresselhaus, G. Dresselhaus, and P. C. Eklund, *Science of Fullerenes and Carbon Nanotubes*, Academic Press, New York, 1996.
133. P. Lambin, A. A. Lucas, and J.-C. Charlier, *J. Phys. Chem. Solids*, 58, 1833, 1997.
134. J.-C. Charlier, P. Lambin, and T. W. Ebbesen, *Phys. Rev. B*, 54, R8377, 1996.
135. H. Hiura, T. W. Ebbesen, J. Fujita, et al., *Nature*, 367, 148, 1994.
136. K. Ray and R. L. McCreery, *Analyt. Chem.*, 69, 4680, 1997.
137. A. M. Rao, E. Richter, S. Bandow, et al., *Science*, 275, 187, 1997.
138. E. T. Mickelson, I. W. Chiang, J. L. Zimmerman, et al., *J. Phys. Chem. B*, 103, 4318, 1999.
139. J. Chen, M. A. Hamon, H. Hu, et al., *Science*, 282, 95, 1998.
140. C. Zhao, L. Ji, H. Liu, et al., *J. Solid State Chem.*, 177, 4394, 2004.
141. H. Hu, B. Zhao, M. A. Hamon, et al., *J. Am. Chem. Soc.*, 125, 14893, 2003.
142. H. Kong, C. Gao, and D. Yan, *J. Am. Chem. Soc.*, 126, 412, 2004.
143. G. Viswanathan, N. Chakrapani, H. Yang, et al., *J. Am. Chem. Soc.*, 125, 9258, 2003.
144. Y. Xu, J. W. Zondlo, H. O. Finklea, et al., *Fuel Process Technol.*, 68, 189, 2000.
145. J. Ma, C. Park, N. M. Rodriguez, et al., *J. Phys. Chem B*, 105, 11994, 2001.
146. M. Endo, Y. A. Kim, M. Ezaka, et al., *Nano Lett.*, 3, 723, 2003.
147. A. Chambers, C. Park, R. T. K. Baker, et al., *J. Phys. Chem. B*, 102, 4253, 1998.
148. R. Andrews and M. C. Weisenberger, *Curr. Opinion Solid State Mater. Sci.*, 8, 31, 2004.
149. V. Datsyuk, C. Guerret-Piecourt, S. Dagreou, et al., *Carbon*, 43, 873, 2005.
150. J. Han, *Chem. Phys. Lett.*, 282, 187, 1998.
151. K. Moriguchi, S. Munetoh, M. Abe, et al., *J. Appl. Phys.*, 88, 6369, 2000.

4 Inorganic Nanotubes and Fullerene-Like Materials of Metal Dichalcogenide and Related Layered Compounds

R. Tenne
Department of Materials and Interfaces,
Weizmann Institute, Rehovot, Israel

CONTENTS

4.1 Preface 135
4.2 Synthesis of Inorganic Nanotubes 138
4.3 Inorganic Nanotubes and Fullerene-Like Structures Studied by
     Computational Methods 144
4.4 Study of the Properties of Inorganic Nanotubes in Relation to
     Their Applications 148
4.5 Conclusions 150
Acknowledgments 150
References 150

4.1 PREFACE

The discovery of carbon fullerenes [1] and later carbon nanotubes [2] stimulated research on carbonaceous nanomaterials and also the search for nanotubes and fullerene-like structures from other layered compounds. Numerous inorganic compounds possess layer (2-D) structure, including metal dichalcogenides (sulfides, selenides, and tellurides), metal dihalides (chlorides, bromides, and iodides), metal oxides, and numerous ternary or quaternary compounds, resembling thereby graphite. The metal dichalcogenides, MX_2 (M = Mo, W, Nb, Ta, Hf, Ti, Zr, Re; X = S, Se) contain a metal layer sandwiched between two chalcogen layers with the metal in a trigonal pyramidal or octahedral coordination mode [3,4]. Analogous to graphite (Figure 4.1a), weak van der Waals forces are responsible for the stacking of the MX_2 layers along the c-axis (Figure 4.1b). The molecular sheets of inorganic layered compounds consist of multiple layers of different atoms chemically bonded together.

The stimulus for the formation of fullerenes and nanotubes can be understood on the basis of the structure of the parent bulk compound, graphite, which consists of atomically flat carbon (graphene) sheets. Here, each carbon atom is bonded to its three nearest-neighboring carbon atoms via sp^2

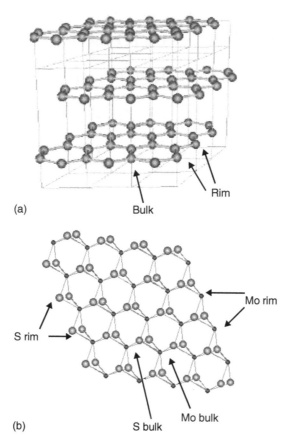

FIGURE 4.1 (a) Schematic structure of graphite. (b) Schematic presentation of MoS_2 layer with both bulk and rim atoms delineated.

hybridization, forming a hexagonal (honeycomb) network. The layers are stacked together via weak van der Waals forces. Carbon atoms that are disposed on the rim of the sheet are only twofold-bonded, each one thus having a dangling bond. When the graphene sheet size shrinks, the ratio of rim/bulk atoms increases as 2^n, where n is the number of cutting cycles of the sheet into smaller pieces. The large increase in the ratio of rim/bulk atoms with shrinking size of the graphene sheet makes the planar nanostructure unstable. This instability results in folding of the graphite layers and the formation of carbon fullerenes and nanotubes, which are seamless structures. While the nanotubes involve winding of the graphene nanosheet in one axis, the fullerenes are obtained by folding the nanosheet along two axes. Since the latter process requires much higher elastic energy, the folding is realized by disposing 12 pentagons into the hexagonal carbon network. It is important to realize that folding of the graphene nanosheet involves overcoming an elastic energy activation barrier, which is provided by heating or irradiation. The invested energy is more than compensated once the fullerene is closed and the dangling bonds at the periphery of the nanosheet disappear.

In analogy to that, one notes that the chalcogen atoms of MX_2 compounds on the basal (0 0 0 1) plane are fully bonded and therefore nonreactive. In contrast the metal and chalcogen atoms on the rim of the nanosheet are not fully bonded. While the metal atom within the MX_2 layer is sixfold-coordinated to the nearest chalcogen atoms, it is only fourfold-coordinated at the edge of the MS_2 nanocluster. Similarly, the chalcogen atom in the rim is only twofold-bonded to the metal atoms instead of being threefold-coordinated as in the bulk (Figure 4.1b). Therefore, planar nanosheets of the dichalcogenide compounds become unstable, forming structures reminiscent of nested carbon fullerenes and nanotubes (Figure 4.2), referred hereafter as inorganic fullerene-like (IF) structures

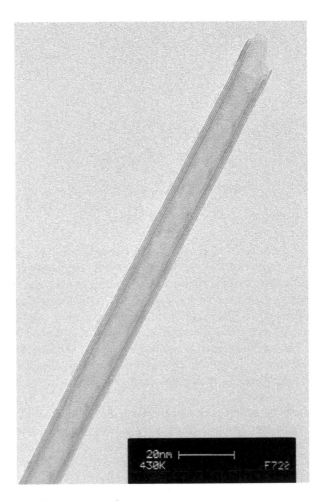

FIGURE 4.2   TEM image of a WS_2 nanotube.

[5,6]. Generally, the fullerene-like structures are less regular in their shape than the respective nanotubes and contain many defects (see below). Other related structures are nanoscrolls, which are obtained by winding molecular sheets several times around their axial direction. In the energy landscape, nanoscrolls can be considered to be an intermediate stage between a perfectly crystalline the stable nanotubular structure and the unstable flat nanosheets. Nanoscrolls are preferably produced by "chemie douce" (low temperature) processes, like sol–gel, intercalation–exfoliation, hydrothermal synthesis, etc., in which the thermal energy is not sufficiently large to induce growth of the ultimately more stable nanotubular structure.

Over the past few years, considerable progress has been achieved in the synthesis of nanotubes from layered metal dichalcogenides and several other layered compounds. Nanotubes of various layered transition metal compounds, including halides like $NiCl_2$ [7], and various layered-type oxides like V_2O_5 [8,9] and $H_2Ti_3O_7$ [10], have been synthesized using different methodologies. The advent of the synthesis, structural characterization, properties, and applications of inorganic nanotubes have been covered by a number of recent review articles, see, e.g., [11,12].

Perhaps, the most well-known example of an early tube-like structure with diameters in the submicrometer range is formed by the asbestos minerals, like chrysotile, kaolinite, etc. The folding of kaolinite sheet was studied by Pauling [13], and was attributed to an asymmetry of the layered structure, which consists of fused silica tetrahedra and alumina octahedra (Figure 4.3). The larger alumina octahedra occupy the outer face of the scrolling sheet, whereas the smaller silica tetrahdra are found

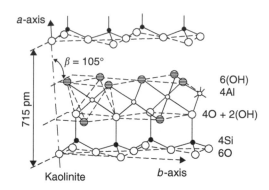

FIGURE 4.3 Schematic representation of the mineral kaolinite.

at the inner face of the asbestos sheet. More recently, a systematic effort to synthesize "misfit" compounds, in which cubic crystals like PbS are intercalated between layers of, e.g., NbS_2, forming thereby an artificial superlattice, is well documented [14]. Owing to lattice mismatch between the two compounds, an asymmetry arises in the "misfit" lattice, which induces folding of the layers and the formation of tubular structures [15]. The synthesis of mesoporous silica with well-defined nanopores in the range 2 to 20 nm was reported by Beck and Kresge and their coworkers [16]. The syntheses strategy involved the self-assembly of liquid crystalline templates. The pore size in zeolitic and other inorganic porous solids is varied by a suitable choice of the template. In contrast with the IF phases, which consist of isolated nanoparticles, mesoporous materials form an interconnected and extended lattice.

Using various templates, recent efforts have resulted in the synthesis of nanotubes from isotropic 3-D (nonlayered) compounds also, further widening the scope of inorganic nanotubes. Even folding of sheets of 3-D compounds, like silica or AlN into nanotues with below 100 nm diameter requires very large amounts of elastic energy and is therefore prohibitive. Even higher elastic energy is required for the formation of quasi-spherical polyhedral (fullerene-like) nanostructures from 3-D compounds. One way for a 3-D compound to overcome this energy barrier is to introduce grain boundaries and dislocations, leading thereby to polycrystalline nanotubes. Nanotubes of 3-D compounds can be obtained through a templated growth using an artificial scaffold. Earlier, amorphous silica nanotubes were obtained by hydrolysis of tetraethylorthosilicate (TEOS) in a mixture of water, ammonia, ethanol, and tartaric acid [17,18]. Later on, polycrystalline silica, alumina, and vanadia nanotubes were obtained by using carbon nanotubes as a template and subsequent calcination of the product at elevated temperatures [19]. Recently, however, highly faceted and crystalline nanotubes of isotropic (3-D) compounds were obtained by growth along an easy axis, as demonstrated for AlN [20] (Figure 4.4) and for microtubules of Ga_2O_3 (Figure 4.5) [21]. Here, the c-axis $\langle 0\ 0\ 0\ 1 \rangle$ serves as the fast growth axis for this wurzite-type lattice, with $(1\ 0\ 1\ 0)$ and $(1\ 1\ \overline{2}\ 0)$ faces making the facets of the tubes. The atomic structure of the edges between two such facets has not been studied in detail so far. Nanotubes of 3-D compounds are inherently unstable and their surface is generally very reactive, a property which can be very useful for catalyzing chemical reactions.

4.2 SYNTHESIS OF INORGANIC NANOTUBES

Even before IF materials were recognized as a universal phenomenon, tubular and fullerene-like forms of layered compounds were reported. Perhaps the earliest among them were nanoscrolls of MoS_2 obtained by reacting $MoCl_4$ with Li_2S in tetrahydrofurane (THF) solution, and subsequent annealing at 400°C [22]. In another study, NiS–MoS_2 core–shell structures, in which closed MoS_2 layers enfold NiS nanoparticles, were reported [23]. Over the past few years, a number of synthetic strategies have been successfully used for the production of nanotubes from layered (2-D) as well

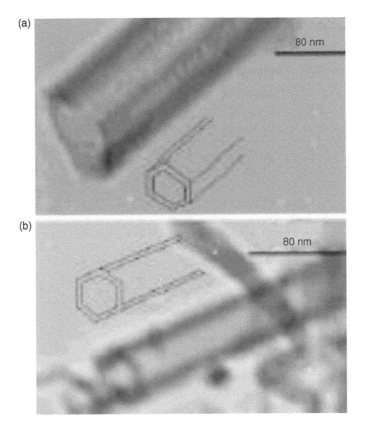

FIGURE 4.4   SEM image of faceted AlN nanotubes. (Adapted from Wu, Q. et al., *J. Am. Chem. Soc.*, 125, 10176, 2003.)

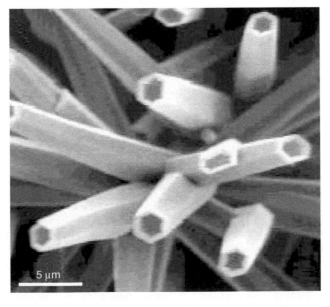

FIGURE 4.5   Faceted microtubules of Ga_2O_3. (Adapted from Sharma, S. and Sunkara, M.K., *J. Am. Chem. Soc.*, 124, 12288, 2002.) Note the clear hexagonal cross section of the tubes in both cases.

TABLE 4.1

Summary of the Inorganic Nanotubes Reported in the Literature and the Synthetic Procedures Used for Their Production

Compound	Synthetic Approach	References	Comments
MS_2 (M=W, Mo) ReS_2	Reaction with the respective oxide at elevated temperature	[5,24–26]	ReS_2 is a 2-D compound with Re-Re bonds
MS_2 (M=W, Mo)	Chemical vapor transport (CVT)	[27,28]	
Single wall MoS_2 nanotubes	$CVT(I_2)+C_{60}$ catalyst	[29]	
MoS_2	Ammonium thiometallate (ATM) solution in porous alumina template, reaction with H_2S	[30,31]	Not perfectly crystalline
MS_2 (M=Nb, Ta, Ti, Zr, Hf)	Decomposition of MS_3 at elevated temperature	[32,33]	
MoS_2	Heating MoS_2 powder in a closed Mo crucible	[34]	
NbS_2, ReS_2	Depositing $NbCl_2$ ($ReCl_3$) from solution onto carbon nanotubes; firing in H_2S at elevated temperature	[35,36]	
$NbSe_2$	Direct reaction of the elements at 800°C	[37]	
MoS_2	Hydrothermal reaction of ATM; crystallization from acetone; firing in H_2S at 600°C	[38]	Nonperfect crystallinity
WS_2	Firing of ATM in thiophene/hydrogen at 360–450°C	[39]	
TiS_2	$CVT(I_2)$	[40]	
MoS_2	Firing of ATM in H_2	[41]	
MoS_2	Firing of MoO_3 nanobelts in the presence of sulfur	[42]	
WS_2	Hydrothermal synthesis with organic amines and a cationic surfactant and firing at 850°C	[43]	
WS_2	Growth of WO_{3-x} nanowhiskers; annealing in H_2S/H_2 atmosphere at elevated temperature	[44,45]	
NbS_2/C	Deposition of $NbCl_4$ on carbon nanotubes template/firing in H_2S	[46]	
MX_2 (M=Mo, Nb; X=S, Se, Te)	Electron beam irradiation of MX_2 powder	[47–50]	
MS_2	Arc discharge of submerged electrodes	[51]	
MoS_2, WS_2	Microwave plasma	[52]	
SnS_2/SnS (misfit)	Laser ablation of SnS_2	[54]	
InS	Reacting t-Bu_3In with H_2S in aprotic solvent at 203°C in the presence of benzenethiol	[88]	Metastable phase

as nonlayered (3-D) compounds. Table 4.1 brings a concise summary of the various metal dichalcogenide nanotubes that have been reported over the last few years, and the synthetic procedures which have been used for their production. This summary shows the large versatility of the chemical apparatus that has been developed for the synthesis of these nanotubes. It further suggests that these nanostructures are thermodynamically stable, with the sole constraint that their diameter is smaller than approximately 100 nm. Most commonly, the synthesis of metal dichalcogenide nanotubes is done at elevated temperatures. "Chemie douce" (low-temperature) processes like hydrothermal synthesis, which are very useful for the synthesis of metal-oxide nanotubes, have not been developed so far for chalcogenide nanotubes. Heating is not only important for accelerating the reaction kinetics, but also to endow sufficient thermal energy to the nanoclusters. This thermal energy induces structural fluctuations in the lamella, leading to folding of the planar nanosheet into a crystalline nanotubular structure. Obviously, ionizing or nonionizing irradiation can serve as a source of energy to elicit the chemical reaction and the structural transformation [47–49]. This effect was first recognized in the electron beam synthesis of fullerene-like MoS_2 nanoparticles [50]. However, such a procedure is not likely to become a useful way to grow macroscopic amounts of inorganic nanotubes. Arc-discharge of MoS_2 electrode submerged in water led to the synthesis of fullerene-like nanoparticles of this compound [51]. Microwave plasma has also been used to synthesize fullerene-like nanoparticles of WS_2 and MoS_2 [52]. Sonochemical reaction has been used to synthesize nanoscrolls of the layered compound $GaO(OH)$ [53].

Laser ablation has been successfully used for the growth of carbon nanotubes and nanowires of Si, GaAs, and similar compounds. Recently, very short nanotubes and nested fullerne-like nanoparticles of "misfit" compound SnS_2/SnS have been obtained by laser ablation of an SnS_2 pellet [54]. Here an ordered superstructure with one or a pair of SnS layers sheathed between two SnS_2 layers was observed. The formation of these unusual nanostructures can be attributed to the volatility of sulfur during the laser ablation process. Unfortunately, the "misfit" nanotubes are rather short (<1 μm) and cannot be produced in large amounts. Laser ablation of MoS_2 and WS_2 targets yielded nanoparticles with fullerene-like structure [55,56]. Growth of $NiCl_2$ nanotubes by a laser ablation process was demonstrated [57]. The nanotubes were shown to grow in a vapor–liquid–solid (VLS) mechanism. However, little control over the growth mode could be established, and the yield of the nanotubes was extremely low.

Hexagonal boron nitride (BN) crystallizes in a graphite-like structure. Here, the iso-electronic B–N pair replaces a C–C pair in the graphene sheet. Partial or entire replacement of the C–C pairs by the B–N pairs in the hexagonal network of graphite leads to the formation of a wide array of two-dimensional phases, like BC_2N. BN tubes with turbostratic structure were reported by Hamilton et al. [58]. The first synthesis of crytalline BN nanotubes was reported shortly afterwards [59]. BN nanotube phases have been produced by several synthetic approaches [60–62]. Double-wall BN nanotubes were obtained by plasma arc reaction using boron anodes impregnated with nickel and cobalt catalysts [63]. Likewise, BC_xN nanotubes were synthesized by reacting aligned CN_x nanotubes with B_2O_3 in the presence of CuO in N_2 atmosphere and at 1800°C [64]. Sheated carbon/BN nanotubes were synthesized as well [65]. Here a graphite cathode was arced with an HfB_2 anode in a nitrogen atmosphere. In this configuration, the three constituents have different sources: the anode for boron, the cathode for carbon, and the chamber atmosphere for nitrogen. In addition, hafnium present in the plasma may be acting as a catalyst. Reactive magnetron sputtering of graphite target in the presence of nitrogen atmosphere led to the synthesis of carbon fullerenes with 20% N content. The aza-fullerene $C_{48}N_{12}$ (Figure 4.6) is considered to be the most stable cluster belonging to the CN_x family of compounds [66]. Each of the 12 pentagons is decorated by a single nitrogen atom.

In summary, the plurality of methods used to synthesize $B_xC_yN_z$ nanotubes suggests that these nanostructures, which are derived from macroscopically stable 2-D compounds are themselves thermodynamically stable phases, given the sole constraint that their diameter is smaller than approximately 0.1 μm.

Systematic attempts to grow nanotubular structures from quasi-isotropic (3-D) oxide and chalcogenide compounds have been undertaken [19,67]. By coating carbon nanotubes (CNTs) with the respective metal-oxide gels and subsequently burning the carbon, polycrystalline nanotubes and nanowires of a variety of metal oxides, including ZrO_2, SiO_2, and MoO_3 have been obtained [19]. Using a surfactant (Triton 100-X) as scaffold at low temperatures, polycrystalline CdSe and CdS nanotubes were obtained after firing at 300 to 400°C [67].

The synthesis of numerous kinds of oxide nanotubes through "chemie douce" processes have been recently documented. Sol–gel synthesis of oxide nanotubes is also possible in the pores of alumina membranes [68]. An interesting case is that of the titania nanotubes, which are hypothesized to exhibit excellent photocatalytic and photoelectrochemical properties for self-cleaning surfaces or

FIGURE 4.6 Schematic representation of $C_{48}N_{12}$ aza-fullerene. (Adapted from Hultman, L., et al., *Phys. Rev. Lett.*, 87, 225503, 2001.)

dye-sensitized solar cells. Titania nanotubes have been obtained by treating anatase or rutile structure TiO_2 with NaOH at 120°C, and subsequent washing with HCl solution [69,70]. The resulting titania nanotubes (nanoscrolls) are 50 to 200 nm long and their diameter is about 10 nm. TEM images of such nanotubes reveal lattice fringes with 0.78 nm spacing. Frequently, the number of molecular layers is not the same on both sides of the nanotube, indicating that the nanotube is made of a continuous molecular sheet which is wound like a folded carpet or a scroll, and hence is termed nanoscroll (Figure 4.7). Indeed, a recent series of works [16,71] concluded that the NaOH treatment leads to recrystallization of the anatase and exfoliation of individual $Na_xH_{2-x}Ti_3O_7$ molecular sheets, which fold into an open-ended nanoscroll. The driving force for the exfoliation of the titanate sheets was attributed to the NaOH, which extracts protons from the exposed surface. This process produces a charge imbalance between the two sides of the solution-exposed $H_2Ti_3O_7$ nanosheet. Consequently, detachment of the nanosheet from the crystal surface and subsequent folding leads to the formation of the nanoscroll [72]. Similarly, nanotubes (nanocrolls) of the lamellar compound $K_4Nb_6O_{17}$ were obtained by exfoliation of the parent crystallites with tetra (n-butyl) ammonium hydroxide [73]. The radius of the nanoscrolls is determined by a very delicate balance between the van der Waals interactions and the elastic energy, which explains the size uniformity of these self-assembled structures.

Nanotubes from various layered metal-silicates were recently prepared via hydrothermal synthesis, i.e., under moderate temperatures (generally < 200°C) and pressure of a few MPa [74]. These nanotubes were shown to exhibit high surface area, appreciable hydrogen storage capacity, and when loaded with Pd metal, high catalytic reactivity toward CO and C_2H_6 oxidation. Nanotubes of δ-MnO_2 were synthesized by hydrothermal conversion of the layered α-$NaMnO_2$ compound [75]. The importance of this advance is in the possibility of using such nanotubular phases as cathodes in high performance Li intercalation batteries.

Nanotubular and fullerene-like structures of various rare-earth hydroxides have been prepared by hydrothermal synthesis [76]. Furthermore, dehydration and subsequent reactions led to the formation of nanotubular structures of rare earth oxides, oxisulfides and oxifluorides. Nanotubes of the ferroelectric materials $BaTiO_3$ and $SrTiO_3$ with perovskite structure were obtained by a two-step hydrothermal synthesis [77]. In the preliminary step, TiO_2 (or titanate) nanotubes were prepared by hydrothermal synthesis with NaOH and subsequent wash with HCl solution. In the following step, the $BaTiO_3$ and $SrTiO_3$ were synthesized by hydrothermal reaction between the titanate nanotubes

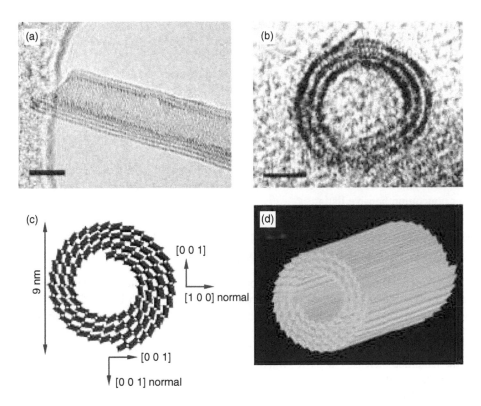

FIGURE 4.7 TEM image and schematic representation of $H_2Ti_3O_7$ nanotubes: (a) axial direction; (b) cross section showing clearly the winding of the sheets into nanoscroll structure; (c) schematic rendering of the cross section; and (d) schematic rendering of the winding of the nanosheet into a nanoscroll. (Adapted from Chen, Q. et al., *Adv. Mater.*, 14, 1208, 2002.)

and $Ba(OH)_2$ and $SrCl_2$, respectively. Although the XRD and TEM analyses suggest that the nanotubes are crystalline, a more thorough analysis is needed in order to understand the morphology of the nanotubes. Nested fullerene-like nanoparticles of the layered compound Tl_2O were synthesized from $TlCl_3$ solutions using a sonochemical method [78]. Tl_2O possesses the relatively rare anti-$CdCl_2$ structure. Here, each molecular sheet consists of an oxygen layer in the center, sandwiched between two outer thalium layers. Numerous metal hydroxides afford a layered structure and can therefore be transformed into nanoscrolls. Thus, $Cu(OH)_2$ nanoscrolls were obtained by simply dipping the metal foil in a basic ammonical solution [79].

Various oxide nanotubes have also been obtained via high-temperature syntheses. Nanotubes of the cubic compound In_2O_3 were synthesized by heating In and In_2O_3 powders under vacuum in an induction furnace at 1300°C [80]. The nanotubes were crystalline, with one of their tips closed and much of their hollow core filled with indium. $Mg_3B_2O_6$ nanotubes with polyhedral cross sections were synthesized by heating Si wafer coated with thin film of boron in the presence of magnesium vapor under mixed oxygen–argon atmosphere [81]. Magnesium borate is a thermoluminescing material, which also possesses excellent tribological properties as an additive to lubricants.

Microtubes and nanotubes of pure elements like Te [82,83] and Se [84], have been synthesized as well. A variety of techniques, including wet [82,84] and simple thermal evaporation [83], were successfully used for this purpose. Bi nanotubes were synthesized by solvothermal reduction of Bi_2O_3 with ethylene glycol [85], or alternatively by reduction of $BiCl_3$ with Zn at room temperature [86]. Similar processes have been used by the same group for the synthesis of Sb nanotubes [87].

An interesting case is that of InS nanotubes [88], which have been synthesized by a "chemie douce" process. While the stable crystalline structure of InSe and GaS is lamellar, the more ionic InS compound crystallizes in an orthorhombic lattice structure. Using a low-temperature solution-based synthesis with benzenethiol (C_6H_5SH) as a catalyst, InS nanotubes were obtained. This work shows that kinetic control could lead to the preferential growth of an otherwise metastable layered structure, which does not exist in the bulk InS material.

Nanotubes of various III–V compounds with zincblende structure, like GaN and AlN nanotubes, were recently reported [20,89]. Clearly, simple folding of a molecular sheet is not possible in this case. Here uniaxial growth along an easy axis is promoted resulting in a highly faceted nanotube morphology. The faceted faces of the nanotube are chemically very reactive. Therefore, protection of the facets' surfaces with respect to the ambient atmosphere remains a great challenge for various applications like field emission, etc.

In summary, over the past few years, a large chemical apparatus has been conceived in order to synthesize new nanotubular and fullerene-like structures; part of this effort has been summarized in this rather noncomprehensive listing.

4.3   INORGANIC NANOTUBES AND FULLERENE-LIKE STRUCTURES STUDIED BY COMPUTATIONAL METHODS

Computational methods play a major role in the progress of materials science in general, and nanomaterials in particular. Pioneering efforts in this direction were undertaken by the Cohen–Louie group, who studied various $B_xC_yN_z$ nanotubes [90–92] (see also Table 4.2). Following this preliminary work, Seifert and coworkers [94–96,101] undertook a systematic approach to elucidate the structural aspects and the physical behavior of metal-dichalcogenide nanotubes [102]. These studies indicated that the strain in the nanotube increases like $1/R^2$, where R is the radius of the nanoubes. Furthermore, due to the bulky structure of the multiple atom S–M–S layer, the strain energy of an MS_2 nanotube was found to be about an order of magnitude larger than that of carbon nanotubes with the same diameter. These calculations clearly indicated that in agreement with the experiment, multiwall WS_2 (MoS_2) nanotubes consisting of 4 to 7 layers, with an inner radius of 6 to 12 nm, are the most stable moities [101].

Generally, the formation of a polyhedral nanoparticle by folding the molecular sheet in two directions is more demanding in terms of elastic energy than folding the lamella in one direction to obtain a nanotube. One manifestation of this conjecture is the fact that the number of defects and irregularities in nested fullerene-like nanoparticles of WS_2 is appreciably higher than those observed in WS_2 nanotubes. Furthermore, while the cross section of WS_2 nanotubes is in general circular, fullerene-like nanoparticles of WS_2 have been shown to exhibit a phase transformation from an evenly curved surface into polyhedral (faceted) configuration, when the number of molecular layers (i.e., the thickness) exceeds a critical number (3 to 4 layers) [125]. This principle has been confirmed also in the case of black phosphorous, where a stable configuration could be calculated for the nanotubes, while fullerene-like nanoparticles of similar dimensions have been found to be unstable [117–119].

The electronic structure of MS_2 nanotubes have been studied in considerable detail. It was found that semiconducting materials, like MoS_2 and WS_2 [94,95], form nanotubes with forbidden gaps, which shrink with decreasing diameter of the nanotubes. It is clear then that in contrast to the ubiquitous semiconducting quantum dots and nanowires, the quantum size effect in MS_2 nanotubes is not particularly large. The diminutive nature of the quantum size effect in IF nanoparticles can be attributed to the confinement of their electronic states within the molecular sheet. On the other hand, the lattice distortion induced by the bending of the molecular layer, which increases with the shrinkage of the nanoparticle radius, leads to an opposite effect, i.e., reduction of the bandgap. Many of the semiconducting layered compounds possess an indirect transition and therefore exhibit a weak luminescence, if any. While (n,n) armchair nanotubes also possess an indirect gap, ($n,0$) zigzag nanotubes were shown to be direct bandgap semiconductors, suggesting a variety of optical effects in such nanotubes. Nanotubes of materials with metallic character, like NbS_2, remain metallic in nature, irrespective of their diameter and chirality [96,98]. The high density of states near the Fermi level suggests that such nanotubes may become superconductors at low temperatures.

TABLE 4.2
Summary of First-Principle Calculations for Inorganic Nanotubes

Compound	Method Used	References	Comment
BN	DFT–LDA	[90]	Insulator
BC_2N	DFT–LDA	[91]	Chiral; metals or semiconductors
BC_3	DFT–LDA	[92]	
GaSe	DFT–LDA	[93]	Semiconductor
WS_2	DFT–TB	[94]	Semiconductor — Z Z, direct; AC indirect; bandgap shrinks with decreasing diameter
MoS_2	DFT–TB	[95]	Like WS_2
NbS_2	DFT–TB	[96]	Metal, possibly superconductor
TiS_2	DFT–TB	[97]	
$NbSe_2$	TB	[98]	Metal, possibly superconductor
ZrS_2	TB	[99]	Semiconductor—Z Z, direct; AC, indirect; bandgap shrinks with decreasing diameter stability: ZZ>AC
$MoS_2I_{1/3}$ bundles	DFT	[100]	Stable mostly as bundles; metallic
WS_2, MoS_2	DFT–TB	[101]	Stability of the nanotubes was confirmed
WS_2, MoS_2	DFT–TB	[102]	Bandgap of the nanotubes as a function of diameter was determined
$(MS_2)_{64}$ (M=Mo, Nb, Zr, S) nanooctahedra	TB	[103]	Stable nanooctahedra; metallic
$Ca(AlSi)_2$ and $Sr(GaSi)_2$	TB	[104]	Metallic, possibly superconducting;
$CaSi_2$	LDA–DFT	[105]	metallic
V_2O_5	TB	[106]	Semiconductors; bandgap shrinks with decreasing diameter
ZrNCl	TB	[107]	ZZ — metal; AC — semiconductor (becomes metal upon removal of 30% chlorine)
B	HF–SCF	[108–110]	Nanotubes are stable nanostructures, metallic character
AlB_2	HF–SCF	[111]	Metallic
Bi	DFT	[112]	Semiconductors
MB_2 (M=Mg,Be,Sc,Ca, Al,Ti,,Cr,Mo,Ta,Zr)	TB	[113–115]	Metallic, possibly superconducting
V_2O_5	TB	[116]	Semiconductors; bandgap shrinks with decreasing diameter — effect is stronger with ZZ as compared with AC nanotubes
Black-P, As	DFT–TB	[117–119]	Nanotubes — stable; fullerenes — unstable
GaN	DFT	[120]	Folded graphite-like GaN; bandgap decreases with shrinking diameter; ZZ — direct gap; AC — indirect gap
AlN	DFT	[121–123]	Folded graphite-like AlN; bandgap decreases with shrinking diameter; ZZ — direct gap; AC — indirect gap. Fullerene-like structures — stable
InP	SCF–MO	[124]	

TB, tight binding; DFT, density functional theory; HF-SCF, Hartree-Fock self-consistent field; LDA, local density approximation; SCF-MO, Self-consistent–field-molecular orbital; ZZ, zigzag; AC, armchair.

The structure of the stable polyhedral clusters of metal dichalcogenide and other layered compounds was first discussed in [6]. It was hypothesized that instead of pentagons, which occur in carbon fullerenes and carbon nanotubes, rhombi and triangles would be the stable elements among IF nanoparticles. Later work provided the first evidence for the existence of nanooctahedra (six rhombi) and nanotetrahedra (four triangles) in MoS_2 [126]. More decisive evidence in support of the existence and stability of MoS_2 nanooctahedra was obtained by laser ablation of an MoS_2 target [55]. Figure 4.8 shows an HRTEM image of one such MoS_2 nanooctahedron (left) and its Fourier transform filtered image (right). These images clearly reveal the precise arrangement of the Mo atoms in the cluster. Recently, the IF nanooctahedra of metal dichalcogenide compounds have been theoretically analyzed and their atomic and electronic structure evaluated [103,127]. Figure 4.9 shows the structure of an $(MoS_2)_{64}$ nanooctahedron calculated by tight binding (TB) theory. The structures of similar nanooctahedra of Nb, Zr, and Sn disulfides were also investigated. Most interestingly, all the nanooctahedra studied were found to be metallic, irrespective of the electronic structure of the parent compound. This conclusion, which remains to be verified experimentally, shows the importance of elucidating the structure of IF nanoparticles and their properties. Also, more accurate first-principle calculations have to be carried out in order to verify the proposed stable structure of such nanooctahedra.

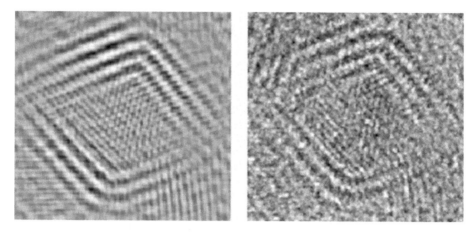

FIGURE 4.8 HRTEM image (right) and its Fourier filtered image (left) of an MoS_2 nanooctahedron prepared by laser ablation of an MoS_2 target. (Courtesy of M. Bar-Sadan and S.Y. Hong.)

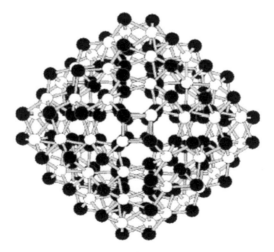

FIGURE 4.9 Schematic representation of the structure of $(MoS_2)_{64}$. (Adapted from Enyashin, A.N. et al., *Inorg. Mater.*, 40, 395, 2004.)

The structure, stability, and properties of GaN and AlN nanotubes were studied as well [120–122]. In analogy to the semiconducting MS_2 nanotubes, and in sharp contrast to carbon nanotubes, the bandgap shrinks with a decreasing diameter of the nanotubes. Furthermore, while (n,n) armchair nanotubes were found to exhibit an indirect transition, direct transition was seen for $(n,0)$ zigzag nanotubes.

An interesting case is presented by nanotubes of the layered compound β-ZrNCl (see Figure 4.10a). This compound is a semiconductor with a bandgap of 2 eV in the bulk. It was converted into a superconductor (T_c = 13 to 15 K) by intercalating such electron donors as alkali atoms, or by deintercalation of some of the lattice chlorine atoms. The structure and electronic properties of β-ZrNCl nanotubes were investigated in detail in [107] (Figure 4.10b). The (n,n) armchair nanotubes were shown to be semiconductors, but $(n,0)$ zigzag nanotubes with diameter larger than 2.8 nm are found to be all metallic. Furthermore, while the (29,0) nanotube is metallic, its unfolded stripe is a semiconductor with a bandgap of 1.2 eV. As in semiconducting MS_2 nanotubes [93,94], the bandgap

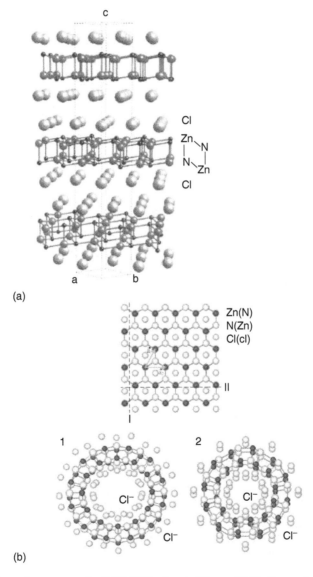

(a)

(b)

FIGURE 4.10 (a) Crystal structure of a β-ZrNCl lattice; 1, Zigzag (12,0); 2, armchair (12,12) ZrNCl nanotube. (Adapted from Enyashin, A.N. et al., *Chem. Phys. Lett.*, 387, 85, 2004.)

of the armchair nanotubes shrinks with decreasing nanotube diameter. Moreover, intrinsic or extrinsic doping may lead to significant variation of the electronic properties of nanotubes. Thus, deintercalation of 30% of chlorine atoms endows metallic character to the ZrNCl nanotubes [107], suggesting that they may become superconductors at sufficiently low temperatures.

In conclusion, the structure and electronic properties of various inorganic nanotubes have been investigated in some detail. While some of the properties could be confirmed by experiment, much more work is needed to elucidate the electronic and optical properties of these novel nanomaterials.

4.4   STUDY OF THE PROPERTIES OF INORGANIC NANOTUBES IN RELATION TO THEIR APPLICATIONS

The properties of inorganic nanotubes and IF nanoparticles have been studied only recently (see Table 4.3). In spite of that, a major application area for the IF nanoparticles in solid lubrication has emerged from these studies. It was initially hypothesized that the quasi-spherical nanoparticles

TABLE 4.3
Summary of Potential Applications of Inorganic Nanotubes and Fullerene-Like Nanoparticles

Compound	Properties and Proposed Applications	References	Comment
MoS_2, WS_2	Tribology: additives to lubricating fluids	[51,128,129]	Oils, greases, machining fluids, etc.
MoS_2, WS_2	Tribological coatings (metallic, polymers)	[129–131]	Medical applications
MoS_2, WS_2	Tribology: self-lubricating porous structures	[132]	Sliding bearings
MoS_2, WS_2	Impact resistance	[133]	
MS_2 (M=W, Mo)	Mechanical properties (nanocomposites, etc.)	[134–136]	
MS_2 (M=Mo,Ti)	Rechargeable batteries	[40,137–139]	
MS_2 (M=Mo, Ti)	Hydrogen storage	[40,41]	
MoS_2	Catalysis	[140]	$CO+3H_2 \rightarrow CH_4+H_2O$
MoS_2I_y	Field emission	[141]	
BN	Mechanical properties	[142,143]	Young's modulus of 700–1200 GPa
BN	H_2 storage	[144]	Up to 4.2 wt%
BCN	Field emission	[145]	Comparable to multiwall carbon nanotubes
VO_x	Cathodes for rechargeable Li batteries	[146–148]	100 cycles; >200 mAh/g
VO_x	Optical limiting	[149]	
$Mg_3Si_2O_5(OH)_4$ $CuSiO_3 \cdot 2H_2O$	H_2 storage; catalyst support for CO oxidation	[150]	
Rare earth oxisulfides	Non-linear optics	[76]	
Titanates (e.g., $H_2Ti_3O_7$)	Dye sensitized solar cells	[151]	Solar to electrical efficiency of 5%
GaN	Optical nanodevices	[89]	
Y_2O_3:Eu	Optical nanodevices	[152]	
Silicate	Catalysis, H_2 stoarge	[74]	Catalytic CO and C_2H_6 oxidation

would exhibit facile rolling and sliding. Coupled with its mechanical stability and chemical inertness it was thought that IF-WS_2 (MoS_2) nanoparticles could become very efficient nanoball bearings, especially under high loads, where lubricating fluids tend to be squeezed out of the contact region between the two moving pieces. This hypothesis was invariably confirmed [51,128]. Apparently, the favorable role played by IF nanoparticles in alleviating friction and wear is a much more complex phenomenon than initially believed. A number of studies have indicated that during friction tests under load, the IF nanoparticles gradually deform and exfoliate, depositing molecular WS_2 (MoS_2) nanosheets on the underlying surfaces. These transferred molecular sheets (third body) provide easy shear for the mating metal pieces, further alleviating friction [129]. Recent technical tests of IF-WS_2 in various industrial environments demonstrated the large potential of such nanoparticles for numerous tribological applications in the automotive, aerospace, electronics, and many other industries. Figure 4.11 shows the detrimental effect of addition of a minute amount of IF-WS_2 nanoparticles on the friction coefficient between a ceramic pair, i.e., a Si_3N_4 ball and alumina flat. The effect of the IF-WS_2 nanoparticles on the wear rate of the ball was similar. The large-scale application of IF-WS_2 (MoS_2) would not be possible, should scaling of the production and its cost effectiveness would not be proved. Indeed, a fluidized-bed reactor with a capacity of more than 100 g/day of a pure IF-WS_2 phase was realized sometime ago [153], and the construction of a mock-up system for full-scale production is in progress. The feedstocks for the present process are quite cheap and notwithstanding the small amount of the nanomaterials needed to alleviate both friction and wear, the cost effectiveness of the IF-WS_2 nanolubricant seems to have been realized.

Many of the proposed applications of inorganic nanotubes are in their infancy stage and thus they are not expected to reach the marketplace before a decade or so. Recent work has indicated the large impact resistance of WS_2 nanotubes to shockwaves, suggesting numerous applications for nanocomposites formulated with such nanoparticles [133].

One of the most appealing applications of inorganic nanotubes is in the field of cathode material for rechargeable lithium batteries. Two of the most demanding requirements for host material are a large capacity to accommodate significant amounts of the foreign lithium atoms and the reversibility with respect to charge/discharge cycles. Intercalation of the lithium and hydrogen atoms have been accomplished within the van der Waals gap between two layers of the nanotube [138,139]. An extra capacity of the alkali atoms may exist also in the cylindrical hollow core of the nanotube, but the mechanism of charge transfer in this case may be quite different from that of the alkali atoms intercalated between each two layers. The metal atoms in the core may not be in direct contact with the wall

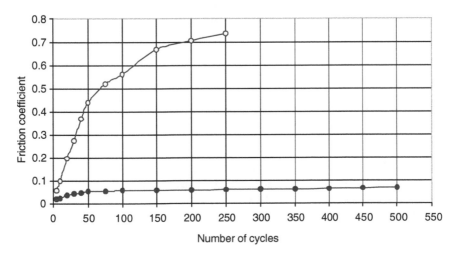

FIGURE 4.11 Friction coefficient vs. number of cycles for alumina Al_2O_3 plat — Si_3N_4 ball pair. Load $P = 0.75$ N. Open circles — non-lubricated part; full circles — IF lubricated pair.

of the nanotube to permit a fast charge transfer from the alkali atoms to the host, and take part in the electrochemical reaction effectively. The other property that makes inorganic nanotubes an intriguing cathode material for rechargeable batteries is their inherent stability, which in general is not common to nanostructured materials. The inherent stability of inorganic nanotubes coupled with their large internal surfaces accessible to small molecules, also suggests that inorganic nanotubes can serve as efficient and selective catalysts, which so far was realized in a few cases only [74,140].

4.5 CONCLUSIONS

Building upon the analogy between the structure of graphite and inorganic layered compounds, nanotubes and fullerene-like nanoparticles of various compounds have been synthesized and investigated in some detail. The versatility in choosing the starting material coupled with the structural complexity of inorganic layered compounds led to numerous intriguing observations, and some important potential applications. Many fundamental issues remain to be addressed; most fundamental among them is the elucidation of the detailed structure of inorganic nanotubes and fullerene-like particles. In order to study the physical properties and chemically modify these nanotubes, macroscopic amounts of the nanotubes must first be synthesized, which in some cases has proved to be an illusive goal. Finally, by careful characterization of these nanophases, new applications may emerge with time.

ACKNOWLEDGMENTS

R.T. is supported by the Helen and Martin Kimmel Center for Nanoscale Science and holds the Drake Family Chair in Nanotechnology. The support of the Israel Science Foundation and "Nanomaterials Ltd." is acknowledged.

REFERENCES

1. Kroto, H.W., Heath, J.R., O'Brien, S.C., Curl, R.F., and Smalley, R.E., C_{60}: Buckminsterfullerene, *Nature*, 318, 162, 1985.
2. Iijima, S., Helical microtubules of graphitic carbon, *Nature*, 354, 56, 1991.
3. Wilson, J.A. and Yoffe A.D., Transition metal dichalcogenides discussion and interpretation of observed optical, electrical and structural properties, *Adv. Phys.* 269, 193–335, 1969.
4. Lévy, F., *Structural Chemistry of Layer-Type Phases,* Vol. 5, D. Reidel, Dordrecht, 1976.
5. Tenne, R., Margulis, L., Genut, M., and Hodes, G., Polyhedral and cylindrical structures of tungsten disulphide, *Nature*, 360, 444, 1992.
6. Margulis, L., Salitra, G., Tenne, R., and Talianker, M., Nested fullerene-like structures, *Nature*, 365, 113, 1993.
7. Rosenfeld, H.Y. et al., Cage structures and nanotubes of $NiCl_2$, *Nature*, 395, 336, 1998.
8. Ajayan, P.M. et al., Carbon nanotubes as removable templates for metal oxide nanocomposites and nanostructures, *Nature*, 375, 564, 1995.
9. Patzke, G.R., Krumeich, F., and Nesper, R., Oxidic nanotubes and nanorods — anisotropic modules for a future nanotechnology, *Angew. Chem. Int. Ed.* 41, 2446, 2002.
10. Chen, Q. et al., Tritanate nanotubes made via a single alkali treatment, *Adv. Mater.*, 14, 1208, 2002.
11. Tenne, R., Advances in the synthesis of inorganic nanotubes and fullerene-like nanoparticles, *Angew. Chem. Int. Ed.*, 42, 5124, 2003.
12. Nath, M. and Rao, C.N.R., Inorganic nanotubes, *Dalton Trans.*, 1, 1, 2003.
13. Pauling, L., The structure of the chlorites, *Proc. Natl. Acad. Sci.*, 16, 578, 1930.
14. Wiegers, G.A. and Meerschut, A., Structures of misfit layer compounds $(MS)_nTS_2$ (MSn, Pb, Bi, rare earth metals; TNb, Ta, Ti, V, Cr; $1.08 < n < 1.23$), *J. Alloys Comp.*, 178, 351, 1992.

15. Suzuki, K., Enoki,T., and Imaeda, K., Synthesis, characterization and physical properties of incommensurate layered compounds/(RES)$_x$TaS$_2$ (RE=Rare earth metal), *Solid State Commun.* 78, 73, 1991.

16. Kresge, C.T. et al., Control of metal radial profiles in alumina supports by carbon-dioxide, *Nature*, 259, 710, 1992.

17. Stöber, W., Fink, A., and Bohn, E., Controlled growth of monodisperse silica spheres in micron size range, *J. Colloid Interf. Sci.*, 26, 62, 1968.

18. Nakamura, M. and Matsui, Y., Silica gel nanotubes obtained by the sol–gel method, *J. Am. Chem. Soc.*, 117, 2651, 1995.

19. Satishkumar, B.C. et al., Oxide nanotubes prepared using carbon nanotubes as templates, *J. Mater. Res.*, 12, 604, 1997.

20. Wu, Q. et al., Synthesis and characterization of faceted hexagonal aluminum nitride nanotubes, *J. Am. Chem. Soc.*, 125, 10176, 2003.

21. Sharma, S. and Sunkara, M.K., Direct synthesis of gallium oxide tubes, nanowires, and nanopaint-brushes, *J. Am. Chem. Soc.*, 124, 12288, 2002.

22. Chianelli, R.R. et al., Molybdenum disuflide in the poorly crystalline "Rag" structure, *Science*, 203, 1105, 1979.

23. Sanders, J.V., Structure of catalytic particles, *Ultramicroscopy*, 20, 33, 1986; Sanders, J.V., Transmission electron-microscopy of catalysts, *J. Electron Microsc. Tech.*, 3, 67, 1986.

24. Feldman, Y. et al., High rate gas phase growth of MoS$_2$ nested inorganic fullerene-like and nanotubes, *Science*, 267, 222, 1995.

25. Rosentsveig, R., et al., WS$_2$ nanotube bundles and foils, *Chem. Mater.*, 14, 471, 2002.

26. Coleman, K.S., The formation of ReS$_2$ inorganic fullerene-like structures containing Re$_4$ parallelogram units and metal–metal bonds, *J. Am. Chem. Soc.*, 124, 11580, 2002.

27. Remskar, M. et al., MoS$_2$ as micro tubes, *Appl. Phys. Lett.*, 69, 351, 1996.

28. Remskar, M. et al., New crystal structures of WS$_2$: microtubes, ribbons, and ropes, *Adv. Mater.*, 10, 246, 1998.

29. Remskar, M. et al., Self-assembly of subnanometer-diameter single-wall MoS$_2$ nanotubes, *Science*, 292, 479, 2001.

30. Zelenski, C.M. and Dorhout, P.K., Template synthesis of near-monodisperse microscale nanofibers and nanotubules of MoS$_2$, *J. Am. Chem. Soc.*, 120, 734, 1998.

31. Santiago, P., et al., Synthesis and structural determination of twisted MoS$_2$ nanotubes, *Appl. Phys. A.*, 78, 513, 2004.

32. Nath, M. and Rao, C.N.R., New metal disulfide naotubes, *J. Am. Chem. Soc.*, 123, 4841, 2001.

33. Nath, M. and Rao, C.N.R., Nanotubes of group 4 metal disulfides, *Angew. Chem. Int. Ed.*, 41, 3451, 2002.

34. Hsu, W.K. et al., An alternative route to molybdenum disulfide nanotubes, *J. Am. Chem. Soc.*, 122, 10155, 2000.

35. Zhu, Y.Q. et al., Carbon nanotube template promoted growth of NbS$_2$ nanotubes/nanorods, *Chem. Commun.*, 2184, 2001.

36. Brorson, M., Hansen, T.W., Jacobsen, C.J.H., Rhenium(IV) sulfide nanotubes, *J. Am. Chem. Soc.*, 124, 11582, 2002.

37. Tsuneta, T. et al., Formation of metallic NbSe$_2$ nanotubes and nanofibers, *Curr. Appl. Phys.*, 3, 473, 2003.

38. Afansiev, P. et al., Molybdenum polysulfide hollow microtubules grown at room temperature from solution, *Chem. Comm.*, 1001, 2000.

39. Chen, J. et al., Synthesis and characterization of WS$_2$ nanotubes, *Chem. Mater.*, 15, 1012, 2003.

40. Chen, J. et al., Titanium disulfide nanotubes as hydrogen-storage materials, *J. Am. Chem. Soc.*, 125, 5284, 2003.

41. Chen J. et al., Electrochemical hydrogen storage in MoS$_2$ nanotubes, *J. Am. Chem. Soc.*, 123, 11813, 2001.

42. Li, X.L. and Li, Y.D., Formation of MoS$_2$ inorganic fullerenes (IFs) by the reaction of MoO$_3$ nanobelts and S, *Chem. Eur. J.*, 9, 2726, 2003.

43. Li., Y.D. et al., Artificial lamellar mesostructures to WS$_2$ nanotubes, *J. Am. Chem. Soc.*, 124, 1411, 2002.

44. Zhu, Y.Q. et. al., Production of WS$_2$ nanotubes, *Chem. Mater.*, 12, 1190, 2000.
45. Rothschild, A., Sloan, J., and Tenne, R., The growth of WS$_2$ nanotubes phases, *J. Am. Chem. Soc.*, 122, 5169, 2000.
46. Zhu, Y.Q. et al., An alternative route to NbS$_2$ nanotubes, *J. Phys. Chem. B*, 106, 7623, 2002.
47. Galvan D.H., Rangel, R. and Adem, E., *Fullerenes Nanotubes Carbon Nanostruct*, 10, 127, 2002.
48. Flores, E. et al., Optimization of the electron irradiation in the production of MoS$_2$ nanotubes, *Fullerene Sci. Tech.*, 9, 9, 2001.
49. Galvan, D.H. et al., Effect of electronic irradiation in the production of NbSe$_2$ nanotubes, *Fullerene Sci. Tech.*, 9, 225, 2001.
50. Yacaman, M.J. et al., Studies of MoS$_2$ structures produced by electron irradiation, *Appl. Phys. Lett.*, 69, 1065, 1996.
51. Chhowalla, M. and Amaratunga, G.A.J., Thin films of fullerene-like MoS$_2$ nanoparticles with ultra-low friction and wear, *Nature*, 407, 164, 2000.
52. Vollath, D. and Szabo, D.V., Nanoparticles from compounds with layered structures, *Acta Mater.*, 48, 953, 2000.
53. Avivi, S. et al., Sonochemical hydrolysis of Ga^{3+} ions: synthesis of scroll-like cylindrical nanoparticles of gallium oxide hydroxide, *J. Am. Chem. Soc.*, 121, 4196, 1999.
54. Hong, S.Y. et al., Synthesis of SnS$_2$/SnS fullerene-like nanoparticles: a superlattice with polyhedral shape, *J. Am. Chem. Soc.*, 125, 10470, 2003.
55. Parilla, P.A. et al., The first inorganic fullerenes, *Nature*, 397, 114, 1999.
56. Sen, R. et al., Encapsulated and hollow closed-cage structures of WS$_2$ and MoS$_2$ prepared by laser ablation at 450–1050°C, *Chem. Phys. Lett.*, 340, 242, 2001.
57. Rosenfeld, H.Y. et al., Synthesis of NiCl$_2$ nanotubes and fullerene-like structures by laser ablation *Phys. Chem. Phys.*, 5, 1644, 2003.
58. Hamilton, E.J.M. et al., Preparation of amorphous boron-nitride and its conversion to a turbostratic, tubular form, *Science*, 260, 659, 1993.
59. Chopra, N.G. et al., Boron-nitride nanotubes, *Science*, 269, 966, 1995.
60. Gleize, P. et al., Growth of tubular boron-nitride filaments, *J. Mater. Sci.*, 29, 1575, 1994.
61. Lourie, O.R. et al., CVD growth of boron nitride nanotubes, *Chem. Mater.*, 12, 1808, 2000.
62. Ma, R., Bando, Y., and Sato, T., CVD synthesis of boron nitride nanotubes without metal catalysts, *Chem. Phys. Lett.*, 337, 61, 2001.
63. Cumings, J. and Zettl, A., Mass-production of boron nitride double-wall nanotubes and nanococoons, *Chem. Phys. Lett.*, 316, 211, 2000.
64. Terrones, M. et al., Production and state-of-the-art characterization of aligned nanotubes with homogeneous BC$_x$N ($1 <= x <= 5$) compositions, *Adv. Mater.*, 15, 1899, 2003.
65. Suenaga, K. et al., Synthesis of nanoparticles and nanotubes with well-separated layers of boron nitride and carbon, *Science*, 278, 653, 1997.
66. Hultman, L. et al., Cross-linked nano-onions of carbon nitride in the solid phase: existence of a novel C$_{48}$N$_{12}$ aza-fullerene, *Phys. Rev. Lett.*, 87, 225503, 2001.
67. Rao, C.N.R. et al., Surfactant-assisted synthesis of semiconductor nanotubes and nanowires, *Appl. Phys Lett.*, 78, 1853, 2001.
68. Imai, H. et al., Direct preparation of anatase TiO$_2$ nanotubes in porous alumina membranes, *J. Mater. Chem.*, 9, 2971, 1999.
69. Kasuga, T. et al., Formation of titanium oxide nanotube, *Langmuir*, 14, 3160, 1998.
70. Kasuga, T. et al., Titania nanotubes prepared by chemical processing, *Adv. Mater.*, 11, 1307, 1999.
71. Sun, X. and Li, Y.D., Synthesis and characterization of ion-exchangeable titanate nanotubes, *Chem. Eur. J.*, 9, 2229, 2003.
72. Zhang, S. et al., Formation mechanism of H$_2$Ti$_3$O$_7$ nanotubes, *Phys. Rev. Lett.*, 91, 256103, 2003.
73. Saupe, G.B. et al., Nanoscale tubules formed by exfoliation of potassium hexaniobate, *Chem. Mater.*, 12, 1556, 2000.
74. Wang, X. et al., Thermally stable silicate nanotubes, *Angew. Chem. Int. Ed.*, 43, 2017, 2004.
75. Wang, X., and Li, Y.D., Rational synthetic strategy. From layered structure to MnO$_2$ nanotubes, *Chem. Lett.*, 33, 48, 2004.

76. Wang, X. and Li, Y.D., Rare-earth-compound nanowires, nanotubes, and fullerene-like nanoparticles: synthesis, characterization, and properties, *Chem. Eur. J.*, 9, 5627, 2003.

77. Mao, Y., Banerjee, S. and Wong, S.S., Hydrothermal synthesis of perovskite nanotubes, *Chem. Commun.*, 408, 2003.

78. Avivi, S., Mastai, Y. and Gednaken, A., A new fullerene-like inorganic compound fabricated by the sonolysis of an aqueous solution of $TlCl_3$, *J. Am. Chem. Soc.*, 2000, 122, 4331.

79. Zhang, W., Single-crystalline scroll-type nanotube arrays of copper hydroxide synthesized at room temperature, *Adv. Mater.*, 15, 822, 2003.

80. Li, Y., Bando, Y. and Golberg, D., Single-crystalline In_2O_3 nanotubes filled with In, *Adv. Mater.*, 15, 581, 2003.

81. Ma, R. et al., Magnesium borate nanotubes, *Angew. Chem. Int. Ed.*, 42, 1836, 2003.

82. Mayers, B. and Xia, Y., Formation of tellurium nanotubes through concentration depletion at the surfaces of seeds, *Adv. Mater.*, 14, 279, 2002.

83. Li, X.-Y. et al., Synthesis and magnetoresistance measurement of tellurium microtubes, *J. Mater. Chem.*, 14, 244, 2004.

84. Zhang, H. et al., Selenium nanotubes synthesized by a novel solution phase approach, *J. Phys. Chem. B*, 108, 1179, 2004.

85. Liu, X. et al., Novel bismuth nanotube arrays synthesized by solvothermal method, *Chem. Phys. Lett.*, 374, 348, 2003; Li, Y.D. et al., Bismuth nanotubes: a rational low-temperature synthetic route, *J. Am. Chem. Soc.*, 123, 9904, 2001.

86. Yang, B. et al., A room-temperature route to bismuth nanotube arrays, *Eur. J. Inorg. Chem.*, 3699, 2003.

87. Hu, H. et al., A rational complexing-reduction route to antimony nanotubes, *New J. Chem.*, 27, 1161, 2003.

88. Hollingsworth, J.A. et al., Catalyzed growth of a metastable InS crystal structure as colloidal crystals, *J. Am. Chem. Soc.*, 122, 3562, 2000.

89. Goldberger, J. et al., Single-crystal gallium nitride nanotubes, *Nature*, 422, 529, 2003.

90. Blasé, X. et al., Stability and band-gap constancy of boron-nitride nanotubes, *Europhys. Lett.*, 28, 335, 1994.

91. Miyamoto, Y. et al., Chiral tubules of hexagonal BC_2N, *Phys. Rev. B*, 50, 4976, 1994.

92. Miyamoto, Y. et al., Electronic-properties of tubule forms of hexagonal BC_3, *Phys. Rev. B*, 50, 18630, 1994.

93. Cote, M., Cohen., M.L., and Chadi, D.J., Theoretical study of the structural and electronic properties of GaSe nanotubes, *Phys. Rev. B*, 58, R4277, 1998.

94. Seifert, G. et al., On the electronic structure of WS_2 nanotubes, *Solid State Commun.*, 114, 245, 2000.

95. Seifert, G. et al., structure and electronic properties of MoS_2 nanotubes, *Phys. Rev. Lett.*, 85, 146, 2000.

96. Seifert, G. et al., Novel NbS_2 metallic nanotubes, *Solid State Commun.*, 115, 635, 2000.

97. Ivanovskaya, V.V. and Seifert, G., Tubular structures of titanium disulfide TiS_2, *Solid State Commun.*, 130, 175, 2004.

98. Ivanovskaya, V.V. et al., Electronic properties of superconducting $NbSe_2$ nanotubes, *Phys. Stat. Sol.*, 238, R1, 2003a.

99. Ivanovskya, V.V. et al., Computational studies of electronic properties of ZrS_2 nanotubes, *Internet Electron J. Mol. Des.*, 2, 499, 2003b.

100. Verstraete, M. and Charlier, J.-C., *Ab initio* study of MoS_2 nanotube bundles, *Phys. Rev. B*, 68, 045423, 2003.

101. Seifert, G., Köhller, T. and Tenne, R, Stability of metal chalcogenide nanotubes, *J. Phys. Chem. B*, 106, 2497, 2002.

102. Scheffer, L. et al., Scanning tunneling microscopy study of WS_2 nanotubes, *Phys. Chem. Chem. Phys.*, 4, 2095, 2002.

103. Enyashin, A.N. et al., Structure and electronic spectrum of fullerene-like nanoclusters based on Mo, Nb, Zr, and Sn disulfides, *Inorg. Mater.*, 40, 395, 2004.

104. Shein, I.R., et al., Electronic properties of new $Ca(Al_xSi_{1-x})_2$ and $Sr(Ga_xSi_{1-x})_2$ superconductors in crystalline and nanotubular states, *JETP Lett.*, 76, 189, 2002.

105. Gemming, S. and Seifert, G., Nanotube bundles from calcium disilicide: a density functional theory study, *Phys. Rev. B*, 68, 75416, 2003.

106. Ivanovskya, V.V. et al., Electronic properties of single-walled V_2O_5 nanotubes, *Solid State Commun.*, 126, 489, 2003.

107. Enyashin, A.N., Makurin, Yu.N. and Ivanovskii, A.L., Electronic band structure of β-ZrNCl-based nanotubes, *Chem. Phys. Lett.*, 387, 85, 2004.

108. Boustani, I. and Quandt, A., Nanotubules of bare boron clusters: *Ab initio* and density functional study *Europhys. Lett.*, 39, 527, 1997.

109. Boustani, I. et al., New boron based nanostructured materials, *J. Chem. Phys.*, 110, 3176, 1999.

110. Gindulyte, A., Lipscomb, W.N., and Massa, L., Proposed boron nanotubes, *Inorg. Chem.*, 37, 6544, 1998.

111. Quandt, A., Liu, A.Y., and Boustani, I., Density-functional calculations for prototype metal-boron nanotubes, *Phys. Rev. B*, 64, 125422, 2001.

112. Su, C., Liu, H.T. and Li, J.M., Bismuth nanotubes: potential semiconducting nanomaterials, *Nanotechnology*, 13, 746, 2002.

113. Ivanovskii, A.L., Band structure and properties of superconducting MgB_2 and related compounds (a Review), *Phys. Solid State*, 45, 1829, 2003.

114. Chernozatonskii, L.A., Diboridebifullerenes and binano tubes, *JETP Lett.*, 74, 335, 2001.

115. Guerini, S. and Piquini, P., Theoretical investigation of TiB_2 nanotubes, *Microelectron. J.*, 34, 495, 2003.

116. Ivanovskaya, V.V. et al., Electronic properties of single-walled V_2O_5 nanotubes, *Solid State Commun.*, 126, 489, 2003.

117. Seifert, G. and Hernandez, E., Theoretical prediction of phosphorus nanotubes, *Chem. Phys. Lett.*, 318, 355, 2000.

118. Seifert, G., Heine, T. and Fowler, P.W., Inorganic nanotubes and fullerenes. Structure and properties of hypothetical phosphorus fullerenes, *Eur. Phys. J. D*, 16, 341, 2001.

119. Cabria, I. and Mintmire, J.W., Stability and electronic structure of phosphorus nanotubes, *Euro Phys. Lett.*, 65, 82, 2004.

120. Lee, S.M. et al., Stability and electronic structure of GaN nanotubes from density-functional calculations, *Phys. Rev. B*, 60, 7788, 1999.

121. Zhao, M., et al., Stability and electronic structure of AlN nanotubes, *Phys. Rev., B*, 68, 235415, 2003.

122. Chang, Ch. et al., Computational evidence for stable fullerene-like structures of ceramic and semiconductor materials, *Chem. Phys. Lett.*, 350, 399, 2001.

123. Zhang, D. and Zhang, R.Q., Theoretical predictions on inorganic nanotubes, *Chem. Phys. Lett.*, 371, 426, 2003.

124. Erkoc, S., Semi-empirical SCF-MO calculations for the structural and electronic properties of single-wall InP nanotubes, *J. Mol. Struct. (Thermochem.)*, 676, 109, 2004.

125. Srolovitz, D.J. et al., Relaxed curvature elasticity and morphology of nested fullerenes, *Phys. Rev. Lett.*, 74, 1779, 1995.

126. Tenne, R., Doped and heteroatom fullerene-like structures and nanotubes, *Adv. Mater.*, 7, 965, 1995.

127. Parilla, P.A., et al., Formation of nanooctahedra in molybdenum disulfide and molybdenum diselenide using pulsed laser vaporization, *J. Phys. Chem., B*, 108, 6197, 2004.

128. Rapoport, L. et al., Hollow nanoparticles of WS_2 as potential solid-state lubricants, *Nature*, 387, 791, 1997.

129. Rapoport, L., Fleishcer, N., and Tenne, R., Fullerene-like WS_2 nanoparticles: superior lubricants for harsh conditions, *Adv. Mater.*, 15, 651, 2003.

130. Chen, W.X. et al., Wear and friction of NiP electroless composite coating including inorganic fullerene-like WS_2 nanoparticles, *Adv. Eng. Mater.*, 4, 686, 2002.

131. Rapoport, L. et al., Polymer nanocomposites with fullerene-like solid lubricant, *Adv. Eng. Mater.*, 6, 44, 2004.

132. Rapoport, L. et al., Slow release of fullerene-like WS_2 nanoparticles from Fe–Ni–graphite matrix: a self-lubricating nanocomposite, *Nanoletters*, 1, 137, 2001.

133. Zhu, Y.Q. et al., WS_2 nanotubes: shockwave resistance, *J. Am. Chem. Soc.*, 125, 1329, 2003.

134. Zhang, W. et al., Use of functionalized WS_2 nanotubes to produce new polystyrene/polymethylmethacrylate nanocomposites, *Polymer.*, 44, 2109, 2003.

135. Kaplan-Ashiri, I. et al., Mechanical behavior of WS_2 nanotubes, *J. Mater. Res.*, 19, 454, 2004.
136. Kis, A. et al., Shear and Young's moduli of MoS_2 nanotube ropes, *Adv. Mater.*, 15, 733, 2003.
137. Dominko, R. et al., Electrochemical preparation and characterisation of Li_2MoS_{2-x} nanotubes, *Electrochim. Acta*, 48, 3079, 2003.
138. Chen, J., Tao, Z.L., and Li, S.L., Lithium intercalation in open-ended TiS_2 nanotubes, *Angew. Chim. Int. Ed.*, 42, 2147, 2003.
139. Chen, J., Li, S.L., and Tao, Z.L., Novel hydrogen storage properties of MoS_2 nanotubes, *J. Alloys Comp.*, 356, 413, 2003.
140. Chen, J. et al., Synthesis of open-ended MoS_2 nanotubes and the application as the catalyst of methanation, *Chem. Commun.*, 1722, 2002.
141. Nemanic, V. et al., Field-emission properties of molybdenum disulfide nanotubes, *Appl. Phys. Lett.*, 82, 4573, 2003.
142. Chopra, N.G. and Zettl, A., Measurement of the elastic modulus of a multiwall boron nitride nanotube, *Solid State Commun.*, 105, 297, 1998.
143. Suryavanshi, A.P., et al., Elastic modulus and resonance behavior of boron nitride nanotubes, *Appl. Phys. Lett.*, 84, 2527, 2004.
144. Tang, C., et al., Catalyzed collapse and enhanced hydrogen storage of BN nanotubes, *J. Am. Chem. Soc.*, 124, 14550, 2002.
145. Dorozhkin, P., et al., Field emission from individual B–C–N nanotube rope, *Appl. Phys. Lett.*, 81, 1083, 2002.
146. Spahr, M.E. et al., Vanadium oxide nanotubes a new nanostructured redox-active material for the electrochemical insertion of lithium, *J. Electrochem. Soc.*, 146, 2780, 1999.
147. Nordlinder, S., Edström, K., and Gustafsson, T., The performance of vanadium oxide nanorolls as cathode material in a rechargeable lithium battery, *Electrochem. Solid-State Lett.*, 4, A129, 2001.
148. Manganese vanadium oxide nanotubes: synthesis, characterization, and electrochemistry, *Chem. Mater.*, 13, 4382, 2001.
149. Xu, J.F. et al., Nonlinear optical transmission in VO_x nanotubes and VO_x nanotube composites, *Appl. Phys. Lett.*, 81, 1711, 2002.
150. Wang, X. et al., Thermally stable silicate nanotubes, *Angew. Chem. Int. Ed.*, 43, 2017, 2004.
151. Adachi, M. et al., Formation of titania nanotubes and applications for dye-sensitized solar cells, *J. Electrochem. Soc.*, 150, G488, 2003.
152. Wu, C. et al., Photoluminescence from surfactant-assembled Y_2O_3: Eu nanotubes, *Appl. Phys. Lett.*, 82, 520, 2003.
153. Feldman, Y. et al., New reactor for production of tungsten disulfide onion-like (inorganic fullerene-like) nanoparticles, *Solid State Sci.*, 2, 663, 2000.

5 Boron Nitride Nanotubes: Synthesis and Structure

Hongzhou Zhang and Ying Chen
Department of Electronic Materials Engineering,
Research School of Physical Sciences and Engineering,
The Australian National University, Canberra, Australia

CONTENTS

Abstract 157
5.1 Introduction 157
5.2 Structures of Boron Nitride Nanotubes 158
    5.2.1 Hexagonal Boron Nitride 158
    5.2.2 Boron Nitride Nanotube Structure 159
    5.2.3 Transmission Electron Microscopy Studies of Boron Nitride Nanotube
        Chirality 161
5.3 Synthesis Methods of Boron Nitride Nanotubes 164
    5.3.1 Arc Discharge and Arc Melting 164
    5.3.2 Laser-Assisted Method 166
    5.3.3 Ball Milling and Annealing 168
    5.3.4 Carbon Nanotube Substitution 169
    5.3.5 Chemical Vapor Deposition and Other Thermal Methods 172
5.4 Summary 173
Acknowledgments 174
References 174

ABSTRACT

This chapter discusses the structure and major synthesis methods of boron nitride nanotubes (BNNTs). Structure and chiralities of BNNTs are introduced from the standpoint of curling up hexagonal boron nitride (h-BN) sheets into nanosized cylinders, and are compared with carbon nanotubes (CNTs). Major synthesis methods including arc-discharge, laser, ball milling-annealing, carbon substitutions and chemical vapor deposition (CVD) have been reviewed. The recent successful synthesis of high yield and large quantities of BNNTs will lead to significant application progresses.

5.1 INTRODUCTION

After the discovery of carbon nanotubes (CNTs) in 1991 [1], researchers immediately started to seek other materials that have the same tubular structure. Boron nitride (BN) is one of them, and NTs in BN were first theoretically predicted in 1994 [2], and subsequently synthesized successfully in 1995 by Zettl's group at University of California, Berkeley [3]. Like CNTs, BNNTs too have superstrong mechanical properties, because of their tubular structure and the strong sp^2 bonding within a tube layer. The upper range of theoretical estimates of the elastic modulus of BNNTs is ~850 GPa, about 80% that of CNTs [4,5]. Experimental results demonstrate a distinct mechanical property with the measured modulus in the range of ~722 GPa to 1.22 Tpa [6,7], which is comparable with that of CNTs. In addition to these useful properties, which are similar to those of CNTs, BNNTs have several other useful properties, and in some parameters, are even better than CNTs. For example, theoretical studies suggest that BNNTs are insulating materials with a uniform electronic band gap of about 5.5 eV [2,8], which is independent of either the diameter or chirality of an NT. Therefore, all BNNTs show insulating or semiconducting behavior in electrical conductance measurements [9,10]. On the other hand, most CNT samples synthesized currently consist of both conducting and semiconducting NTs, and it is quite difficult to separate them. Boron nitride nanotubes also exhibit more stable chemical properties — for instance, they have a much stronger resistance to oxidation at elevated temperatures [11,12]. Boron nitride nanotubes synthesized using a ball-milling annealing method can be oxidized only at over 800°C. Some BNNTs with perfect nanocrystalline structures can withstand temperatures of up to 900°C, while CNTs are readily oxidized in air at a temperature of 400°C, and burned completely at 700°C when sufficient oxygen is supplied [12]. Owing to the high resistance to oxidation, BNNTs are ideal for composite material applications.

Though BNNTs have such promising properties when compared with CNTs, several of their properties, such as energy storage [13–15], optical [16], piezoelectricity and electrical-polarization effects [5,16,17], field emission [10], and so on, have not yet been adequately studied. To explore BNNTs' properties and applications, a sufficient amount of high yield BNNT samples is essential. This was not possible until recently because the synthesis methods known till recently did not give the requisite yields. A large number of methods for BNNT synthesis have been documented hitherto, and in recent years, a significant progress has been achieved. This chapter first provides a brief introduction of BNNT structures, which will help in understanding the process of the formation of BNNTs. Following this, it presents a literature review on major synthesis methods reported from 1995 to 2004.

5.2 STRUCTURES OF BORON NITRIDE NANOTUBES

Boron nitride nanotubes can be regarded as wrapped hexagonal BN (h-BN) sheets from the standpoint of their structure — the simplest way to construct a NT. It is thus helpful to start from the discussion of the structures of bulk BN. Bulk BN exhibits polymorphism in its crystalline structures and has at least four different crystallographic forms: hexagonal (h-BN), rhombohedra (r-BN), cubic (c-BN), and wurtzite (w-BN) structures. Interestingly, bulk C also has four similar corresponding structures — this stems from the fact that BN and C are isoelectronic substances, with the same average number of valence electrons per atom. However, in this section, only h-BN will be discussed, because the layered structure of h-BN is the key to understanding BNNT structures. The similar and different aspects of h-BN and carbon graphite structures will also be presented. Some geometric quantities of BNNTs such as chirality will be defined in this section. Finally, typical transmission electron microscopy (TEM) experimental results of BNNT structures will be discussed briefly.

5.2.1 HEXAGONAL BORON NITRIDE

The ball-stick model of h-BN structure is illustrated in Figure 5.1. The model shows four coordinated layers (or sheets) of threefold hexagonal networks stacked together and forming the h-BN structure. The hexagonal layers are generally referred to as basal planes. The flat basal planes consist only of hexagon rings, built with strong sp^2 bonds between B and N atoms. Between basal

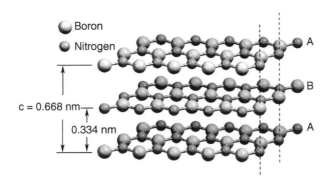

FIGURE 5.1 Crystal structure of h-BN. Large and small balls represent boron and nitrogen atoms, respectively. The lattice constant of c_0 is 0.668 nm. Along the c-direction, B and N atoms alternate in adjacent basal planes, thus forming a stacking sequence of ABAB.

planes, much weaker van der Waals interactions operate to keep them together in certain coordination to form a three-dimensional structure. As shown in Figure 5.1, a boron atom in one layer is directly over an nitrogen atom in the adjacent layers, i.e., along the c-axis, B and N atoms alternate. The adjacent layers are thus distinguished from each other and exhibit an ABAB... stacking sequence. h-BN hence has a P6$_3$mmc symmetry. The distance between adjacent basal planes of h-BN is $d =$ 3.34 Å, while the lattice constant along the c-direction should be $2d = 6.68$ Å.

5.2.2 BORON NITRIDE NANOTUBE STRUCTURE

Single-walled nanotubes (SWNTs) can be regarded as a seamlessly wrapped mono-atomic sheet of hexagonal networks. Multiwalled nanotubes (MWNTs) are several such tubulars concentrically organized and nested into each other. As with CNTs, there are many different ways to wrap BN basal planes and, consequently, the NTs exhibit different diameters and chiralities, which will be discussed in this section.

To construct an SWNT, a sheet of hexagonal network can be curled up so that a selected lattice point in the network is superposed on a predefined origin. For example, the predefined origin (0, 0) is labeled "O" and the other lattice point A (12,6) is selected as shown in Figure 5.2(a). As to the indices of the lattice points, i.e., (*n, m*), a crystallographic convention is adopted. The inter-angle between basis vectors is 120° as shown in Figure 5.2(a). The NT formed by bringing point O to A is shown in Figure 5.2(b). The NT can be named after the indices of lattice point A; thus a (12, 6) BNNT is shown in Figure 5.2(b). The following geometric relation of the tubular structure is obvious but it is still worth pointing out: the length of OA equals the perimeter of the NT, and the axis of the NT is along the direction of OA′.

We do not distinguish h-BN and carbon graphite in the above discussion. This is because the graphite and h-BN have the same symmetry, and the wrapping operation depends solely on the symmetry of the hexagonal network. In the case of graphite, two kinds of carbon atoms exist in a single graphite sheet, with their adjacent atoms in different geometric relationships; for example, the atoms at points O and O′ (see Figure 5.2[a], disregarding the difference of species between B and N atoms, the schematic picture of a BN sheet shown in Figure 5.2[a] is similar to a carbon graphitic sheet). Two adjacent carbon atoms (O and O′) form a pair, and these pairs are fixed on the hexagonal Bravais lattice, which exhibits the required symmetry for the curling-up operation, i.e., these two kinds of carbon atoms can never be superimposed to form a tube. In the case of h-BN, geometrically, two atoms (B and N) play the same roles as the two types of carbon atoms play in a graphite sheet. Therefore, we cannot superimpose a boron atom with a nitrogen atom, which also means that, theoretically, by curling up an h-BN sheet to construct a perfect tubular structure, a homo-atomic bonding (B–B or N–N) is impossible. There are infinite ways to roll up a sheet into

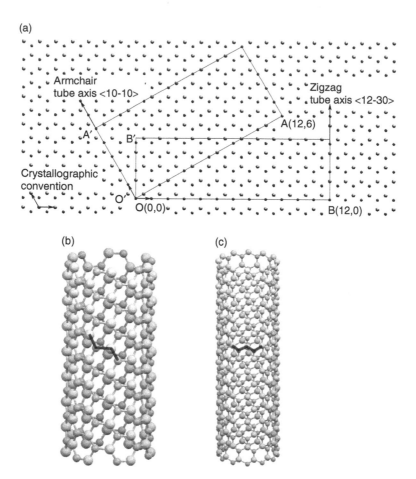

FIGURE 5.2 (a) Atom configuration of h-BN basal plane. The lattice constant of a_0 is 0.252 nm. For crystallographic convention, the inter-angle between the basis vectors is 120°. By wrapping the basal plane and superimposing lattice point A(12,6) with origin O, an armchair tube, as shown in (b), can be constructed. Point B(12,0) hence corresponds to a zigzag tube shown in (c). The black lines in (b) and (c) indicate the armchair and zigzag shapes, respectively.

an NT with different atomic configurations, though some duplicate cases and mirror-symmetric equivalents can be ruled out by symmetry considerations. By defining the chirality angle of an NT as the angle between the zigzag edge (shown in Figure 5.2[a]) and the tube perimeter (OA in Figure 5.2[a]), different ways of rolling result in NTs with different chirality angles. Two configurations of NTs are special: the NT shown in Figure 5.2(b) is the so-called armchair NT, since we can liken the configuration of the atoms (see bold lines) to an armchair. All $(2n, n)$ NTs, where n is a positive integer, are armchair-type NTs. Similarly, NTs with $(n, 0)$ indices are zigzag. A (12,0) zigzag NT is shown in Figure 5.2(c), which corresponds to the lattice point B in Figure 5.2(a) in the sense of curling up the basal plane.

There is no essential difference between BNNTs and CNTs as far as wall structures or configuration are concerned. However, there are differences in their tip morphologies. An SW CNT is generally considered to be capped by a half-fullerene molecule (C60) consisting of six-membered hexagons and five-membered pentagons, and thus CNTs have typical cone-shape tips [18]. Most BNNTs have flat tips as shown in Figure 5.3, which are formed with four-membered squares instead of five-membered pentagons [19]. Theoretical investigations reveal that five-membered pentagons in BN would require energetically unfavorable B–B or N–N bonds that destabilize the structure [20].

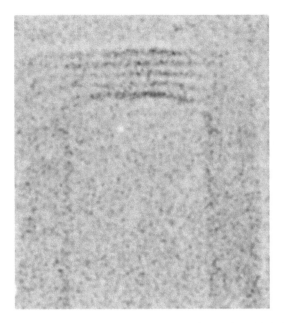

FIGURE 5.3 A typical flat tip of BNNTs. (From Loiseau, A. et al., *Carbon*, 36, 743–752, 1998. Reproduced with permission from *Elsevier Science*.)

In addition to typical straight parallel-walled cylindrical tubes, bamboo and cone-like BNNTs can also be produced [21]. Moreover, BNNTs can also be filled with metals (Fe, Ni, W, …) and buckyballs [22]; BN sheath can also be coated on the surface of CNTs [23].

In the following section, chirality of straight and hollow BNNT will be introduced.

5.2.3 TRANSMISSION ELECTRON MICROSCOPY STUDIES OF BORON NITRIDE NANOTUBE CHIRALITY

From the above discussion, we know that by determining the chirality and diameter of an SWNT, its atomic configuration is determined completely. Transmission electron microscopy is an indispensable tool for measuring the diameter and chirality of NTs. While diameter measurement is straightforward, the determination of BNNT chirality is not a trivial job. Electron diffraction (ED) has been used to elucidate the helical structures of both CNTs and BNNTs [19,24–28]. The other complementary method within TEM for measuring NT chirality is high-resolution TEM [29–31]. In addition to TEM, any measurement that can reach atomic resolution with the scales of NTs can discern the chirality of CNTs — for example, scanning tunneling microscopy or atomic force microscopy [32,33]. In this section, basic knowledge of ED of NTs is first illustrated briefly. Then we will introduce some of the experimental work on BNNT chirality.

The theory of ED, and thorough reviews can be found in some distinguished review articles [22,24,25]. Instead of trying to go through the diffraction theories strictly and systematically, here we have tried to give a simple picture of the ED of thin NTs. For an MWNT, with its tube axis perpendicular to the incidence beam as shown in Figure 5.4(a), the top and bottom walls of the NT are perpendicular to the incidence beam, i.e., the electron beam is along the 0002 direction of these parts of the NT (in the z-axis direction). Hence, the top and bottom basal planes of the NT will result in a diffraction pattern like the two-dimensional reciprocal lattice shown in Figure 5.4(b). If the walls of the NT are aligned to each other, a simple diffraction pattern is expected; otherwise the 10−10 diffraction spots of each wall (even the top and bottom layers of a single wall) may split, which is the case of helical NTs or the walls of MWNTs with different chiralities. On the other hand, the sidewalls of the NT shown in Figure 5.4(a) are parallel to the incidence beam, i.e., the

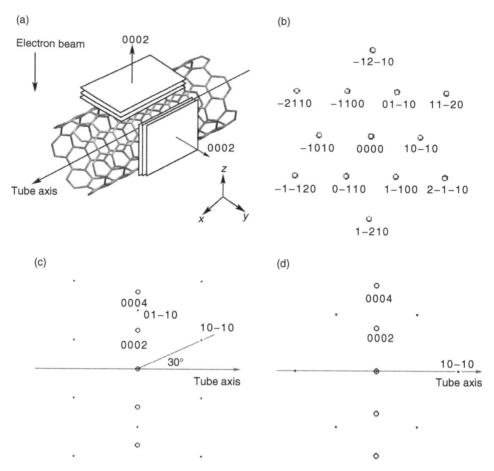

FIGURE 5.4 (a) Electron diffraction of MWNTs. The innermost layer is shown as a hexagonal network, and the outer layers are represented by the flat sheets. Tube walls normal to the incident electron beam generate diffraction patterns as shown in (b), which are also shown as solid dots in (c) and (d). Tube walls oriented edge-on to the incident electron beam display a row of (0002) spots, shown as open circles in (c) and (d). (c) and (d) are diffraction patterns of armchair and zigzag tubes, respectively.

0002 direction of these parts of the NT is along the y-axis direction. The electron beam incidences along some direction in the basal planes. The significant diffraction spots of such a configuration are 0002 spots, which are perpendicular to the tube axis — this is because the stacking of the walls is perpendicular to the tube axis. We can now draw a schematic picture of the diffraction patterns of an armchair and zigzag NTs — shown in Figures 5.4(c) and (d), respectively. The open circles in these pictures are the diffraction spots from the sidewalls, and the small dots are from the top and bottom walls. In detail, the schematic picture of the diffraction patterns is constructed as follows: (1) we draw the 0002 diffraction spots, which can determine the tube axis; (2) along the tube axis, in real space, the crystallographic direction is 10–10 for an armchair NT. In reciprocal space, the corresponding direction of 10–10 is along a direction with a 30° inter-angle between its real space counterparts. The angle between 10–10 diffraction spot and the tube axis is thus 30°, which is an indicator for the identification of an armchair NT. The same analysis applies to zigzag tubes and results in the pattern shown in Figure 5.4(d). The characteristic of a zigzag diffraction pattern is that the 10–10 diffraction spot is along the tube axis, i.e., perpendicular to the direction of the 0002 diffraction spots. It is worth clarifying that if one claims that the tube axis of a zigzag tube is along

10–10, it means that the indices are in the reciprocal space; correspondingly, the very same direction in the real space is 12–30.

In practice, nanobeam diffraction (NBD) technique is usually conducted to investigate the chirality of NTs [34]. The chiralities of BNNTs have been investigated by several research groups. It seems that different chiralities were obtained from BNNTs prepared through different growth methods. For the BNNTs produced by arc-discharge method using HfB_2 as the electrode [19], diffraction patterns from two as-grown NTs suggest either chiral or nonchiral, however, a preference toward nonchiral is suggested. Compared with Figures 5.4(c) and (d), Figure 5.5(A) is an armchair and Figure 5.5(B) is a zigzag [19]. For a given NT, the distribution peak of helicities of its walls is always close to either 0, i.e., armchair, or 30, i.e., zigzag, with a rather small dispersion of ~10.

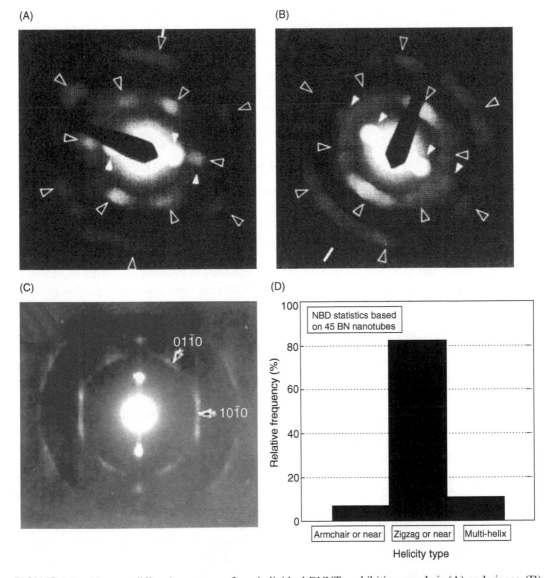

FIGURE 5.5 Electron diffraction patterns from individual BNNTs exhibiting armchair (A) and zigzag (B) configurations. (From Loiseau, A. et al., *Carbon*, 36, 743–752, 1998.) (C) A characteristic NBD pattern taken from a rope of zigzag BNNTs, and the histogram (D) shows the preference of zigzag orientation. (From Golberg, D. et al., *Solid State Commn.*, 116, 1–6, 2000. Reproduced with permission from *Elsevier Science.*)

In Golberg's [27,31] work, the zigzag type of BNNTs is overwhelming. An NBD pattern of the sample is shown in Figure 5.5(C), and the distribution of the NT chirality is shown in Figure 5.5(D). Besides different tip morphologies, it is believed that the second major difference between the structures of BNNTs and CNTs is that dominant BNNT structures are zigzag tubes. Armchair and chiral tubes are fewer, possibly due to the special tip configurations [35], while CNTs with zigzag, armchair, and chiral structures are all particularly abundant.

To conclude the discussion on BNNT structure, the preferences of both their flat tips and the achiral configurations should be stressed, which could be a characteristic of BNNTs. However, the tips of BNNTs could also be cone-like [36], open [37], and flag-like [38]. Although pentagons or heptagons can result in homogeneous bonding (B–B or N–N), which is energy unfavorable [2], these multimorphology tips and the bending of BNNTs indicate the existence of these rings [31]. Different chiralities of BNNTs have been observed, and the preference of specific chirality may depend on the growth technique employed. As pointed by Golberg [26], BNNTs grown directly from the vapor phase frequently have armchair configuration, while those grown from laser-heating and carbon substituted methods, are zigzag-dominant type [26,39]. The formation of achiral BNNTs, either armchair or zigzag, is probably a consequence of the so-called lip–lip interaction during the NT growth, which induces a correlation between the chiralities of adjacent layers and selects the growth of particular pairs of tubes [20]. It is definite that the growth method affects the morphologies of BNNTs. In the following section, we will review the major synthesis methods developed since the discovery of BNNTs.

5.3 SYNTHESIS METHODS OF BORON NITRIDE NANOTUBES

The growth of BNNTs involves the formation of boron and nitrogen hexagon networks curled up to seamless tubular forms as discussed in the previous section. This is directly related to the rearrangement of boron and nitrogen atoms via nitriding chemical reactions. To make the rearrangement possible and efficient, atomic scale boron and nitrogen clusters need to be first generated, which requires a significant amount of energy to be supplied. Different forms of energy can be exploited to fulfill this task, and the growth methods can then be coarsely classified according to the types of energy supplied. In this section, we will discuss most successful growth methods reported so far.

5.3.1 ARC DISCHARGE AND ARC MELTING

Pure BNNTs, like their carbon counterparts [1], were first fabricated by using arc-discharge method [3]. Conventional arc discharge is a method in which the reactants are used as electrodes and vaporized between the two electrodes by electric energy. So far, several variations of the growth technique for BNNTs have been proposed. Figure 5.6(A) is a schematic diagram of an arc-discharge set-up. The basic configuration of the set-up consists of a vacuum chamber, gas-flow controls, and two electrodes with a DC power supply. Depending on the requirements of specific experiments, the ambient condition inside the chamber can be either inert gas (e.g., helium, argon) or some reactive gas, for instance, nitrogen gas in the case of growing BNNTs. During the growth, the pressure inside the chamber is usually a few hundred torr [3,19,36–38,40–42]. The reactant materials are compressed or shaped into rods, which are used as the electrodes. To synthesize CNTs, the electrodes used are graphite rods, while for the growth of BNNTs, the situation is more complicated and is discussed below. The voltage exerted between the electrodes is ~20 to 40 V, and a high current up to 150 A is applied to generate the arc [3,19,36–38,40–42]. For given values of the voltage and current, we can adjust the gap between the electrodes (i.e., cathode and anode) to sustain a stable arc. The duration of the discharge is normally several minutes. During the discharge, the high current can easily increase the temperature of the electrodes to ~4000 K and evaporate the electrodes into clusters at atomic scale. The anode is normally the consumed electrode because a large

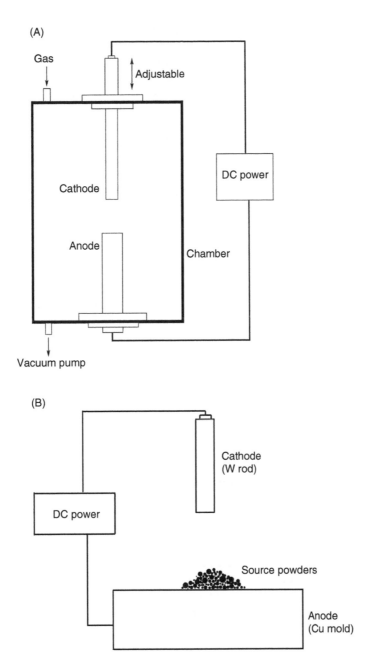

FIGURE 5.6 Schematic diagrams of (A) arc-discharging and (B) arc-melting apparatus.

quantity of electrons from the arc-discharge is accelerated toward the anode and collides with the anodic rod. Deposits can be found on the cathode as well as on the inner wall of the chamber. The deposits consist of many nanosized structures including NTs.

The rod-like electrodes, i.e., the reactants, should exhibit reasonable electrical conductance — otherwise a high current, and thus a stable arc, are not possible. Unfortunately, this criterion was the difficulty encountered when scientists first tried to make BNNTs using the arc-discharge method. Practically, bulk h-BN is an electrical insulator because of its wide band gap of 5.8 eV. Therefore, pure BN electrodes cannot carry such a high current as required in the normal discharge process.

Alternative conductive electrodes need to be considered. To circumvent this problem, Chopra et al. [3] used a hollow tungsten tube stuffed with BN powders as the anode, and water-cooled copper as the cathode. This modification in the preparation of new electrodes made arc discharge the first successful method for pure BNNT synthesis. Multiwalled BNNTs were found in the dark gray cathode soot. The typical length of these MW BNNTs was about 200 nm, with diameters in the range of 6 to 8 nm. It is worth noting that metal particles were found at the ends of all the investigated BNNTs. The composition ratio of B over N was about 1.14, as determined by electron energy loss spectroscopy (EELS). One year later, Terrones et al. [36] tried tantalum tubes filled with BN powders. Stoichiometric BN MWNTs have been fabricated, and the as-grown BN MWNTs can be several micrometers in length. In the two cases above, the starting materials are BN compounds, no chemical reactions are needed to bind B and N together, although the break and reconnection of B–N bonds may happen. The formation process is very similar to the case of CNTs. In both cases, metal particles from the anodes were found in the BNNT samples and they accidentally played the role of catalysts.

Some groups have used conducting metal borides as both the cathode and anode. Hot-pressed HfB_2 [19,37] or ZrB_2 [38,40] electrodes have been demonstrated to be successful so far. In the case of boride electrodes, nitriding reactions definitely occur during the arcing process. An nitrogen or ammonia ambient is thus usually established as the nitrogen source for the nitriding reaction. As well as MW BNNTs, some SW BNNTs have been observed, which was first reported by Loiseau et al. [19,37]. The image of an SW BNNT is reproduced as Figure 5.7(A). The aforementioned modifications have resulted in the successful synthesis of BNNTs, and verified the theoretical prediction proposed several years ago [2]. However, unlike the case of CNTs, in which the arc-discharge is already a routine method for high yield synthesis, the yield of BNNTs is very low through the use of these methods. To improve the yield and, also to overcome the conductance problem of electrodes, Cumings [41] and Altoe [42] have developed another technique within the framework of arc discharge. Metal addictives (e.g., Ni and Co) and boron powders were mixed, melted and then cooled to form an ingot. The ingot was shaped into electrodes. This method can improve the NT yield. Mass quantity production of the order of a milligram has also been realized. An interesting feature of the as-grown BNNTs is that double-walled BNNTs (see Figure 5.7[B]) are the dominant morphology of the products, as can be seen from the histogram of the wall numbers as shown in the inset of Figure 5.7(C).

Arc melting is similar to the conventional arc-discharge method but with a simpler setup configuration and experimental procedure. Instead of shaping the reactants into rod-like electrodes, the reactant powders are put on a Cu mold (anode), as shown in Figure 5.6(B). The cathode is a tungsten gun. When the voltage and current (e.g., 200 V, 125 A) are applied between the electrodes, an arc is generated and the powders on the mold are melted [45,46]. Together with the contribution from the ambient gas, which may be one of the reactants, the evaporated clusters from the melted sources form an ion gas. The growth of nanostructures can then occur in the ion gas by forming quasi-liquid particles first. The presence of a catalyst, especially transition metal borides exemplified by LaB_6 [43], are necessary in this method. The as-grown BNNTs are MW, with an appropriate stoichiometry. The effects of different catalysts have been examined as well [44]. This method is also called plasma-jet evaporation [45,46]. Only MW BNNTs were synthesized using arc-discharge method, and SW BNNTs were not reported. This is different from CNTs, and might suggest different growth conditions between BNNTs and CNTs. Arc-discharge method was the first technique in successful synthesis of CNTs and BNNTs, but complications in the scale-up process keep it as a laboratory technique.

5.3.2   LASER-ASSISTED METHOD

In addition to electrical energy, photonic energy can also be converted into heat, which evaporates the starting materials into ion gas instantly. Nowadays, the outputs of some advanced lasers have a very high-energy density in either a continuous wave mode or at kHz-repetition rates. For example,

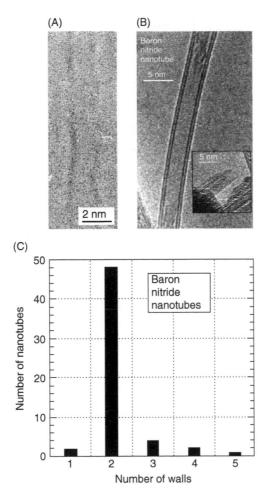

FIGURE 5.7 Boron nitride nanotube produced by arc-discharge method (A) SW BNNT. (From Loiseau, A. et al., *Carbon*, 36, 743–752, 1998.) (B) Double-walled BNNT. (C) Histogram of BNNT walls. (From Cumings, J. and Zettl, A., *Chem. Phys. Lett.*, 316, 211–216, 2000. Reproduced with permission from *Elsevier Science*.)

at 10.6 μm, the output power of a CO_2 laser-emitting infrared radiation can reach as high as 1 kW, and can be further focused to a spot of several millimeters. When the light, either continuous beam or discrete pulses, is focused on the target of the source materials with such small spot size, the energy provided by the incident light elevates the temperature of the irradiated zone to several thousand Kelvin within a very short period of time. If the temperature is above the sublimation temperature of the target material, local explosions may occur and the source materials may effuse from the surface. The ion gas of atomic scale reactants is thus generated. This process is called laser ablation [47–49]. On the other hand, if the target temperature is below the sublimation point under the laser radiation, the radiation then only increases the temperature of the target without obvious ejection of materials from it. This process is referred to as laser heating [50,51]. Both these processes have been attempted for the fabrication of BNNTs in recent years.

A schematic diagram of the laser-ablation setup is shown in Figure 5.8. Air-cooled metallic trap and filer are used to collect the ablation products, while the products of laser heating are obviously located on the surface and around the irradiated zone of the target. As shown in Figure 5.8, a tube furnace can be used as an additional energy supply to maintain the temperature of the target [47]. A diamond anvil cell under high N_2 pressure (~GPa) was also used for this purpose by Golberg et al. [50] in laser-heating process. Targets of laser-assisted processes are usually pure BN or B, and

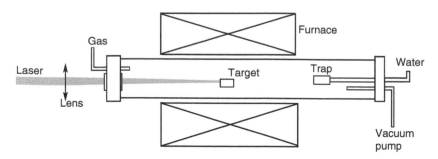

FIGURE 5.8   Schematic setup of Laser-assisted method for the growth of BNNTs.

therefore the products are metal catalyst free. It is worth mentioning that both laser-heating [51] and laser-ablation [48] methods produce BNNTs in macroscopic quantity (up to 1 g) with stoichiometric ratio of B over N.

By ablating a rotating BN target in a flowing N_2 ambient (100 ml/s) under pressure of 1 bar, Lee et al. [48] have synthesized SW-dominant BNNTs in gram quantities (0.6 g/h). The SWNTs are self-organized into bundles as shown in Figure 5.9(A). When Ni/Co are used as catalyst, SWNTs can also be found in the products (see Figure 5.9(B)) of laser ablation [47]. Similar to the arc-discharge method, the laser method involves a significant amount of energy and very high temperatures (several thousands Kelvin). The formation processes of BNNTs are probably very similar. Both methods produce BNNTs with almost the same small size range and perfect cylindrical structures with minimum defects. Scaling-up of the process is required to produce much larger quantity and high yield of SWNTs, which is very important for property and application studies.

5.3.3   BALL MILLING AND ANNEALING

This is a two-step method involving first a ball-milling process at room temperature, and a subsequent annealing at relatively low temperatures. These two processes actually correspond to separate nucleation and growth processes, respectively. As mentioned in the preceding section, atomic scaled B and N clusters are expected for the growth of BNNTs. The starting materials used in the previous methods are usually powders or rods that have bulk properties. Generally, the surface to volume ratio of bulk materials is low. To increase the effectiveness of the effusion, a large surface to volume ratio of the starting material is helpful. Regarding the chemical reaction involved in the formation of BN, when one of the reactants is bulk boron, high temperature is required to achieve a practical rate of the reaction. They are the reasons for exploiting high-temperature routines such as with the assistance of an electric arc or a high-energy laser for the growth of BNNTs. A different type of energy is used to achieve this reaction at lower temperatures. Chen et al. [52–54], in 1999, were the first to fabricate BNNTs using high-energy ball milling (HEBM) to pretreat boron or boron nitride powders.

Repeated high-energetic milling impacts provide mechanical energy to material powders during ball-milling process [55]. The grinding energy of HEBM is at least a thousand times higher than that from conventional ones. Therefore, HEBM can create structural defects and induce structural changes as well as chemical reactions at room temperature. There are several types of high-energy ball mills, such as Spex vibrating mill, planetary ball mill, rotating ball mill, attritors [56]. As an example, a photo and a schematic diagram of a room-temperature vertical-planetary ball mill are shown in Figures 5.10(A) and (B), respectively. For the synthesis of BNNTs, boron (or BN) powders were loaded into a stainless-steel cell with several hardened steel balls. The milling chamber was then filled with NH_3 or N_2 up to a pressure of 300 kPa. An external magnet applies pulling forces on the balls and thus increases milling energy. When the milling chamber rotates, the balls drop to the bottom of the chamber due to both gravity and magnetic force, as they reach the top of

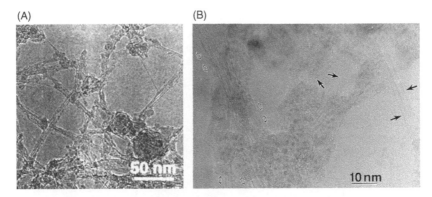

FIGURE 5.9 (A) BNNTs produced by laser-assisted method organized in bundles. (Lee, R.S. et al., *Phys. Rev. B*, 64, 121405, 2001. Reproduced with permission from *American Physical Society*.) (B) SW BNNTs produced by laser-assisted method. (From Yu, D.P. et al., *Appl. Phys. Lett.*, 72, 1966–1968, 1998. Reproduced with permission from *American Institute of Physics*.)

the chamber. This impact movement of the balls generates a strong energetic collision toward the boron/BN powders on the bottom of the chamber. The impact energy can be controlled by adjusting the position of the magnet and the rotation frequency of the chamber. The milling process usually lasts over 100 h to ensure complete structural changes.

The milling process embraces a complex mixture of fracturing, grinding, high-speed plastic deformation, cold welding, thermal shock, intimate mixing, etc. of the materials. Because the structural changes and chemical reactions are induced by mechanical energy rather than thermal energy, reactions are possible at low temperatures (room temperature). In the case of growth of BNNTs, a nitriding reaction between boron and atomic nitrogen inside the milling chamber was realized at room temperature. Boron nitride was evidenced in the as-milled powders by X-ray diffraction (XRD) as shown in Figure 5.11. XRD results also indicate that the as-milled boron powders are either amorphous or consist of extremely small, highly disordered crystallites. The ball-milling treatment creates a precursor.

The as-milled boron powders were then put into a tube furnace and annealed in nitrogen or ammonia at a temperature of 1000°C or higher for several hours. Large quantities of BN MWNTs have been found in the products. The high yield can be observed in the scanning electron microscope (SEM) image shown is Figure 5.12. The yield of the growth is up to 85%, and due to the capability of the ball milling, a kg-quantity production is possible. Therefore, this method provides an opportunity to synthesize BNNTs in an industrial-compatible quantity. This method is very different from the other methods, and special formation mechanisms have been observed. In the case where BN compounds are used as the starting materials, because the growth temperature during thermal annealing (1200°C) is far below the melting point of BN phase, no vapor could possibly be generated as in arc-discharge and laser-ablation cases. The BNNT formation is a solid-state process. Transmission electron microscopy investigation has revealed that crystal growth driven by surface diffusion is a possible mechanism under such low-temperature growth conditions [53,54,57,58]. By varying growth conditions, morphologies such as hollow tubes, and bamboo-like structures, (see Figure 5.13) can be produced separately [21,52,59]. This method has been used by several groups and demonstrated to be a very efficient method for production of BNNTs in large quantities [60,61]. Actually, the first commercial source for BNNTs was established based on this synthesis method.

5.3.4 CARBON NANOTUBE SUBSTITUTION

Several one-dimensional structures have been prepared by using CNTs as templates. The functionalities of CNTs in these syntheses might be (1) their open hollow structures being filled by other materials through capillary effects, (2) their fine cylinders being coated by some materials,

(A)

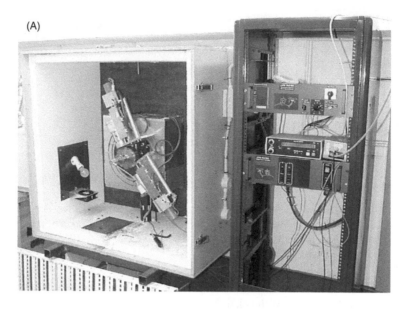

(B)

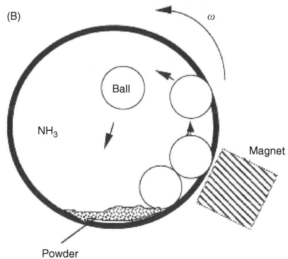

FIGURE 5.10 Photo (A) and schematic diagram (B) of an HEBM setup.

and (3) the hexagonal-carbon network in the tubular network reacting with other materials. The last functionality may result in either solid nanowire structures or hollow NTs. If the product structure reserves the framework of CNTs, the reaction is named CNT-substituted reaction and has been used for the synthesis of BNNTs. Bando's group in Japan covered B_2O_3 powders with carbon MWNTs and annealed them at 1773 K in flowing nitrogen for half an hour [26,62]. The substitution reaction is

$$B_2O_3 + 3C(NTs) + N_2(g) \rightarrow 2BN(NTs) + 3CO(g)$$

B_2O_3 exhibits a much lower melting point of 450°C compared with B (2076°C) and BN (~3000°C). Therefore, reasonable amount of B_2O_3 vapor can flow up and react with CNTs.

The general morphologies of BNNTs and CNTs are much alike. However, on an average, as-grown BNNTs have a smaller number of walls than those of the CNT templates. The EELS

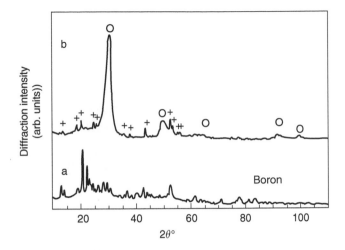

FIGURE 5.11 XRD patterns of ball-milled and annealed samples. (Chen, Y. et al., *Chem. Phys. Lett.*, 299, 260–264, 1999. Reproduced with permission from *Elsevier Science.*)

FIGURE 5.12 Scanning electron microscope image of ball-milling and annealing products showing large quantity and high yield of BNNTs.

shows that the composition of the products can be pure BN without obvious trace of carbon. A striking increase in BN MWNT yield and self-assemblage of tubes into ropes have also been seen when MoO_3 or V_2O_5 is introduced as an oxidizing-promoting agent during the substituted reaction [27,63–65]. In addition to the improved yield, this method takes advantage of the alignment of carbon nitride nanotube (CNNT) arrays, i.e., by starting from aligned CNNTs, the as-grown BNNTs are also aligned [66]. Pure BNNTs can be fabricated by this method, but it has been noted that these attempts frequently resulted in C traces in the products and there were insurmountable difficulties in preparing 100% pure BN nanomaterials. In fact, similar growth techniques, but with a lower temperature (1373 K) and a longer anneal time (4 h), are used to fabricate boron doped CNTs [67,68]. The boron carbon nitride nanotubes (BCN NTs) are easier to get than pure BNNTs, but it is difficult to control the B level. To purify the BCN NTs from carbon contaminations so as to transform BCN NTs into pure BNNTs, Han et al. [69] burned BCN NTs in air at about 700°C for 30 min,

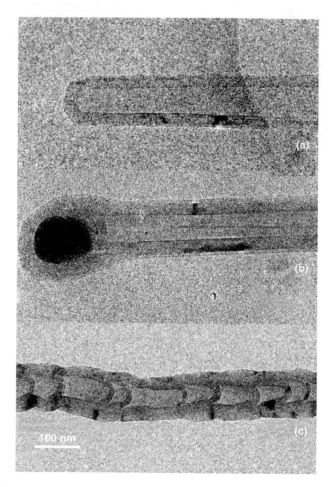

FIGURE 5.13 Different morphologies of BNNTs fabricated from the ball-milling and annealing method. (From Chen, L. et al., *J. Mater. Res.*, 17, 1896–1899, 2002. Reproduced with permission from *Material Research Society.*)

and found that up to 60% of the BCN NTs could be transformed. The quantity of BNNTs produced using this method depends on that of CNTs.

5.3.5 CHEMICAL VAPOR DEPOSITION AND OTHER THERMAL METHODS

In this section, we review several methods that, from the standpoint of energy provided, use thermal energy only. During heat treatment, the formation of BNNTs may be through different kinds of chemical reactions, depending on the starting materials that are used in the process. This has resulted in several terminologies concerning these processes.

When amorphous boron (or mixed with h-BN) was heated in ammonia gas with the presence of Li vapor for 10 to 20 h [70,71], boron nitride MWNTs with a typical diameter of 10 nm were formed. In addition to Li vapor, oxides such as Fe_2O_3, MgO, Ga_2O_3, and SiO_2 have also been co-heated with amorphous B, and NH_3 was usually exploited for the nitriding reaction as well. Using various oxides, Tang [72] reported that straight concentric pure BNNTs with a large diameter distribution from several nm to 70 nm were fabricated in large scale. The conversion rate was ~40%. The oxides may act as media, reacting with boron and generating an intermediate product of

B_2O_2, which is in vapor phase at the reaction temperature. B_2O_2 vapor and ammonia react and thus facilitate the formation of BNNTs. Taking MgO as an example, the possible reactions are

$$2B(s) + 2MgO(s) \rightarrow B_2O_2(g) + 2Mg(g)$$

$$B_2O_2(g) + 2NH_3(g) \rightarrow 2BN(s) + 2H_2O(g) + H_2(g)$$

Heating temperature and B/oxides ratio have noticeable effects on the NT formation [73,74]. Boron oxides (B_2O_3) can also be heated directly either as discussed in the carbon-substituted reaction [26,62], or as described in Bartnitskaya et al. [75]. Boron nitride powders have also been heated at high temperature (1750 to 2000°C) in nitrogen ambient to fabricate BNNTs [76]. Local three-dimensional order was first observed in the as-grown products.

Chemical vapor deposition (CVD) has also been attempted for fabrication of BNNTs. A CVD process is a process of chemically reacting a volatile compound of a material to be deposited, with other gases, to produce a nonvolatile solid that deposits atomistically on a suitably placed substrate. The volatile compound of the first CVD routine for BNNTs was $B_3N_3H_6$ [77], and it was carried by N_2 to the substrate of Si covered with catalysts (Co, Ni, and NiB). The temperature of the substrate was 1000 to 1100°C. $B_3N_3H_6$ vapor was decomposed and BNNTs were deposited on the Si wafer. Changing the volatile materials, Ma et al. [78–81] have developed several CVD routines for the fabrication of BNNTs. Other compounds such as NH_4BF_4 and KBH_4 are also used as precursor materials, and heated in a nitrogen-containing atmosphere to synthesize BNNTs [82]. The yield of the CVD synthesis for BNNTs is not clear, though the CVD process itself is believed to be a low-cost and high-yield method. Integrated into the CVD method, nanoporous templates have become widely used to confined grow 1-D nanostructures. Anodic alumina membranes (AAMs) are one of the most popular templates. The pores of AAM with adjustable diameters in the range of 5 to 200 nm are perpendicular to the surface of the AAM and are organized into arrays. The density of the pores may be up to $10^{11}/cm^2$. The nanopores are filled with desired materials (e.g., BN) via CVD process. After the filling, the template itself can be removed by proper chemical solutions. Shelimov has used this technique to grow aligned BNNTs. Compared with other growth methods, the temperature of the deposition is very low (750°C) [83]. However, because of the low temperature, the as-grown BNNTs are polycrystalline in nature.

5.4   SUMMARY

BNNTs have a similar tubular structure as CNTs with different capes and possibly different chiralities. More detailed research on a large number of BNNT samples is needed to fully clarify the structure. BNNT synthesis is, however, much more complicated than that of CNTs, even using the same processing techniques. This is probably because two different elements, B and N, are involved rather than the single C, and consequently B–N pairs have to be created via a chemical reaction prior to (or simultaneous to) the growth of tubular nanostructure. According to supplied energy classification of various synthesis methods, most successful synthesis methods reviewed previously do not rely on traditional thermal energy as general chemical reactions and crystal growths normally require. Electric (arc discharge), photonic (laser), mechanical (ball milling), and chemical (CNT substitution) energies are more successful than thermal energy (pure-furnace heating). This implies that the nitriding-reaction and NT-formation processes are thermo-dynamical nonequilibrium processes. Compared with bulk h-BN, the curved BN basal planes in NTs at nanometer scale are metastable structures, which are difficult to be produced with pure thermal energy under equilibrium conditions. Similar to the CNT, the formation process needs to be detailed fully. Nevertheless, the growth techniques of BNNTs still require significant improvement, and new synthesis methods with better control over the growth directions and locations are needed. There is only one paper [48] describing the BN SWNT dominant fabrication, while, to the best of our knowledge, study on the alignment growth and selective growth of BNNTs is still absent. We believe that the research and application of BNNTs will be extensively developed, and a fundamental breakthrough of the growth technique can be achieved.

ACKNOWLEDGMENTS

We wish to acknowledge the permission given by Dr. Dmitri Golberg, Dr. Francois Willaime, Dr. John Cumings, and Dr. Dapeng Yu for using their published images of BNNTs. This research has been supported by the Australian Research Council under the Centre of Excellent program (ARC Centre of Nanofunctional Materials).

REFERENCES

1. Iijima, S., Helical microtubules of graphitic carbon, *Nature*, 354, 56–58, 1991.
2. Rubio, A., Corkill, J.L., and Cohen, M.L., Theory of graphitic boron-nitride nanotubes, *Phys. Rev.*, B 49, 5081–5084, 1994.
3. Chopra, N.G., Luyken, R.J., Cherrey, K., Crespi, V.H., Cohen, M.L., Louie, S.G., and Zettl, A., Boron-nitride nanotubes, *Science*, 269, 966–967, 1995.
4. Kudin, K.N., Scuseria, G.E., and Yakobson, B.I., C2F, BN, and C nanoshell elasticity from ab initio computations, *Phys. Rev.*, B 64, 235406, 2001.
5. Mele, E.J. and Kral, P., Electric polarization of heteropolar nanotubes as a geometric phase, *Phys. Rev. Lett.*, 88, 56803, 2002.
6. Chopra, N.G. and Zettl, A., Measurement of the elastic modulus of a multiwall boron nitride nanotube, *Solid State Commn.*, 105, 297–300, 1998.
7. Suryavanshi, A.P., Yu, M.F., Wen, J.G., Tang, C.C., and Bando, Y., Elastic modulus and resonance behavior of boron nitride nanotubes, *Appl. Phys. Lett.*, 84, 2527–2529, 2004.
8. Blase, X., Rubio, A., Louie, S.G., and Cohen, M.L., Stability and band-gap constancy of boron-nitride nanotubes, *Europhys. Lett.*, 28, 335–340, 1994.
9. Radosavljevic, M., Appenzeller, J., Derycke, V., Martel, R., Avouris, P., Loiseau, A., Cochon, J.L., and Pigache, D., Electrical properties and transport in boron nitride nanotubes, *Appl. Phys. Lett.*, 82, 4131–4133, 2003.
10. Cumings, J. and Zettl, A., Field emission and current-voltage properties of boron nitride nanotubes, *Solid State Commn.*, 129, 661–664, 2004.
11. Golberg, D., Bando, Y., Kurashima, K., and Sato, T., Synthesis and characterization of ropes made of BN multiwalled nanotubes, *Scr. Mater.*, 44, 1561–1565, 2001.
12. Chen, Y., Zou, J., Campbell, S.J., and Le Caer, G., Boron nitride nanotubes: pronounced resistance to oxidation, *Appl. Phys. Lett.*, 84, 2430–2432, 2004.
13. Oku, T. and Kuno, M., Synthesis, argon/hydrogen storage and magnetic properties of boron nitride nanotubes and nanocapsules, *Diamond Relat. Mater.*, 12, 840–845, 2003.
14. Seayad, A.M. and Antonelli, D.M., Recent advances in hydrogen storage in metal-containing inorganic nanostructures and related materials, *Adv. Mater.*, 16, 765–777, 2004.
15. Ma, R.Z., Bando, Y., Zhu, H.W., Sato, T., Xu, C.L., and Wu, D.H., Hydrogen uptake in boron nitride nanotubes at room temperature, *J. Am. Chem. Soc.*, 124, 7672–7673, 2002.
16. Wu, J., Han, W.Q., Walukiewicz, W., Ager, J.W., Shan, W., Haller, E.E., and Zettl, A., Raman spectroscopy and time-resolved photoluminescence of BN and BxCyNz nanotubes, *Nano Lett.*, 4, 647–650, 2004.
17. Kral, P., Mele, E.J., and Tomanek, D., Photogalvanic effects in heteropolar nanotubes, *Phys. Rev. Lett.*, 85, 1512–1515, 2000.
18. Iijima, S., Ichihashi, T., and Ando, Y., Pentagons, heptagons and negative curvature in graphite microtubule growth, *Nature*, 356, 776–778, 1992.
19. Loiseau, A., Willaime, F., Demoncy, N., Schramchenko, N., Hug, G., Colliex, C., and Pascard, H., Boron nitride nanotubes, *Carbon*, 36, 743–752, 1998.
20. Charlier, J.C., Blase, X., De Vita, A., and Car, R., Microscopic growth mechanisms for carbon and boron-nitride nanotubes, *Appl. Phys. a—Mater. Sci. Process.*, 68, 267–273, 1999.

21. Chen, Y., Conway, M., Williams, J.S., and Zou, J., Large-quantity production of high-yield boron nitride nanotubes, *J. Mater. Res.*, 17, 1896–1899, 2002.
22. Golberg, D. and Bando, Y., Electron microscopy of boron nitride nanotubes, in *Electron Microscopy of Nanotubes*, 1st ed., Wang, Z.L. and Hui, C., Eds., Tsinghus University Press, Beijing, 2004, pp. 221–250.
23. Chen, L., Ye, H., and Gogotsi, Y., Synthesis of boron nitride coating on carbon nanotubes, *J. Am. Ceram. Soc.*, 87, 147–151, 2004.
24. Qin, L.C., Ichihashi, T., and Iijima, S., On the measurement of helicity of carbon nanotubes, *Ultramicroscopy*, 67, 181–189, 1997.
25. Lucas, A.A., Bruyninckx, V., Lambin, P., Bernaerts, D., Amelinckx, S., Landuyt, J.V., and Tendeloo, G.V., Electron diffraction by carbon nanotubes, *Scanning Microsc.*, 12, 415–436, 1998.
26. Golberg, D., Han, W., Bando, Y., Bourgeois, L., Kurashima, K., and Sato, T., Fine structure of boron nitride nanotubes produced from carbon nanotubes by a substitution reaction, *J. Appl. Phys.*, 86, 2364–2366, 1999.
27. Golberg, D., Bando, Y., Kurashima, K., and Sato, T., Ropes of BN multi-walled nanotubes, *Solid State Commn.*, 116, 1–6, 2000.
28. Saito, Y., Maida, M., and Matsumoto, T., Structures of boron nitride nanotubes with single-layer and multilayers produced by arc discharge, *Japanese J. Appl. Phys. Part 1-Regular Pap. Short Notes Rev. Pap.*, 38, 159–163, 1999.
29. Demczyk, B.G., Cumings, J., Zettl, A., and Ritchie, R.O., Structure of boron nitride nanotubules, *Appl. Phys. Lett.*, 78, 2772–2774, 2001.
30. Narita, I. and Oku, T., Atomic structure of boron nitride nanotubes with an armchair-type structure studied by HREM, *Solid State Commn.*, 129, 415–419, 2004.
31. Golberg, D., Bando, Y., Bourgeois, L., Kurashima, K., and Sato, T., Insights into the structure of BN nanotubes, *Appl. Phys. Lett.*, 77, 1979–1981, 2000.
32. Wildoer, J.W.G., Venema, L.C., Rinzler, A.G., Smalley, R.E., and Dekker, C., Electronic structure of atomically resolved carbon nanotubes, *Nature*, 391, 59–62, 1998.
33. Odom, T.W., Huang, J.L., Kim, P., and Lieber, C.M., Atomic structure and electronic properties of single-walled carbon nanotubes, *Nature*, 391, 62–64, 1998.
34. Cowley, J.M., *Nanodiffraction of Carbon Nanotubes*, 1st ed., Tsinghua University Press, Beijing, 2004.
35. Menon, M. and Srivastava, D., Structure of boron nitride nanotubes: tube closing versus chirality, *Chem. Phys. Lett.*, 307, 407–412, 1999.
36. Terrones, M., Hsu, W.K., Terrones, H., Zhang, J.P., Ramos, S., Hare, J.P., Castillo, R., Prassides, K., Cheetham, A.K., Kroto, H.W., and Walton, D.R.M., Metal particle catalysed production of nanoscale BN structures, *Chem. Phys. Lett.*, 259, 568–573, 1996.
37. Loiseau, A., Willaime, F., Demoncy, N., Hug, G., and Pascard, H., Boron nitride nanotubes with reduced numbers of layers synthesized by arc discharge, *Phys. Rev. Lett.*, 76, 4737–4740, 1996.
38. Saito, K., Maida, M., and Matsumoto, T., Structures of boron nitride nanotubes with single-layer and multilayers produced by arc discharge, *Japanese J. Appl. Phys. Part 1-Regular Pap. Short Notes Rev. Pap.*, 38, 159–163, 1999.
39. Golberg, D., Bando, Y., Eremets, M., Takemura, K., Kurashima, K., Tamiya, K., and Yusa, H., Boron nitride nanotube growth defects and their annealing-out under electron irradiation, *Chem. Phys. Lett.*, 279, 191–196, 1997.
40. Saito, Y. and Maida, M., Square, pentagon, and heptagon rings at BN nanotube tips, *J. Phys. Chem. A*, 103, 1291–1293, 1999.
41. Cumings, J. and Zettl, A., Mass-production of boron nitride double-wall nanotubes and nanococoons, *Chem. Phys. Lett.*, 316, 211–216, 2000.
42. Altoe, M.V.P., Sprunck, J.P., Gabriel, J.C.P., and Bradley, K., Nanococoon seeds for BN nanotube growth, *J. Mater. Sci.*, 38, 4805–4810, 2003.
43. Kuno, M., Oku, T., and Suganuma, K., Synthesis of boron nitride nanotubes and nanocapsules with LaB6, *Diamond Relat. Mater.*, 10, 1231–1234, 2001.
44. Narita, I. and Oku, T., Synthesis of boron nitride nanotubes by using NbB2, YB6 and YB6/Ni powders, *Diamond Relat. Mater.*, 12, 1912–1917, 2003.
45. Shimizu, Y., Moriyoshi, Y., Komatsu, S., Ikegami, T., Ishigaki, T., Sato, T., and Bando, Y., Concurrent preparation of carbon, boron nitride and composite nanotubes of carbon with boron nitride by a plasma evaporation method, *Thin Solid Films*, 316, 178–184, 1998.

46. Shimizu, Y., Moriyoshi, Y., Tanaka, H., and Komatsu, S., Boron nitride nanotubes, webs, and coexisting amorphous phase formed by the plasma jet method, *Appl. Phys. Lett.*, 75, 929–931, 1999.

47. Yu, D.P., Sun, X.S., Lee, C.S., Bello, I., Lee, S.T., Gu, H.D., Leung, K.M., Zhou, G.W., Dong, Z.F., and Zhang, Z., Synthesis of boron nitride nanotubes by means of excimer laser ablation at high temperature, *Appl. Phys. Lett.*, 72, 1966–1968, 1998.

48. Lee, R.S., Gavillet, J., de la Chapelle, M.L., Loiseau, A., Cochon, J.L., Pigache, D., Thibault, J., and Willaime, F., Catalyst-free synthesis of boron nitride single-wall nanotubes with a preferred zigzag configuration, *Phys. Rev. B*, 64, 121405, 2001.

49. Golberg, D., Rode, A., Bando, Y., Mitome, M., Gamaly, E., and Luther-Davies, B., Boron nitride nanostructures formed by ultra-high-repetition rate laser ablation, *Diamond Relat. Mater.*, 12, 1269–1274, 2003.

50. Golberg, D., Bando, Y., Eremets, M., Takemura, K., Kurashima, K., and Yusa, H., Nanotubes in boron nitride laser heated at high pressure, *Appl. Phys. Lett.*, 69, 2045–2047, 1996.

51. Laude, T., Matsui, Y., Marraud, A., and Jouffrey, B., Long ropes of boron nitride nanotubes grown by a continuous laser heating, *Appl. Phys. Lett.*, 76, 3239–3241, 2000.

52. Chen, Y., Fitz Gerald, J.D., Williams, J.S., and Bulcock, S., Synthesis of boron nitride nanotubes at low temperatures using reactive ball milling, *Chem. Phys. Lett.*, 299, 260–264, 1999.

53. Chen, Y., Chadderton, L.T., FitzGerald, J., and Williams, J.S., A solid-state process for formation of boron nitride nanotubes, *Appl. Phys. Lett.*, 74, 2960–2962, 1999.

54. Chen, Y., Chadderton, L.T., Williams, J.S., and Fitz Gerald, J.D., Solid-state formation of carbon and boron nitride nanotubes, *Metastable, Mech. Alloyed Nanocryst. Mater., Parts 1 and 2*, 343, 63–67, 2000.

55. Benjamin, J.S., Dispersion strengthened superalloys by mechanical alloying, *Metall. Trans.*, 1, 2943, 1970.

56. Suryanarayana, C., Nanocrystalline materials, *Int. Mater. Rev.*, 40, 41, 1995.

57. Chadderton, L.T. and Chen, Y., A model for the growth of bamboo and skeletal nanotubes: catalytic capillarity, *J. Cryst. Growth*, 240, 164–169, 2002.

58. Chadderton, L.T. and Chen, Y., Nanotube growth by surface diffusion, *Phys. Lett. A*, 263, 401–405, 1999.

59. Chen, Y., Halstead, T., and Williams, J.S., Influence of milling temperature and atmosphere on the synthesis of iron nitrides by ball milling, *Mater. Sci. Eng. A*, 206, 24–29, 1996.

60. Tang, C.C., Bando, Y., and Sato, T., Synthesis and morphology of boron nitride nanotubes and nanohorns, *Appl. Phys. a—Mater. Sci. Process.*, 75, 681–685, 2002.

61. Bae, S.Y., Seo, H.W., Park, J., Choi, Y.S., Park, J.C., and Lee, S.Y., Boron nitride nanotubes synthesized in the temperature range 1000–1200°C, *Chem. Phys. Lett.*, 374, 534–541, 2003.

62. Han, W.Q., Bando, Y., Kurashima, K., and Sato, T., Synthesis of boron nitride nanotubes from carbon nanotubes by a substitution reaction, *Appl. Phys. Lett.*, 73, 3085–3087, 1998.

63. Golberg, D., Bando, Y., Kurashima, K., and Sato, T., Synthesis, HRTEM and electron diffraction studies of B/N-doped C and BN nanotubes, *Diamond Relat. Mater.*, 10, 63–67, 2001.

64. Golberg, D. and Bando, Y., Unique morphologies of boron nitride nanotubes, *Appl. Phys. Lett.*, 79, 415–417, 2001.

65. Golberg, D., Bando, Y., Mitome, M., Kurashima, K., Sato, T., Grobert, N., Reyes-Reyes, M., Terrones, H., and Terrones, M., Preparation of aligned multiwalled BN and B/C/N nanotubular arrays and their characterization using HRTEM, EELS and energy-filtered TEM, *Physica B-Condens. Matter*, 323, 60–66, 2002.

66. Deepak, F.L., Vinod, C.P., Mukhopadhyay, K., Govindaraj, A., and Rao, C.N.R., Boron nitride nanotubes and nanowires, *Chem. Phys. Lett.*, 353, 345–352, 2002.

67. Han, W.Q., Bando, Y., Kurashima, K., and Sato, T., Boron-doped carbon nanotubes prepared through a substitution reaction, *Chem. Phys. Lett.*, 299, 368–373, 1999.

68. Golberg, D., Bando, Y., Bourgeois, L., Kurashima, K., and Sato, T., Large-scale synthesis and HRTEM analysis of single-walled B- and N-doped carbon nanotube bundles, *Carbon*, 38, 2017–2027, 2000.

69. Han, W.Q., Mickelson, W., Cumings, J., and Zettl, A., Transformation of BxCyNz nanotubes to pure BN nanotubes, *Appl. Phys. Lett.*, 81, 1110–1112, 2002.

70. Terauchi, M., Tanaka, M., Matsuda, H., Takeda, M., and Kimura, K., Helical nanotubes of hexagonal boron nitride, *J. Electron Microsc.*, 46, 75–78, 1997.

71. Terauchi, M., Tanaka, M., Suzuki, K., Ogino, A., and Kimura, K., Production of zigzag-type BN nanotubes and BN cones by thermal annealing, *Chem. Phy. Lett.*, 324, 359–364, 2000.
72. Tang, C., Bando, Y., Sato, T., and Kurashima, K., A novel precursor for synthesis of pure boron nitride nanotubes, *Chem. Commn.*, (12), 1290–1291, 2002.
73. Tang, C.C., de la Chapelle, M.L., Li, P., Liu, Y.M., Dang, H.Y., and Fan, S.S., Catalytic growth of nanotube and nanobamboo structures of boron nitride, *Chem. Phys. Lett.*, 342, 492–496, 2001.
74. Tang, C.C., Fan, S.S., Li, P., Liu, Y.M., and Dang, H.Y., Synthesis of boron nitride in tubular form, *Mater. Lett.*, 51, 315–319, 2001.
75. Bartnitskaya, T.S., Oleinik, G.S., Pokropivnyi, A.V., and Pokropivnyi, V.V., Synthesis, structure, and formation mechanism of boron nitride nanotubes, *JETP Lett.*, 69, 163–168, 1999.
76. Bourgeois, L., Bando, Y., and Sato, T., Tubes of rhombohedral boron nitride, *J. Phys. D-Appl. Phys.*, 33, 1902–1908, 2000.
77. Lourie, O.R., Jones, C.R., Bartlett, B.M., Gibbons, P.C., Ruoff, R.S., and Buhro, W.E., CVD growth of boron nitride nanotubes, *Chem. Mater.*, 12, 1808, 2000.
78. Ma, R., Bando, Y., and Sato, T., CVD synthesis of boron nitride nanotubes without metal catalysts, *Chem. Phys. Lett.*, 337, 61–64, 2001.
79. Ma, R.Z., Bando, Y., Sato, T., and Kurashima, K., Growth, morphology, and structure of boron nitride nanotubes, *Chem. Mater.*, 13, 2965–2971, 2001.
80. Ma, R.Z., Bando, Y., Sato, T., and Kurashima, K., Thin boron nitride nanotubes with unusual large inner diameters, *Chem. Phys. Lett.*, 350, 434–440, 2001.
81. Ma, R.Z., Bando, Y., and Sato, T., Controlled synthesis of BN nanotubes, nanobamboos, and nanocables, *Adv. Mater.*, 14, 366–368, 2002.
82. Xu, L.Q., Peng, Y.Y., Meng, Z.Y., Yu, W.C., Zhang, S.Y., Liu, X.M., and Qian, Y.T., A co-pyrolysis method to boron nitride nanotubes at relative low temperature, *Chem. Mater.*, 15, 2675–2680, 2003.
83. Shelimov, K.B. and Moskovits, M., Composite nanostructures based on template-crown boron nitride nanotubules, *Chem. Mater.*, 12, 250–254, 2000.

6 Nanotubes in Multifunctional Polymer Nanocomposites

Fangming Du
Department of Chemical and Biomolecular Engineering,
University of Pennsylvania, Philadelphia, Pennsylvania

Karen I. Winey
Department of Materials Science and Engineering,
University of Pennsylvania, Philadelphia, Pennsylvania

CONTENTS

6.1   Introduction 179
6.2   Nanocomposite Fabrication and Nanotube Alignment 181
6.3   Mechanical Properties 185
6.4   Thermal and Rheological Properties 187
6.5   Electrical Conductivity 190
6.6   Thermal Conductivity and Flammability 192
6.7   Conclusions 193
Acknowledgments 195
References 195

6.1   INTRODUCTION

Polymer nanocomposites have attracted great attention due to the unique properties introduced by nanofillers, which typically refer to carbon blacks, silicas, clays, or carbon nanotubes (CNT). The polymer matrix acts as a supporting medium and the improvement in the properties of the nanocomposites generally originates from the nature of these nanofillers. Compared to other nanofillers, the unique structures of CNT potentially provide superior mechanical, electrical, and thermal properties. There are two types of CNT with high structural perfection, single-walled nanotubes (SWNT) and multi-walled nanotubes (MWNT) (Figure 6.1). SWNT can be considered as a single graphite sheet seamlessly wrapped into a cylindrical tube. MWNT comprise an array of such nanotubes that are concentrically nested. Theoretical and experimental results[1] on individual CNT have shown extremely high elastic modulus, greater than 1TPa, and their reported strength is many times higher than the strongest steel at a fraction of the weight. The fiber-like structure of CNT with low density makes them particularly attractive for reinforcement of composite materials. The nearly one-dimensional (1D) electronic structure of CNT allows electrons to be transported along the nanotubes without scattering, enabling them to carry high current with essentially no heating. Phonons also propagate easily along the nanotubes, such that the measured room temperature thermal conductivity for an individual MWNT[2] (>3000 W/m K) is greater than that of natural diamond (~2000 W/m K). The thermal

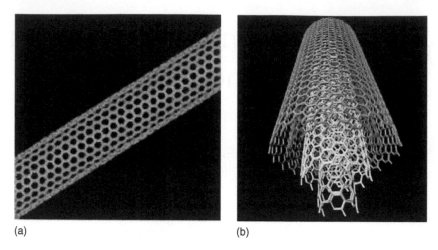

(a) (b)

FIGURE 6.1 (a) Image of SWNT. (Copyright, Dr. Smalley Group, Rice University, 2004.) (b) Image of MWNT. (Copyright, Rochefort, Nano-CERCA, University of Montreal, Canada, 2004.)

conductivity of an individual SWNT is expected to be even higher.[3] The high aspect ratio (length/diameter) of CNT suggests lower percolation thresholds for both electrical and thermal conductivity in their nanocomposites. Therefore, CNT have considerable potential in multifunctional polymer nanocomposites for structural, electrical, and thermal applications.

To date this potential has not yet been fully realized, although many researchers are devoting much attention to developing CNT/polymer nanocomposites. One significant challenge is to obtain uniform nanotube dispersion within the polymer matrix. CNT have diameters on the nanoscale and substantial van der Waals attractions between them, so CNT tend to agglomerate. As shown in Figure 6.2, these SWNT (synthesized by a high-pressure carbon monoxide method at Rice University)[4] are highly entangled and the diameter of the nanotube bundles is tens of nanometers, indicating hundreds of individual nanotubes in one bundle. Unfortunately, the nanotube bundles have significantly lower aspect ratios and inferior mechanical (due to slipping of nanotubes inside the bundles)[5] and electrical properties. Lower aspect ratios are likely to lead to higher percolation thresholds for both electrical and thermal conductivity in these nanocomposites. Therefore, uniformly dispersed nanotubes within the polymer matrix is critical for improved properties at the minimal CNT content. Various nanotube processing procedures and nanocomposite fabrication methods have been used toward this goal.[6–8] The second challenge is to understand the effect of nanotube alignment on nanocomposite properties because the nanotubes have asymmetric structures and properties. For example, nanotube alignment increases the elastic modulus[6] and increases the electrical conductivity[9] of the nanocomposites along the nanotube alignment direction. The third challenge is to create strong physical or chemical bonds between the nanotubes and the polymer matrix. These bonds can efficiently transfer load from the polymer matrix to the nanotubes, which is important for mechanical reinforcement. Covalent bonds can also benefit phonon transferring between the nanotubes and the polymer matrix, which is a key factor for improving thermal conductivity of the nanocomposites. Therefore, various surfactants and chemical functionalization procedures have been adopted to modify the surface of nanotubes, so as to enable more adhesion to the polymer matrix.

It is to be noted that nanotubes are synthesized by a variety of methods that give rise to different mean lengths, mean diameters, distributions in lengths and diameters, relative amounts of conductive nanotubes, and concentrations and types of impurities.[10] Furthermore, we have found important differences between different batches of nanotubes produced by seemingly equivalent synthetic processes. Added to this complexity are the observations that nanotube distribution and alignment significantly influence properties and is expected that the physical properties of

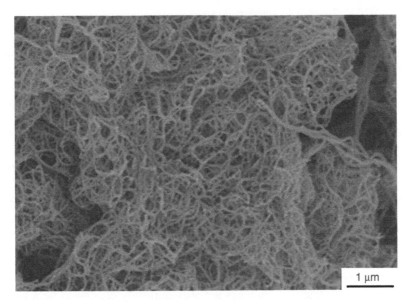

1 µm

FIGURE 6.2   SEM image of as-produced SWNT, showing that nanotube bundles are highly entangled. (Copyright, Winey group.)

nanotube/polymer composites will vary dramatically in the published literature. Thus, this chapter focuses on trends rather than absolute values for elastic modulus, electrical percolation, and so forth, because the nanotubes and composite processing methods employed by different researchers are substantially different. After reviewing various methods for producing nanotube/polymer nanocomposites, the mechanical, rheological, electrical, and thermal conductivities will be discussed separately.

6.2   NANOCOMPOSITE FABRICATION AND NANOTUBE ALIGNMENT

Solvent casting and melt mixing are two common fabrication methods for the nanotube-based nanocomposites. Solvent casting involves preparing a suspension of nanotubes in a polymer solution and then allowing the solvent to evaporate to produce a nanotube/polymer nanocomposite. Qian et al.[11] cast a MWNT/polystyrene (PS)/toluene suspension after sonication into a dish to produce the nanocomposites with enhanced elastic modulus and break stress. Benoit et al.[7] obtained electrically conductive nanocomposites by dispersing SWNT and poly(methyl methacrylate) (PMMA) in toluene, followed by drop casting the mixture on substrates. Nanocomposites with other thermoplastic matrices with enhanced properties have also been fabricated by solvent casting,[12] but nanotubes tend to agglomerate during solvent evaporation, which leads to inhomogeneous nanotube distribution in the polymer matrix. Unlike solvent casting methods, in which nanotubes are typically dispersed by sonication, melt mixing uses elevated temperatures and high shear forces to disrupt the nanotube bundles. Bhattacharyya et al.[13] made a 1 wt% SWNT/polypropylene (PP) nanocomposite by melt mixing, but found that melt mixing alone did not provide uniform nanotube dispersion. A subsequent processing step, namely passing the 1 wt% nanocomposite through a stainless-steel filter at 240°C, removed the larger SWNT agglomerates and reduced the nanotube loading to 0.8 wt%.

Some researchers use combined methods, such as solvent casting in conjunction with sonication, followed by melt mixing. For example, Haggenmueller et al.[6] first demonstrated the combination of solvent casting and melt mixing for SWNT/PMMA composites with considerable improvement in nanotube dispersion. They dissolved PMMA in dimethylformamide that was also used to disperse SWNT by sonication and the resulting suspension was then cast into a dish. After drying, the nanocomposite film exhibited heterogeneous nanotube dispersion when subjected to optical microscopy. These SWNT/PMMA films prepared by solvent casting alone contained micron-scale nanotube agglomerates due to reaggregation during solvent evaporation. Thus, the nanocomposite films were broken into pieces, stacked together, and then hot pressed. This hot pressing procedure was repeated for up to 20 times and the state of dispersion steadily improved with each additional melt mixing cycle. This combination of solvent casting and melt processing produced superior SWNT dispersion.

As seen above, solvent casting can allow CNT to agglomerate during solvent evaporation. There have been some attempts to reduce the solvent evaporation time by spin casting or drop casting.[7,14,15] Spin casting, also called spin coating, involves putting a nanotube/polymer suspension on a rotating substrate, so that the solvent evaporates within seconds. Drop casting involves dropping a nanotube/polymer suspension on a hot substrate, so that the solvent can evaporate very quickly. To better eliminate nanotube reagglomeration, Du et al.[8] developed a coagulation method to produce nanotube-based nanocomposites with uniform dispersion. Optical microscopy (Figure 6.3a) shows that there are no obvious agglomerations of the nanotubes in a 1 wt% SWNT/PMMA nanocomposite fabricated by the coagulation method, indicating that the nanotubes are uniformly distributed within the polymer matrix. Scanning electron microscopy (SEM) of a 7 wt% nanocomposite (Figure 6.3b) shows that the nanotubes are also uniformly distributed on a sub-micrometer scale. This coagulation method begins with a nanotube/polymer suspension in which the nanotubes have been well dispersed by sonication, and then poured into an excess of a nonsolvent, causing the polymer to precipitate and entrap the nanotube bundles. This precipitation process is rapid, so the nanotubes apparently do not aggregate or flocculate during the coagulation process.[16] Therefore, the coagulation method provides better nanotube dispersion.

In addition to the solvent casting, melt mixing, and coagulation methods, which combine nanotubes with high molecular weight polymers, *in situ* polymerization method have also been used to make nanotube-based nanocomposites starting with nanotubes and monomers. The most common *in situ* polymerization methods involve epoxy in which the resins (monomers) and hardeners are combined with SWNT or MWNT prior to curing (polymerizing).[17,18] Epoxy is the most commonly used thermoset polymer. Its resin can be crosslinked after mixing with a hardener of an appropriate ratio (resin/hardener) based on the number of functional groups on them. In addition to thermoset matrixes, a few examples of thermoplastic nanocomposites have been prepared by *in situ* polymerization. Park et al.[19] performed *in situ* polymerization of polyimide while sonicating to keep the nanotubes dispersed in the reaction medium. The role of sonication during polymerization is to disperse the SWNT in the low-viscosity monomer solution. As polymerization progresses, the

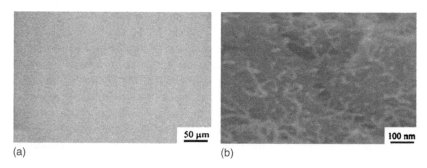

(a)   (b)

FIGURE 6.3 (a) Optical micrograph of a 1wt% SWNT/PMMA nanocomposite thin film illustrating the absence of nanotube agglomerates; (b) SEM image of a 7wt% SWNT/PMMA nanocomposite. These nanocomposites were fabricated by the coagulation method and show uniform nanotube distribution. (Adapted from Du, F. et al.[8] *J. Poly. Sci.: Part B: Poly. Phys.* 41, 3333–3338, 2003.)

high-molecular-weight polymer increases the solution viscosity, restricts Brownian motion and sedimentation of the nanotubes, and stabilizes the nanotube dispersion against agglomeration.

Many of the fabrication methods described above require nanotubes to be well dispersed in solvents. The nanotube dispersion in the polymer matrix largely depends on the state of nanotube dispersion in a solvent, assuming that the following processing procedures effectively avoid nanotube flocculation. However, the chemical structure of CNT makes dissolving long CNT in common solvents to form true solutions virtually impossible. Most nanotube/solvent systems are suspensions of nanotube bundles where the size and concentration of the bundles dictate how long the suspensions persist before flocculation. To date, large fractions of individual nanotubes have only been achieved either by functionalizing the nanotubes or by surrounding the nanotubes with dispersing agents, such as surfactants and polymers. The improved nanotube suspensions resulting from functionalization or dispersing agents can be employed in many of the methods described above to make nanotube nanocomposites with improved nanotube dispersion. Furthermore, functionalized CNT might also allow covalent bonds between the nanotubes and the polymer matrix.

Nanotube functionalization typically begins with oxidative conditions, such as refluxing in nitric acid, to attach carbonoxylic acid moieties to the defect sites on the nanotubes surface.[20,21] These acid moieties can be further transformed to more reactive groups, like –COCl, or –CORNH$_2$. These reactive groups allow CNT to chemically bond with the polymer chains through certain chemical reactions. Hill et al.[22] showed that SWNT and MWNT could be chemically bonded to a PS copolymer based on esterification. Lin et al.[23] also reported that the carbonoxylic acid on the surface of MWNT could attach aminopolymers via the formation of amide linkages. As functionalization chemistries advance, those interested in nanocomposites will continue to adapt these methods for the preparation of nanotube/polymer nanocomposites. One caveat in this regard is the degradation of properties, particularly aspect ratio and electrical conductivity, as the bonding of the CNT is disrupted by attaching functional groups to the nanotubes. There is likely to be an optimal level and type of nanotube functionalization that captures the advantages of better nanotube dispersion while minimizing the disadvantages of reduced nanotube performance.

The addition of dispersing agents, particularly surfactants, to nanotube/solvent systems has also shown improved dispersion.[24] Islam et al.[24] reported that sodium dodecylbenzene sulfonate was an effective surfactant to solubilize high-molecular-weight SWNT fractions in water by sonication. Their atomic force microscopy (AFM) showed that ~63 ± 5% of SWNT bundles exfoliated into single tubes even at 20 mg(SWNT)/ml(H$_2$O). Barrau et al.[25] used palmitic acid as a surfactant to disperse SWNT into the epoxy resin, followed by a curing procedure in the presence of the hardener. In comparison with nanocomposites not using palmitic acid, the use of palmitic acid lowers the threshold of the electrical conductivity percolation, indicating better SWNT dispersion in the epoxy. Polymers can also solubilize nanotubes in water by a noncovalent association. O'Connell et al.[26] first reported improved nanotube dispersion in water associated with modest concentration of polyvinyl pyrrolidone and also found that the nanotubes can be disassociated by changing the solvent system. Star et al.[27] discovered that starch improves the dispersion of SWNT in water and Barisci et al.[28] found that DNA could disperse SWNT well in water. Many groups investgating the interactions between nanotubes and polymers describe the mechanism by which nanotubes show improved dispersion as "polymer wrapping."[26,27] While there might be a few examples in which a polymer chain finds it energetically favorable to assume a periodic helical conformation about an isolated nanotube, it is far more likely that the improved nanotube solubility is accomplished with a more random association between the polymers and nanotubes. Regardless of the polymer conformations and mechanisms, improved nanotube solubility with the addition of surfactants and polymers provides a means to improve the nanotube dispersion in nanocomposites by improving the initial state of dispersion and by hindering the reaggregation during solvent casting and other fabrication methods.

Nanotubes can be aligned by magnetic force[29–31] (Figure 6.4a) and by polymer flow intro-duced by mechanical stretching,[32] spin casting,[33] or melt fiber spinning,[6,8] (Figure 6.4b). Magnetic fields have been used to align nanotubes in low-viscosity suspensions based on the anisotropic magnetic susceptibility of nanotubes. While magnetic alignment has shown some promise in the preparation of nanotube samples or buckypapers, this method is less appealing in nanotube/polymer composites due to the higher viscosities of the nanotube/polymer suspensions. Alternatively, polymer flow elongates and orients polymer molecules along the flow direction and the frictional forces of this process align nanotubes within the polymer matrix. Of the mechanical methods used to align nanotubes in a polymer matrix, melt fiber spinning methods exhibit the greatest degree of alignment.

The characterization of dispersion and alignment requires special attention in nanotube/polymer nanocomposites, because the physical properties are quite sensitive to these morphological attributes. Transmission electron microscopy (TEM) is a direct method to characterize MWNT/polymer composites, although there has been little success with SWNT/polymer composites due to the small size of the filler. Figure 6.5 is an aligned 5 wt% MWNT/PS nanocomposite film made by melt mixing and the MWNT were further aligned by drawing the nanocomposite in the molten state.[34] It shows that the nanocomposite has quite good dispersion and the nanotubes are well aligned along the polymer flow direction. In contrast to the qualitative measurements by TEM, polarized Raman spectroscopy and x-ray scattering provide quantitative descriptions of the degree of nanotube alignment in the nanocomposites. Besides detecting nanotube alignment, Raman spec-troscopy can also be used to measure quantitatively nanotube dispersion in the polymer matrix.[16]

Nanotubes show a resonance-enhanced Raman scattering effect when a visible or near-infrared laser is used as the excitation source, while most polymers do not. By taking advantage of this scattering effect, polarized Raman spectra are recorded for the nanocomposites with the nanotube alignment direc-tion at some angles with respect to the incident polarization axis. In the recorded spectra, the intensity

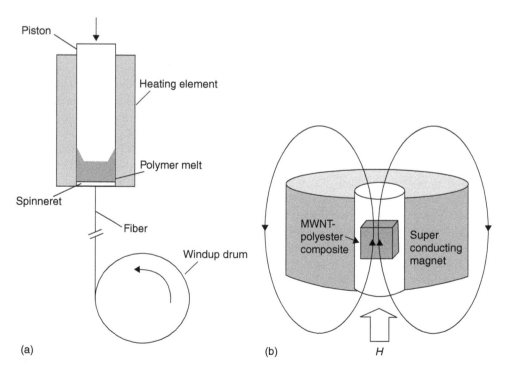

(a)   (b)   H

FIGURE 6.4 Schematics for aligning nanotubes in nanocomposites (a) by melt fiber spinning using extensional polymer flow and (b) by a magnetic field. (Adapted from Kimura, T.et al., *Adv. Mater.*, 14, 1380–1383, 2002.)

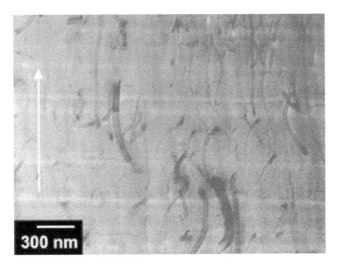

FIGURE 6.5   TEM image of a 5 wt% MWNT/PS nanocomposite film showing strong nanotube alignment after mechanical stretching. The arrow indicates the stretch direction. (Adapted from Thostenson, E.T. et al., *Compos. Sci. Technol.*, 61, 1899–1912, 2001.)

of the unique nanotube scattering mode as a function of angle can be fitted by a Lorentzian or a Gaussian function whose full width at half maximum (FWHM) can quantitatively determine the extent of nanotube alignment. For example, Haggenmueller et al.[35] obtained the degree of nanotube alignment for their SWNT/polyethylene (PE) fibers aligned by melt spinning with polarized Raman spectroscopy.

However, Raman spectroscopy is surface sensitive because the penetration depth of the incident beam is generally no more than several micrometers, depending on the nanotube loading. This problem can be solved by x-ray scattering, which is a bulk measurement for the nanocomposite in transmission and can describe the degree of nanotube alignment. Jin et al.[32] aligned nanotubes in a 50% MWNT/polyhydroxyaminoether nanocomposite film by mechanical stretching. Wide-angle x-ray scattering patterns that were dominated by a strong Bragg peak resulting from the intershell of MWNT ((002) peak) were collected from both unstretched and stretched samples. The 2D scattering intensity, I, was then integrated along the 2θ axis and plotted as I vs. azimuth, f, after subtracting the background intensity, as shown in Figure 6.6 for the unstretched and stretched samples, respectively. The scattering intensity from the unstretched sample is essentially a constant over the whole f range, as shown in Figure 6.6a, indicating random nanotube orientation. The spectrum in Figure 6.6b was fitted by 2D Lorentzian function with a FWHM of 46.4°, corresponding to a mosaic angle of ±23.2° around the stretching direction. Du et al.[8] used small-angle x-ray scattering to characterize the degree of nanotube alignment in the SWNT/PMMA nanocomposites, in which SWNT were aligned by melt spinning. X-ray intensity from SWNT nanocomposites in the small-angle range ($0.01\,\text{Å} < q < 0.1\,\text{Å}$) was mainly from the form factor scattering of SWNT and SWNT bundles. They observed that the aligned 1 wt% SWNT/PMMA had a FWHM of 16°, indicating that SWNT were highly aligned.

6.3   MECHANICAL PROPERTIES

Nanotubes were first considered as reinforcing fillers in polymer matrices due to their fiber-like structure and their exceptionally high axial strengths and axial Young's moduli. To date, the mechanical properties of various polymer matrices have shown only moderate improvements with the addition of either SWNT or MWNT.

Qian et al.[11] measured mechanical properties of MWNT/PS nanocomposites. With the addition of only 1 wt% MWNT the nanocomposite exhibited a 36–42% increase in the elastic stiffness and

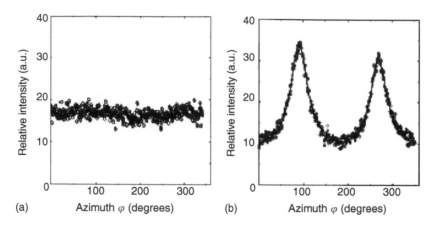

FIGURE 6.6 Integrated x-ray intensity (along the 2θ axis) vs. the azimuth for (a) an as-cast 50% MWNT/polyhydroxyaminoether nanocomposite film, and (b) the nanocomposite after being mechanically stretched. The solid line is a fit to the data by two Lorentzian functions and a constant background. (Adapted from Jin, L. et al., *Appl. Phys. Lett.*, 84, 2660–2669, 2002.)

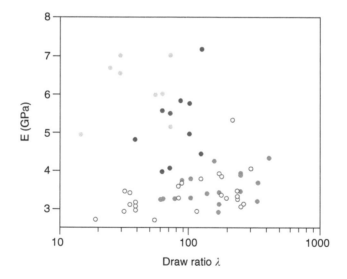

FIGURE 6.7 Elastic modulus as a function of draw ratio λ for the SWNT/PMMA nanocomposites melt spun fibers containing 0 wt% (white), 1 wt% (blue), 5 wt% (red), or 8 wt% (green) of SWNT. (Adapted from Haggenmueller, R. et al., *Chem. Phys. Lett.*, 330, 219–225, 2000.)

a 25% increase in the tensile strength. Haggenmueler et al.[6] investigated the effect of both nanotube loading and nanotube alignment on the elastic modulus of SWNT/PMMA nanocomposite fibers. Nanotube alignment was achieved by melt fiber spinning and characterized by polarized resonant Raman spectroscopy. The elastic modulus of the nanocomposite fibers increased with nanotube loading and nanotube alignment, as shown in Figure 6.7. Note that the x-axis in Figure 6.7 is the draw ratio, where higher draw ratios correspond to better nanotube alignment. Andrews et al.[36] also reported that the tensile strength and modulus of a 5 wt% SWNT/pitch nanocomposite fiber were enhanced by ~90 and ~150%, respectively, as compared to pure pitch fiber.

To enhance mechanical properties, mechanical load has to be efficiently transferred from the polymer matrix to the nanotubes. Lordi and Yao[37] suggest that the strength of the interface might result from molecular-level entanglement of pristine nanotubes and polymer chains. Thus, many researchers have investigated the interactions and load transfer between nanotubes and polymer matrices in nanocomposites. Lourie et al.[38] collected TEM images showing good epoxy-nanotube wetting and significant nanotube-epoxy interfacial adhesion, and concluded SWNT-epoxy interfacial affinity. This group also performed *in situ* deformation studies within the TEM and followed the rupture of SWNT, another indication of strong nanotube/epoxy interfaces. Wagner and co-workers[39] performed reproducible nano-pullout experiments using AFM to measure the force required to separate a CNT from poly(ethylene-co-butene) and found a separation stress of 47 MPa. In comparison, composite materials containing fiber reinforcements with weak fiber/polymer interactions typically have separation stresses of <10 MPa.[39] Simulating nanotube pullout in a CNT/PS nanocomposite, Liao and Li[40] obtained an interfacial shear stress of ~160 MPa, corresponding to a strong interface. These results from TEM, AFM, and simulation indicate that CNT can form strong interfaces with various polymer matrices.

Stress transfer can be most directly investigated by Raman spectroscopy, because the second-order (disorder-induced) Raman peak for nanotubes shifts with applied strain. If nanotubes within a nanocomposite carry part of the mechanical strain, then the Raman peak shifts with increasing applied strain. Cooper et al.[41] observed that when their SWNT/epoxy nanocomposite was mechanically strained the G' Raman band (2610 cm^{-1}) shifted to a lower wave number. This shift in the G' Raman band corresponds to strain in the nanotube graphite structure, indicating stress transfer between the epoxy matrix and nanotubes, and hence reinforcement by the nanotubes.

In addition to the publications showing improvements in mechanical properties along with good interfacial adhesion and load transfer, other reports have been less promising and highlight the sensitivity of nanotube/polymer composites to all aspects of materials and fabrication. Fisher et al.[42] used a combined finite element and micromechanical approach and found that the nanotube waviness significantly reduced the effective reinforcement when compared to straight nanotubes. Cooper et al.[43] investigated the detachment of MWNT from an epoxy matrix using a pullout test for individual MWNT. Observed values for the interfacial shear strength ranged from 35 to 376 MPa, and the authors attribute this variability to differences in structure, morphology, or surface properties of the nanotubes. This structural variety in nanotubes occurs during their synthesis, but can also be introduced during purification and other processing procedures.[10] Defects to the nanotube structure are expected to reduce significantly the mechanical properties of nanotubes.[44] Some methods of handling nanotubes, including acid treatments and high- as well as low-power sonication for extended periods of time, are known to shorten nanotubes.[10] Shorter nanotubes and thereby reduced aspect ratios are detrimental to mechanical properties. In contrast, some purification and processing steps might also introduce functional groups, such as carboxylic acid to the nanotube surface that subsequently form secondary bonds, such as hydrogen bonds, between the nanotubes and polymer matrices. Such functional groups might improve interfacial strength and promote nanotube compatibility and composite properties. Given the complexity of the starting materials (raw nanotubes), the range of chemical and physical processing methods and the breadth of polymer matrices being explored, variability in the mechanical properties in nanotube/polymer composites will persist.

Although improvements in mechanical properties have been reported for various nanotube/polymer nanocomposites, the gains are modest and fall far below simple estimates. Haggenmueller et al.[35] applied the Halpin–Tsai composite theory to their SWNT nanocomposites and found their experimental elastic modulus more than an order of magnitude smaller than predicted. They attribute this large difference mainly to the lack of perfect load transfer from the nanotube to the matrix, but other sources of uncertainty in the model include the aspect ratio and modulus of the nanotubes.

6.4 THERMAL AND RHEOLOGICAL PROPERTIES

The glass transition temperature (T_g) is a measure of the thermal energy required to allow polymer motion involving 10 to 15 monomeric units and corresponds to the softening of a polymer. Park et al.[19] reported that T_g did not change for their *in situ* polymerized SWNT/polyimide nanocomposites. The SWNT/PMMA nanocomposites produced by the coagulation method have the same T_g over a wide range of nanotube loadings.[16] The addition of nanotubes does not change the glass transition temperature in nanotube/polymer nanocomposites, because in the absence of strong interfacial bonds and at low nanotube loadings, the majority of polymers are locally constrained only by other polymers.

In contrast, at larger length scales, nanotubes do impede the motion of polymer molecules as measured by rheology. Rheological (or dynamic mechanical) measurements at low frequencies probe the longest relaxation times of polymers that correspond to time required for an entire polymer molecule to change conformation. Du et al.[16] found that although it has little effect on polymer motion at the length scales comparable to or less than an entanglement length, the presence of nanotubes has a substantial influence at large length scales corresponding to an entire polymer chain. The storage modulus, G', at low frequencies becomes almost independent of the frequency as nanotube loading increases (Figure 6.8a). These data show a transition from liquid-like behavior (short relaxation

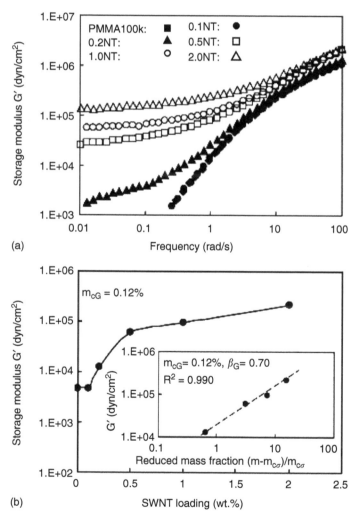

FIGURE 6.8 (a) Storage modulus (G') vs. shear frequency for SWNT/PMMA nanocomposites with various nanotube loadings; (b) G' as a function of the nanotube loading for SWNT/PMMA nanocomposites at a fixed frequency, 0.5rad/sec. The inset is a power law plot of G' of the nanocomposites vs. reduced mass fraction on a logarithmic scale. (Adapted from Du, F. et al., *Macromolecules*, 37, 9048–9055, 2004.)

times) to solid-like behavior (infinite relaxation times) with increasing nanotube loading. By plotting G' versus nanotube loading and fitting with a power law function, the rheological threshold of these nanocomposites is ~0.12 wt% (Figure 6.8b). This rheological threshold can be attributed to a hydro-dynamic nanotube network that impedes the large-scale motion of polymer molecules. This phenom-enon has previously been reported in polymer nanocomposites filled with nanoclays by Krishnamoorti et al.[45] A network of nanoscale fillers restrains polymer relaxations, leading to solid-like or nontermi-nal rheological behavior. Therefore, any factor that changes the morphology of the nanotube network will influence the low-frequency rheological properties of the nanocomposites.

Du et al.[16] found that better nanotube dispersion, less nanotube alignment, and longer polymer chains result in more restraint on the mobility of the polymer chains; i.e., the onset of solid-like behav-ior occurs at low nanotube concentrations. In addition to the loading, dispersion and alignment of the nanotubes, the size, aspect ratio, and interfacial properties of the nanotubes are expected to influence the rheological response in nanocomposites. For example, at a fixed loading, nanotubes with smaller nanotube diameters and larger aspect ratios will produce a network with smaller mesh size and larger surface area/volume, which might restrain polymer motion to a greater extent. Experimental results support this hypothesis. Lozano et al.[46] observed a rheological threshold of 10–20 wt% in carbon nanofiber/PP nanocomposites in which the diameter of the carbon nanofiber is ~150 nm. The rheo-logical threshold is ~1.5 wt% in MWNT/polycarbonate nanocomposites,[47] and only 0.12 wt% for the SWNT/PMMA system.[16] Although these three systems have different polymer matrices and their states of dispersion are unclear, the diameters of carbon nanofibers, MWNT, and SWNT differ by orders of magnitude. As the filler size decreases, the filler loading required for solid-like behavior increases substantially.

The constraints imposed by nanotubes on polymers in nanocomposites are also evident in the polymer crystallization behavior. Bhattacharyya et al.[13] studied crystallization in 0.8 wt% SWNT/PP nanocomposites using optical microscopy (with cross-polars) and differential scanning calorimetry (DSC) (Figure 6.9). The spherulite size in PP is much larger than in SWNT/PP nanocomposites. DSC results show that upon cooling, the SWNT/PP nanocomposites begins crys-tallizing at ~11°C higher than PP, suggesting that nanotubes act as nucleating sites for PP crystal-lization. They also observed that both melting and crystallization peaks in the nanocomposite are narrower than in pure PP. The authors proposed that higher thermal conductivity of the CNT as compared to that of the polymer, at least in part may be responsible for the sharper but narrower crystallization and melting peaks, as heat will be more evenly distributed in the samples containing the nanotubes.

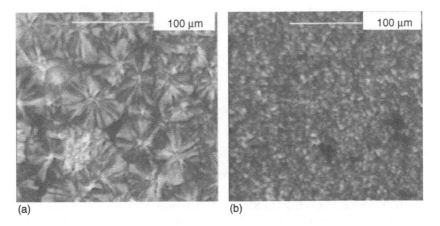

(a)　　　　　　　　　　　　　　(b)

FIGURE 6.9 Optical micrographs using cross-polarizers of (a) pure PP and (b) a 0.8 wt% SWNT/PP nanocomposite, showing that the nanocomposite has much smaller spherulite size compared to pure PP. (Adapted from Bhattacharyya, A.R. et al., *Polymer*, 44, 2373–2377, 2003.)

6.5   ELECTRICAL CONDUCTIVITY

A variety of applications are being pursued using electrically conductive nanotube/polymer nanocomposites including electrostatic dissipation, electromagnetic interference shielding, printable circuit wiring, and transparent conductive coatings. For example, charge buildup can lead to explosions in automotive gas lines and filters, and carbon black is typically added to plastics to dissipate the charge. Nanotube/polymer composites also dissipate charge and have the added advantage that the barrier properties against fuel diffusion are superior to composites made with carbon black.[48] Compared to carbon black's globular shape and micron-scale size, nanotubes are cylindrical in shape and with diameter in nanometer-scale. The larger aspect ratios and smaller diameters of nanotubes led to improved electrical conductivity in polymers at lower filler concentrations as compared to carbon black, so that a polymer's other desired performance aspects, such as mechanical and permeability properties, can be preserved.

Figure 6.10a shows the conductivity of SWNT/PS nanocomposites as a function of nanotube loading measured by Ramasubramaniam et al.[49] The conductivity of the nanocomposite increases sharply between 0.02 and 0.05 wt% SWNT loading, indicating the formation of a conductive nanotube network; this behavior is typical of percolation. According to percolation theory, the conductance should follow the following power law close to the threshold concentration: $\sigma \sim (v-v_c)^\beta$, where v is the volume fraction of the conductive component and β is the critical exponent for the conductivity. Most

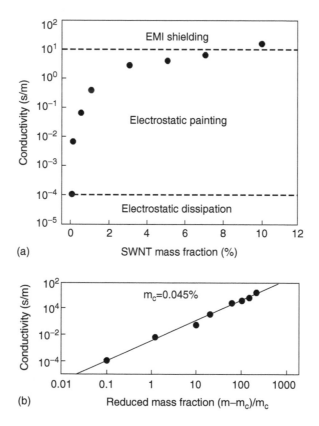

(a)

(b)

FIGURE 6.10   (a) Electrical conductivity of SWNT/PS nanocomposites as a function of nanotube loading, showing a typical percolation behavior. Dashed lines represent the lower limits of electrical conductivity required for the specified applications, (b) Electrical conductivity as a function of reduced mass fraction of nanotubes, showing a threshold of 0.045 wt%. (Adapted from Ramasubramaniam, R. et al., *Appl. Phys. Lett.*, 83, 2928–2930, 2003.)

researchers assume that the nanotube volume fraction is comparable to the nanotube mass fraction, because the densities of nanotubes and polymers are similar. The power law expression for percolation fits the conductivity data for SWNT/PS nanocomposites giving a low percolation threshold, 0.045 wt%. Ounaies et al.[50] also observed percolation behavior for electrical conductivity in SWNT/polyimide nanocomposites with a percolation threshold of 0.1 vol%. Their current–voltage measurements exhibited nonohmic behavior, which is most likely due to a tunneling conduction mechanism. Conduction probably occurs by electron hopping between adjacent nanotubes when their separation distance is small. At concentrations greater than the percolation threshold, conductive paths are formed through the whole nanocomposite, because the distance between the conductive filler (nanotubes or nanotube bundles) is small enough to allow efficient electron hopping. Ounaies et al.[50] developed an analytical model and numerical simulation using high aspect ratio and rigid spherocylinders in a unit cube to mimic SWNT/polymer nanocomposites to aid in understanding these results. The predictions from both the analytical model and the numerical simulation were in good agreement with the experimental results.

As presented earlier, a hydrodynamic nanotube network forms upon increasing nanotube loading and leads to the nonterminal rheological behavior. Similarly, a conductive nanotube network turns polymers from insulating to conducting. Du et al.[16] have compared the hydrodynamic and the conductive nanotube networks. In their SWNT/PMMA nanocomposites, the rheological percolation threshold, 0.12 wt%, is significantly smaller than the percolation threshold for electrical conductivity, 0.39 wt%. They understand this difference in the percolation threshold in terms of the smaller nanotube–nanotube distance required for electrical conductivity as compared to that required to impede polymer mobility. Specifically, the proposed electron hopping mechanism requires tube–tube distance of less than ~5 nm, while for rheological percolation the important length scale is comparable to the size of the polymer chain, which is ~18 nm for PMMA (M = 100 kDa). Thus, they conclude that a less dense nanotube network can restrict polymer motion than can conduct electricity.

Three main factors that influence the percolation threshold for the electrical conductivity are nanotube dispersion, aspect ratio, and alignment. Barrau et al.[25] used palmitic acid as a surfactant to improve the nanotube dispersion in SWNT/epoxy nanocomposites and reduced the threshold concentration for electrical conductivity from ~0.18 to ~0.08 wt%. Bai and Allaoui[51] pretreated MWNT to alter their aspect ratios before preparing MWNT/epoxy nanocomposites and found that the threshold concentration for electrical conductivity varied from 0.5 to >4 wt% with decreasing aspect ratio. This observation is consistent with the predictions from Balberg's model.[52] As the quality of nanotube dispersion improves and the aspect ratios of nanotubes increase, lower nanotube loadings are required to increase the electrical conductivities and these loadings are smaller than the loading required, obtaining comparable conductivities by adding other conductive fillers, like carbon black and graphite.[53]

Owing to their highly anisotropic shape the alignment of nanotubes must be considered when studying the properties, including electrical conductivity of nanotube/polymer nanocomposites. Du et al.[8] found that the electrical conductivity of a 2 wt% SWNT/PMMA nanocomposite decreased significantly (from ~10^{-4} to ~10^{-10}S/cm) when the SWNT were highly aligned (FWHM=20° as measured by x-ray scattering as described above). This decrease in electrical conductivity is the result of fewer contacts between nanotubes when they are highly aligned as compared to having an isotropic orientation. In contrast, Choi et al.[9] observed that nanotube alignment increased the conductivity of a 3 wt% SWNT/epoxy nanocomposite from ~10^{-7} to ~10^{-6} S/cm. Note that nanotubes in the SWNT/PMMA systems[8] were aligned by melt spinning, while they were aligned in the SWNT/epoxy[9] by magnetic force during fabrication. Although Choi et al. did not quantify the degree of alignment, it is reasonable to assume that SWNT are better aligned by the extensional flow of melt fiber spinning. More recently, we found an optimal degree of nanotube alignment that yields a maximum electrical conductivity.[63] The degree of nanotube alignment was varied by controlling the melt fiber spinning conditions and was characterized by x-ray scattering. In all cases, the SWNT/PMMA nanocomposites with isotropic nanotube orientation have greater electrical conductivity than the nanocomposites with highly aligned nanotubes. Furthermore, at low nanotube concentrations, there are intermediate levels of nanotube alignment with higher electrical conductivities than the isotropic condition. We attribute the maximum electrical conductivity at an intermediate nanotube alignment observed in the competition between the number of tube–tube contacts and the distance between these contacts.

The electrical properties of nanocomposites made from electrically conductive conjugated polymers have also been studied.[14,54] Tchmutin et al.[55] compared the electrical conductivity behavior of SWNT/polyanaline (PA) and SWNT/PP nanocomposites produced by the same method. The nanocomposites prepared with PA have a lower threshold for electrical conductivity than SWNT/PP nanocomposites, specifically ~2 vol% compared to ~ 4 vol%. They attribute this lower threshold to a "double percolation," involving both the nanotubes and the conjugated polymer matrix, where the PA becomes conductive when injected with charge carriers from the nanotubes. In SWNT/PA composites, the nanotubes are surrounded by PA domains of higher conductivity that increase the effective volume of conductive nanotubes and thereby reduce the concentrations required for percolation in these nanocomposites.

6.6   THERMAL CONDUCTIVITY AND FLAMMABILITY

Nanotube/polymer nanocomposites are expected to have superior thermal conductivity due to the exceptionally high thermal conductivities reported for nanotubes. Biercuk et al.[56] first reported the improvement in thermal conductivity for SWNT/polymer nanocomposites; their 1 wt% SWNT/epoxy nanocomposite showed a 70% increase in thermal conductivity at 40 K, increasing to 125% at room temperature. Choi et al.[9] reported a 300% increase in thermal conductivity at room temperature with 3 wt% SWNT in epoxy. The thermal conductivity of individual nanotube is approximately four orders of magnitude higher than that of typical polymers (~0.2 W/m K), so these reported enhancements are smaller than expected if one assumes perfect phonon transfer between nanotubes in the composite. A more appropriate benchmark for thermal conductivities in nanotube/polymer composites is the thermal conductivity of nanotube buckypaper (a low-density felt or mat of nanotubes typically made via filtration of nanotube suspensions) so as to account for the thermal resistance between nanotubes. Hone et al.[17] used a comparative method to measure the thermal conductivity of an unaligned SWNT buckypaper at room temperature and found it to be ~30 W/m K. In the absence of thermal resistance between nanotubes, this value would be expected to be less of the order of 10^3 W/m K. Researchers are currently exploring the extent, to which substantial interfacial thermal resistance is intrinsic to nanotubes.

A comparison of the electrical and thermal conductivity behaviors in nanotube-based nanocomposites provides some insight to heat transport. A 2 wt% SWNT/PMMA nanocomposite is well above the threshold concentration for electrical conductivity (~0.4 wt%),[16] but has the thermal conductivity of pure PMMA.[64] In other words, the electrically conductive nanotube network in the 2 wt% SWNT/PMMA is not thermally conductive. This can be explained by the different transport mechanisms for electrons and phonons. As we discussed above, an electron hopping mechanism has been applied to nanotube-based nanocomposites, which requires close proximity (but not direct contact) of the nanotubes or nanotube bundles in the composites. However, heat transport inside nanotube-based nanocomposites proceeds by phonon transfer. The nanotubes and polymer matrix are coupled only by a small number of low-frequency vibrational modes in the absence of covalent bonds at the interface. Thus, thermal energy contained in high-frequency phonon modes within the CNT must first be transferred to low frequencies through phonon–phonon couplings before being exchanged with the surrounding medium.[57] This is the origin of the high interfacial thermal resistance in nanotube/polymer composites. Huxable et al.[57] used picosecond transient absorption to measure the interface thermal conductance of CNT suspended in surfactant micelles in water, a system comparable to polymer-based composites. They estimated that the thermal resistance posed by the nanotube–polymer interface was equivalent to the resistance of a 20-nm-thick layer of polymer. This finding indicates that heat transport in a nanotube-based nanocomposite material will be limited by the exceptionally small interface thermal conductance and is in marked contrast to the findings regarding electrical conduction in these composites. Thus, the thermal conductivity of nanotube/polymer nanocomposite will be much lower than the value estimated from the intrinsic thermal conductivity of nanotubes. Improvements in thermal conductivity will require strategies for overcoming the interfacial thermal resistance, perhaps by interfacial covalent bonding.

In addition to interfacial bonding, nanotube dispersion, alignment, and aspect ratio are expected to influence the thermal conductivity of the nanocomposites. There are a limited number of reports on thermal conductivity of nanotube-based nanocomposites, due at least in part to the difficulty of the experiments. Thus, only an incomplete view of how these parameters influence thermal conductivity is available. Choi et al.[9] showed that the thermal conductivity of the magnetically aligned 3 wt% SWNT/epoxy was 10% higher than that of the unaligned nanocomposite with the same loading.

Like silica[58] and clay,[59,60] nanotubes also improve the thermal stability of polymer matrices in nanocomposites. In these nanocomposites, the filler particles absorb a disproportionate amount of thermal energy, thereby retarding the thermal degradation of the polymer. Du et al.[8] showed that the temperature of the maximum weight loss peak shifts from 300°C for pure PMMA to 370°C for the 0.5 wt% SWNT/PMMA nanocomposite, as determined by thermal gravimetric analysis. Similar results were also found using MWNT in various polymer matrices.[12,61]

Nanotubes have also proven to be effective flame-retardant additives. Kashiwagi et al.[12,31] reported that small amount of MWNT greatly improve flammability of PP, as measured both by cone calorimetery and radiant gasification in nitrogen. Mass loss rates decrease with increasing MWNT concentration (Figure 6.11a) indicating that the addition of MWNT effectively prolongs the burning of PP. The sample residue collected after nitrogen gasification (Figure 6.11b) is nearly the same shape as the original sample and is a low-density, self-supporting structure of nanotubes, a nanotube network. Kashiwagi et al.[62] attribute the improvement in flammability to nanotube networks spanning the nanocomposite that dissipate heat and thereby reduce the external radiant heat transmitted to the PP in the sample. A recent study[62] of SWNT/PMMA nanocomposites also shows similar flame retardancy effectiveness with lower concentrations of SWNT.

Kashiwagi et al.[31] compared MWNT/PP and carbon black/PP nanocomposites. With the same carbon loading, 1 wt%, carbon black is much less flame retardant than MWNT. The residue of the 1 wt% carbon black/PP nanocomposite after nitrogen gasification contains discrete aggregates and granular particles in contrast to the residue from the MWNT/PP composites having a MWNT network. The presence of the nanotube network is critical to improving the flame retardancy, because this morphology provides efficient heat dissipation. A similar network was observed in clay/PP nanocomposites with 5 wt% filler,[59] but the residue was brittle. At only 0.5 wt% loading, intact nanotube networks form in both SWNT and MWNT nanocomposites, indicating that nanotubes are more potent flame-retardant additives than nanosilicates. As with other properties of nanotube/polymer composites, the flame-retardant performance will depend on nanotube dispersion,[62] aspect ratio, and alignment.

6.7   CONCLUSIONS

Polymers are frequently modified to improve their physical properties for specific applications. Among the additives now available are CNT that provide unique opportunities to improve mechanical, electrical, and thermal properties of a variety of polymer matrices. We have summarized the current state of this field, though each month brings new publications that expand the understanding of nanotube/polymer nanocomposites. The published literature, as presented above, is only a fraction of the current activity as companies pursue research and development on nanotube/polymer materials. Upon review of the available literature the following generalizations are evident:

1. In addition to nanotube concentration, the properties of nanotube/polymer composites strongly depend on the nanotube dispersion within the matrix, the aspect ratio of the

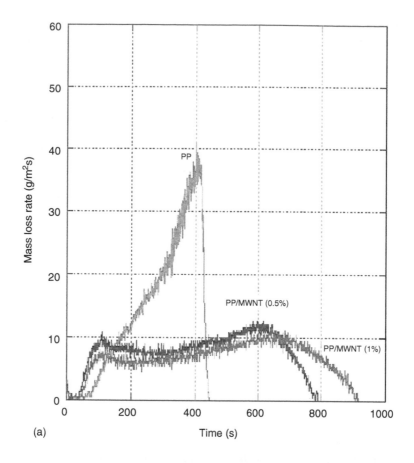

(a)

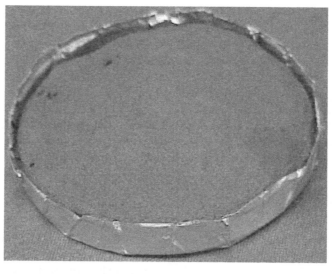

(b)

FIGURE 6.11 (a) Mass loss rate for PP and MWNT/PP nanocomposites in nitrogen gasification, showing that the addition of MWNT slows the thermal degradation of the polymer matrix; (b) highly porous residue of the 1 wt% MWNT/PP nanocomposite after nitrogen gasification suggesting the presence of a nanotube network in the nanocomposite. (Adapted from Kashiwagi et al.[31])

nanotubes and the alignment of the nanotubes within the polymer matrix. These traits, dispersion, aspect ratio, and alignment, are difficult and tedious to determine, and many publications fail even to attempt a qualitative description of these characteristics of their composites. As these traits are not independent (e.g., shorter nanotubes are easier to disperse) control experiments are challenging. Nanotube dispersion in a nanocomposite must be explored over a range of length scales from nm to mm. The aspect ratios of nanotubes depend on the synthetic route of the nanotubes with batch-to-batch variations that are considerable. In addition, postsynthesis processing to increase purity and improve dispersion can alter the distribution of aspect ratios. The high aspect ratios of nanotubes make them prone to alignment by even modest applied forces (e.g., during solvent casting gravity is sufficient to bias the spatial distribution of nanotubes to be preferentially horizontal). These nuances are important to nanotube/polymer composites and their properties and serve to caution researchers in the field.

2. The concept of networks of nanotubes provides strong correlations between nanocomposite morphology and properties. Nanotube networks are structures in which interconnected nanotubes form an open and irregular framework of nanotubes that span across the nanocomposites. Such nanotube networks are readily detected by dynamic mechanical (rheological) or electrical conductivity measurements. In fact, the detection of a nanotube network provides a reliable, though indirect and qualitative, measure of the relative nanotube dispersion and aspect ratio. For example, the presence of solid-like behavior indicates better dispersion than a comparable nanocomposite with the same nanotube loading and aspect ratio having liquid-like behavior. The presence of nanotube networks in nanocomposites provides a framework for discussing rheological, electrical, and flame-retardant properties. This framework is inappropriate for properties such as mechanical and thermal properties in which the interfacial properties of the nanotube/polymer nanocomposite are dominant.

3. Nanotubes are certainly capable of improving the physical properties of polymers. As with any filler, there will be a balance of advantages and disadvantages. The higher electrical conductivity available at low nanotube loadings will be balanced with the onset of solid-like behavior that will make the nanocomposite more difficult to process. The current cost of nanotubes is such that loadings as small as 1 wt% could be prohibitive over the next few years, except in highly selective markets. Because the cost of nanotubes will continue to drop, research with available nanotube materials should continue in earnest. At the moment, improvements in the electrical and thermal properties are perhaps the most promising, although mechanical properties of nanotube/polymer composites will benefit substantially from the emerging availability of all nanotube fibers.

In conclusion, nanotube/polymer nanocomposites have demonstrated their promise as multifunctional materials. Attention to the challenges detailed above will maximize their potential as a new class of nanocomposites for a variety of applications in the coming decades.

ACKNOWLEDGMENTS

The authors thank the Office of Naval Research (N00014-3-1-0890) for funding and Dr. Fischer, Dr. Kashiwagi, Dr. Lukes, R. Haggenmueller, and C. Guthy for helpful discussions.

REFERENCES

1. Thostenson, E.T., Ren, Z., and Chou, T.W., *Comp. Sci. Technol.*, 61, 1899–1912, 2001.
2. Kim, P., Shi, L., Majumdar, A., and McEuen, P.L., *Phys. Rev. Lett.*, 87, 21550201–21550204, 2001.
3. Maruyama, S., *Physica B*, 323, 193–195, 2002.
4. Bronilowski, M.J., Willis, P.A., Colbert, D.T., Smith, K.A., and Smalley, R. E., *J. Vac. Sci. Technol. A*, 19, 1800–1805, 2001.

5. Salvetat, J.P., Andrew, G., Briggs, D., Bonard, J.M., Bacsa, R.R, Kulik, A. J., Stockli, T., Burnham, N.A., and Forro, L., *Phys. Rev. Lett.*, 82, 944–947, 1999.
6. Haggenmueller, R., Commans, H.H., Rinzler, A.G., Fischer, J.E., and Winey, K.I., *Chem. Phys. Lett.*, 330, 219–225, 2000.
7. Benoit, J.M., Corraze, B., Lefrant, S., Blau, W., Bernier, P., and Chauvet, O., *Synth. Met.*, 121, 1215–1216, 2001.
8. Du, F., Fischer, J.E., and Winey, K.I., *J. Polym. Sci.: Part B: Polym. Phys.* 41, 3333–3338, 2003.
9. Choi, E.S., Brooks, J.S., Eaton, D.L., Al-Haik, M.S., Hussaini, M.Y., Garmestani, H., Li, D., and Dahmen, K., *J. Appl. Phys.*, 94, 6034–6039, 2003.
10. Furtado, C.A., Kin, U.J., Gutierrez, H.R., Pan, L., Dickey, E.C., and Eklund, P.C., *J. Am. Chem. Soc.*, 126, 6095–6105, 2004.
11. Qian, D., Dickey, E.C., Andrews, R., and Rantell, T., *Appl. Phys. Lett.*, 76, 2868–2870, 2000.
12. Kashiwagi, T., Grulke, E., Hilding, J., Harris, R., Awad, W., and Douglas, J., *Macromol. Rapid Commun.*, 23, 761–765, 2002.
13. Bhattacharyya, A.R., Sreekumar, T.V., Liu, T., Kumar, S., Ericson, L. M., Hauge, R.H., and Smalley, R.E., *Polymer*, 44, 2373–2377, 2003.
14. Coleman, J.N., Curran, S., Dalton, A.B., Davey, A.,P., McCarthy, B., Blau, W., and Barklie, R.C., *Phys. Rev. B*, 58, 57–60, 1998.
15. Benoit, J.M., Corraze, B., and Chauvet, O., *Phys. Rev. B*, 65, 24140501–24140504, 2002.
16. Du, F., Scogna, R.C., Zhou, W., Brand, S., Fischer, J.E., and Winey, K.I., *Macromolecules*, 37, 9048–9055, 2004.
17. Hone, J., Batlogg, B., Benes, Z., Llaguno, M.C, Nemes, N.M., Johnson, A.T., and Fischer, J.E., *Mater. Res. Soc. Symp. Proc.*, 633, A17.ll.ll–A17.ll.12, 2001.
18. Ajayan, P.M., Schadler, L.S., Giannaris, C., and Rubio, A., *Adv. Mater.*, 12, 750–753, 2000.
19. Park, C., Ounaies, Z., Watson, K.A., Crooks, R.E., Smith J., Jr. Lowther, S.E., Connell, J.W., Siochi, E.J., Harrison, J.S., and Clair, T.L.S., *Chem. Phys. Lett.*, 364, 303–308, 2002.
20. Sun, Y., Fu, K., Lin, Y., and Huang, W., *Acc. Chem. Res.*, 35, 1096–1104, 2002.
21. Chen, J., Hamon, M.A., Hu, H., Chen, Y., Rao, A.M., Eklund, P.C., and Haddon, R.C., *Science*, 282, 95–98, 1998.
22. Hill, D.E., Lin, Y., Rao, A.M., Allard, L.F., and Sun, Y., *Macromolecules*, 35, 9466–9471, 2002.
23. Lin, Y., Rao, A.M., Sadanadan, B., Kenik, E.A., and Sun, Y, *J. Phys. Chem. B*, 106, 1294–1298, 2001.
24. Islam, M.F., Rojas, E., Bergey, D.M., Johnson, A.T., and Yodh, A.G, *Nano Lett.*, 3, 269–273, 2003.
25. Barrau, S., Demont, P., and Perez, E., *Macromolecules*, 36, 9678–9680, 2003.
26. O'Connell, M.J., Boul, P., Ericson, L.M., Huffman, C., Wang, Y., Haroz, E., Kuper, C., Tour, J., Ausman, K.D., and Smalley, R.E., *Chem. Phys. Lett.*, 342, 265–271, 2001.
27. Star, A., Steuerman, D.W., Heath, J.R., and Stoddart, J.F., *Angew. Chem. Int. Ed*, 41, 2508–2512, 2002.
28. Barisci, J.N., Tahhan, M., Wallace, G.G., Badaire, S., Vaugien, T., Maugey, M., and Poulin, P., *Adv. Funct. Mater.*, 14, 133–138, 2004.
29. Kimura, T., Ago, H., Tobita, M., Ohshima, S., Kyotani, M., and Yumura, M., *Adv. Mater.*, 14, 1380–1383, 2002.
30. Walters, D.A., Casavant, M.J., Qin, X.C, Huffman, C.B., Boul, P.J., Ericson, L.M., Haroz, E.H., O'Connell, M.J., Smith, K., Colbert, D.T., and Smalley, R.E, *Chem. Phys. Lett.*, 338, 14–20, 2001.
31. Kashiwagi, T., Grulke, E., Hilding, J., Groth, K., Harris, R., Shields, J., Kathryn, B., Kharchenko, S., and Douglas, J, *Polymer*, 45, 4227–4239, 2004.
32. Jin, L., Bower, C, and Zhou, O., *Appl. Phys. Lett.*, 73, 1197–1199, 1998.
33. Safadi, B., Andrews, R., and Grulke, E.A., *J. Appl. Poly. Sci.*, 84, 2660–2669, 2002.
34. Thostenson, E.T., and Chou, T.-W., *J. Phy. D: Appl. Phy.*, 35, 77–80, 2002.
35. Haggenmueller, R., Zhou, W., Fischer, J.E., and Winey, K.I., *J. Nanosci. Nanotechnol.*, 3, 104–108, 2003.
36. Andrews, R., Jacques, D., Rao, A.M., Rantell, T., Derbyshire, F., Chen, Y., Chen, J., and Haddon, R.C., *Appl. Phys. Lett.*, 75, 1329–1331, 1999.
37. Lordi, N., and Yao, V., *J. Mater. Res.*, 15, 2770–2779, 2000.
38. Lourie, O., and Wagner, H.D., *Appl. Phys. Lett.*, 73, 3527–3529, 1998.
39. Barber, A.H., Cohen, S.R., and Wagner, H.D., *Appl. Phys. Lett.*, 82, 4140–4142, 2003.
40. Liao, K., and Li, S., *Appl. Phys. Lett.*, 79, 4225–4227, 2001.
41. Cooper, C.A., Young, R.J., and Halsall, M., *Comp. Part A: Appl. Sci. Manuf.*, 32, 401–411, 2001.

42. Fisher, F.T., Bradshaw, R.D., and Brinson, L.C., *Appl. Phys. Lett.,* 80, 4647–4649, 2002.
43. Cooper, C.A., Cohen, S.R., Barber, A.H., and Wagner, H.D., *Appl. Phys. Lett.*, 81, 3873–3875, 2002.
44. Wagner, H.D., *Chem. Phys. Lett.*, 361, 57–61, 2002.
45. Krishnamoorti, R., and Giannelis, E.P., *Macromolecules,* 30, 4097–4102, 1997.
46. Lozano, K., Bonilla-Rios, J., and Barrera, E.V., *J. Appl. Poly. Sci.,* 80, 1162–1172, 2000.
47. Cumings, J., and Zettl, A., *Science*, 602, 289–291, 2000.
48. Baughman, R.H., Zakhidov, A.A., and deHeer, W.A., *Science*, 297, 787–792, 2002.
49. Ramasubramaniam, R., Chen, J., and Liu, H. *Appl. Phys. Lett.*, 83, 2928–2930, 2003.
50. Ounaies, Z., Park, C., Wise, K.E., Siochi, E.J., and Harrison, J.S., *Comp. Sci. Technol*, 63, 1637–1646, 2003.
51. Bai, J.B., and Allaoui, A., *Comp. Part A: Appl. Sci. Manuf.*, 34, 689–694, 2003.
52. Balberg, I., and Binenbaum, N., *Phys. Rev., B*, 28, 3799–3812, 1983.
53. Lau, K., Shi, S., and Cheng, H., *Comp. Sci, Technol.*, 63, 1161–1164, 2003.
54. Kymakis, E., Alexandou, I., and Amaratunga, G.A.J., *Synthetic Metals*, 127, 59–62, 2002.
55. Tchmutin, I.A., Ponomarenko, A.T., Krinichnaya, E.P., Kozub, G.I., and Efimov, O.N., *Carbon*, 41, 1391–1395, 2003.
56. Biercuk, M.J., Llaguno, M.C., Radosavljevic, M., Hyun, J.K., Johnson, A.T., and Fischer, J.E., *Appl. Phys. Lett.,* 80, 2767–2739, 2002.
57. Huxtable, S.T., Cahill, D.G., Shenogin, S., Xue, L., Ozisik, R., Barone, P., Usrey, M., Strano, M.S., Siddons, G., Shim, M., and Keblinski, P., *Nat. Mater.*, 2, 731–734, 2003.
58. Morgan, A.B., Antonucci, J.M., VanLandingham, M.R., Harris, R.H., and Kashiwagi, T., *Polym. Mater. Sci. Eng.*, 83, 57–58, 2000.
59. Marosi, G., Marton, A., Szep, A., Csontos, I., Keszei, S., Zimonyi, E., Toth, A., Almeras, X., and Le Bras, M., *Polym. Degrad. Stabil.*, 82, 379–385, 2003.
60. Qin, H., Su, Q., Zhang, S., Zhao, B., and Yang, M., *Polymer*, 44, 7533–7538, 2003.
61. Pötschke, P., Fornes, T.D., and Paul, D.R., *Polymer,* 43, 3247–3255, 2002.
62. Kashiwagi, T., Du, F., Winey, K.I., Groth, K.M., Shields, J.R., Bellayer, S.P., Kim, H., and Douglas, J.F., *Polymer*, 46, 471–481, 2005.
63. Du, F., Fischer, J.E., and Winey, K.I., *Phys. Rev. B*, 72, 12140401–12140404, 2005.
64. Du, F., Guthy, C., Kashiwagi, T., Fischer, J.E., and Winey, K.I., In preparation.

7 One-Dimensional Semiconductor and Oxide Nanostructures

Jonathan E. Spanier
Department of Materials Science and Engineering,
Drexel University, Philadelphia, Pennsylvania

CONTENTS

Abstract 199
7.1 Introduction 200
7.2 Strategies for the Synthesis of 1-D Nanostructures 201
    7.2.1 Metal Nanoclusters: Facilitating 1-D Growth 202
    7.2.2 Laser-Assisted Metal-Catalyzed Nanowire Growth 203
    7.2.3 Metal-Catalyzed Vapor–Liquid–Solid Growth 204
    7.2.4 Vapor–Solid–Solid Growth 206
    7.2.5 Catalyst-Free Vapor-Phase Growth 207
    7.2.6 Chemical Solution-Based Growth 207
    7.2.7 Template-Assisted Growth 209
    7.2.8 Selected Other Methods 211
7.3 Hierarchal Complexity in 1-D Nanostructures 211
    7.3.1 Control of Diameter and Diameter Dispersion 211
    7.3.2 Control of Shape: Novel Topologies 212
    7.3.3 Other Binary Oxide 1-D Nanostructures 213
    7.3.4 Hierarchal 1-D Nanostructures 213
    7.3.5 Axial and Radial Modulation of Composition and Doping 215
7.4 Selected Properties and Applications 220
    7.4.1 Mechanical and Thermal Properties and Phonon Transport 220
    7.4.2 Electronic Properties of Nanowires 221
    7.4.3 Optical Properties of Nanowires 222
7.5 Concluding Remarks 225
Acknowledgments 225
References 225

ABSTRACT

Within the last 10 years, inorganic single-crystalline semiconducting nanowires (NWs), functional-oxide NWs have emerged as one of the most important nanostructured-material platforms. These nanomaterials have enabled fundamental investigations of the size and shape dependence on physical, optical, electronic, magnetic, mechanical, and functional properties. Moreover, these hierarchal nanostructures are the building blocks for the development of a range of new devices and materials applications in several fields including electronics, photonics, sensing, photovoltaics, and thermoelectric power. In this chapter, we review methods of producing these inorganic nanostructures, and highlight some of the widening array of nanowire (NW) and other one-dimensional (1-D) materials. We also discuss some of the recent progress in the chara-cterization of some of the more unique structural, mechanical, thermal, electronic, optical, and other functional properties of semiconducting and oxide NWs and related nanostructures. Finally, we highlight important application areas, and we discuss current levels of understanding with respect to modeling of the size- and shape-dependent properties of these single-component and hierarchal 1-D nanostructures.

7.1   INTRODUCTION

Fundamental understanding, control, and application of the rich variation in the properties of nanoscaled crystalline solids with size and shape remain central themes at the frontier of nanomaterials research. For example, insights into the size-induced depression of melting temperature [1], and on solid–solid phase transitions, temperatures, and pressures [2,3] gained from studies of low-dimensional materials have important implications for the development of new process routes for material synthesis and of new nanostructural topologies. The altered and often unusual behavior of electronic carriers confined to within one dimension in materials with altered or unique electronic-band structures [4] has enabled new classes of electronic logic, memory, and sensing devices. Twelve years following their introduction, the technological and commercial impact of chemically synthesized semiconducting quantum dots (QDs) [5] as "artificial atoms" continues to grow.

The technology sectors in which advances in nanomaterials research are expected to contribute significantly are diverse: drug delivery, medical therapies, medical testing and diagnostics, imaging and sensing, electronics, computing logic and architecture, telecommunications, transportation, information storage, thermal management, energy conversion and storage, and manufacturing. The transfer of nanoscience from the research laboratory to practical applications will be aided in part by (1) progress in bridging length scales between micro- or mesoscopic materials and molecular systems, and an improved understanding of the origin of size evolution of properties, (2) engineering functional and adaptable surfaces for the interfacing of inorganic-functional materials and biological systems at the nanoscale, (3) providing a high level of connectivity within nanostructured systems, and (4) providing addressability of nanoscaled components — whether by mechanical, electrical, optical, or other means. *Inorganic nanowires* and related nanostructures produced by so-called "bottom-up" synthesis methods have emerged as important building blocks within nanotechnology for addressing these challenges. With nanowires (NWs) and segmented NWs, finite-size effects can be realized in the radial and axial dimensions without the loss of connectivity. Considerable progress has been achieved in the development of (1) a large and growing library of nanowire (NW) materials systems, (2) methods that yield structures with control of diameter, length, shape, composition, and hierarchal complexity, (3) improved techniques for probing their properties, (4) effective methods of controlling their assembly, and (5) application of computational methods for predicting properties in nanostructures.

This chapter is organized into five sections. In Section 7.2, we review the methods and likely mechanisms for growth of 1-D single-crystalline inorganic nanostructures, and highlight some members of the array of 1-D nanomaterials that have been produced. In Section 7.3, we discuss recent progress in realizing more advanced nanostructures, including ongoing efforts in controlling the diameter, shape, topology, and spatial modulation of composition in axial and radial directions. We also briefly discuss branched and hierarchal nanostructures, nanobelts, nanosprings, and other nanostructures. In Section 7.4, we survey experimental work to date in characterizing selected properties that are unique to these

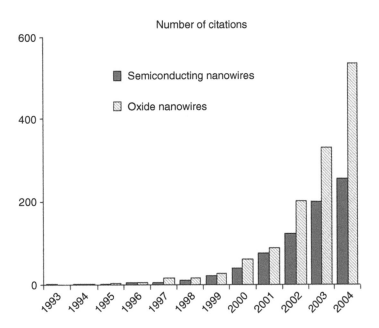

FIGURE 7.1 Annual growth in numbers of 1-D nanostructure-related publications.

1-D nanomaterials. We also identify selected applications. Finally, in Section 7.5, we conclude with a brief outlook for 1-D semiconductor and oxide-nanostructure material technology.

An important clarification is warranted: the phrase "one-dimensional" nanostructure is ambiguous; structures may be regarded as 1-D with respect to the behavior of one or more elementary excitations — e.g., electronic carriers, phonons, or photons — as distinguished from two- or zero-dimensional (0-D) structures. Specifically, this would involve the comparison of a relevant length scale, e.g., the Fermi wavelength, excitonic radius, coherence length, or scattering length, with the NW diameter. Alternatively, an even more stringent use of 1-D is reserved for cases in which fluctuations in an order parameter prevent the long-range order necessary to support the existence of a phase transition. In this chapter, the term 1-D is used to indicate that the diameters of the structures are all <100 nm, and that the length-to-diameter aspect ratio is significantly greater than unity. Given the importance of crystalline quality, for example, on electronic carrier mobility, and its effect on overall electronic device characteristics and performance, and that grain boundaries typically have a deleterious effect on many electronic, photonic, and other functional properties, we focus our discussion here primarily on *single-crystalline* NWs.

The NW research and development landscape is vast: a search of the keywords "nanowires" and "semiconductor" and "nanowires" and "oxide" produced well over 2400 publications over the last 12 years. (Included were synonyms for nanowire, e.g., nanowhiskers, nanofibers, and nanorods.) Figure 7.1 shows the rapid growth in annual number of publications for each. In early 2004, one estimate [6] placed the number of groups actively engaged in research and development of NW materials and devices at ~100. Recently, comprehensive reviews of inorganic NWs have been presented [7,8].

7.2 STRATEGIES FOR THE SYNTHESIS OF 1-D NANOSTRUCTURES

Common to the synthesis or deposition of bulk and nanocrystalline solids — whether from the liquid or vapor phase — are the concepts of nucleation and growth. Nucleation of the solid phase on a surface or in solution, by definition involves the initial formation of clusters of atoms due to supersaturation and subsequent precipitation of a selected elemental species or compound. Viewing synthesis of nanostructures as the early stage of bulk crystal growth, it should be emphasized that in contrast to 0-D (sphere-like) and 2-D growth, the formation of 1-D structures is far less likely to occur in the absence of a symmetry-breaking event, of interfacial strain between the substrate and the nuclei, or of differences among crystallographic surfaces that can lead to strongly anisotropic growth rates.

Some 1-D nanostructured materials may be grown using methods in which there is inherent growth rate anisotropy to the crystalline structure. However, for other technologically important materials (e.g., Si, GaAs) crystalline structure alone often does not typically provide the necessary driving force to overcome more energetically favorable growth modes. Thus, a central challenge in the rational synthesis of 1-D nanostructured materials is to employ growth conditions, which employ a symmetry-breaking feature. Additionally, for some applications and fundamental studies, methods that permit dislocation and twin-free structures or precise control of diameter and length, and size dispersion may be important. In some situations, methods that produce periodic or nonperiodic arrays of aligned nanostructures may be desired. Finally, the need to adapt a process for other material systems or compositions to carry out growths at lower temperatures, to have compatibility with existing processes, and adaptability for multi-component architectures, may also be important considerations.

The methods of synthesizing single-crystalline 1-D nanostructured materials from the bottom-up may be classified into those that employ inherent crystalline anisotropy, those that exploit a metal nanocluster catalyst to facilitate 1-D growth, those that rely on templates, and those that rely on kinetic control and capping agents. Specifically, kinetic control is gained by monomer concentration or by multiple surfactants, bringing about conditions that strongly favor anisotropic growth. Though not discussed here, self- or directed-assembly of 0-D nanostructures into 1-D nanostructures, and size reduction of 1-D structures are additional process routes. Alternatively, one can classify the strategies into chemical solution-based syntheses, and those via vapor condensation routes (e.g., by solid source evaporation), by metallorganic-vapor phase, by thermal means (e.g., sublimation), or by laser ablation.

7.2.1 METAL NANOCLUSTERS: FACILITATING 1-D GROWTH

Vapor–liquid–solid (VLS) growth refers to a deposition route involving the condensation of species from the vapor phase onto a miscible liquid, followed by supersaturation of a species and its subsequent precipitation, forming the solid phase. Though VLS growth is commonly used for the synthesis of NWs, an understanding of the conditions for whisker growth via VLS methods was developed several decades ago by Wagner [9]. For more than two decades, 1-D semiconductor nanostructures were realized by subtractive or top-down methods: narrow mesas were etched from thin films, or structures produced by vicinal growth on selected steps were used in device configurations in which electrostatic potential applied to electrodes was used to provide adjustable degrees of confinement of electronic carriers to narrow channels within a planar 2-D electron or hole gas. Though the top-down approaches are technologically significant and robust, the lower limits of diameter or width are often restricted by the limits imposed by lithographic patterning techniques (e.g., electron-beam lithography, typically ~20 nm), by film removal, or by direct writing techniques (e.g., focused-ion beam). Following the early work of Wagner and Ellis [9], Hiruma et al. [10] produced free-standing, single-crystalline GaAs NWs with diameters as small as 10 nm via metallorganic vapor-phase epitaxy (MOVPE) on SiO_2-patterned substrates. In 1992, the same group showed that metal nanocluster droplets selectively catalyze NW growth, demonstrating the synthesis of InAs NWs [11]. A brief review of early GaAs and InAs nanowhisker growth was also presented by these authors [12]. We note that the first synthesis of silicon nanowires (Si NWs) was reported by Yu et al. [13]; the authors used thermal evaporation, specifically the sublimation of hot-pressed targets containing Si and Fe in flowing Ar at 1200°C at ~100 torr. This relatively simple method produced NWs with diameters of ~15 nm, with lengths reaching tens of microns. However, it was the 1998 publication by Morales and Lieber [14], of a generalizable strategy for synthesizing high-quality single-crystalline NWs via laser ablation-produced nanocluster catalyst and reactant that stimulated enormous interest in the synthesis, characterization and application of semiconductor and oxide NW and other 1-D nanostructures.

Today, two of the most common methods of producing high-quality elemental and binary semiconductor NWs are condensation of laser-ablation-produced reactant and flowing of gaseous precursors (or a combination thereof), with a carrier gas in vacuum and at elevated temperature, such that selective adsorption onto substrate-bound metal nanocluster catalyst particles is promoted. In practice, metal nanoclusters can be in the form of colloidal nanocrystals (NCs) dispersed on the substrate or by deposition of an ultrathin film followed by an elevated thermal treatment in an inert atmosphere, to promote the formation of nanoscale islands. In the case of growth via laser ablation [14], nanocluster catalysts can also be formed by ablating target material that includes the catalyst metal. The requirement for the needed symmetry-breaking at the nanoscale for a given gas-phase species may be satisfied by a combination of appropriately chosen metal nanocluster species, temperature, and partial pressures. Following condensation and supersaturation of the adsorbate in the metal droplets, a nucleation event occurs, producing an interface between the precursor component and the catalyst alloy, and excess atoms from the precursor precipitate from the melt and crystallize as NW material. Under conditions that suppress the thermal decomposition of the precursor and the resulting uncatalyzed deposition, subsequent growth is strongly favored at the interface of the nanodroplet and crystal, forming the basis for the desired shape-anisotropic growth mode. In practice, the NW synthesis via metal-nanocluster catalyst routes can be achieved using some of the same methods that are used to deposit thin films: i.e., via laser ablation of a source target, via decomposition of flowing gaseous precursors (or both), by metallorganic chemical vapor deposition (MOCVD), or thermal decomposition and evaporation of a source material. In general, the selection of metal catalyst is guided by the requirement that it forms a miscible liquid phase with the NW components, and that the metal in the solid phase cannot be more thermodynamically stable than the NW material.

Specifically, the conditions that permit growth of NWs can be understood in terms of the Gibbs–Thomson effect. If the chemical potentials of the relevant bulk-solid phase, NW-solid phase, and vapor-phase precursor component are denoted by μ_{bulk}, μ_{wire}, and μ_{vapor}, the differences in chemical potential $\Delta\mu_{wire} \equiv \mu_{wire} - \mu_{vapor}$ and $\Delta\mu_{bulk} \equiv \mu_{bulk} - \mu_{vapor}$ will determine whether NW growth can ($\Delta\mu_{wire} < 0$) or cannot ($\Delta\mu_{wire} > 0$) occur. In addition, $\Delta\mu_{wire} = \Delta\mu_{bulk} + 4\Omega\alpha/d$, where Ω, α, and d denote the atomic volume, specific surface energy, and the nanocluster diameter, respectively. Two effects should be noted: (1) for a fixed negative value of $\Delta\mu_{bulk}$, there exists a minimum for d in order to satisfy the condition for growth ($\Delta\mu_{wire} < 0$), and (2) this minimum for d can be altered by increasing the partial pressure of the vapor phase during growth, resulting in a larger negative value for $\Delta\mu_{bulk}$. However, an increase in the partial pressure of the precursor gas can lead to defects in the NW product. In addition, heating of nanocluster-catalyst particles can lead to agglomeration, effectively increasing d and lowering the required minimum vapor pressure at which growth can occur.

7.2.2 LASER-ASSISTED METAL-CATALYZED NANOWIRE GROWTH

In laser-assisted metal-catalyzed nanowire growth (LCG) [14], laser ablation of a target containing the desired NW material(s) provides the source; the process is based, in part, on pulsed laser deposition (PLD), a common route for producing a wide variety of different classes of inorganic thin-film materials. The LCG method involves the flowing of an inert or carrier gas through a furnace, and the use of a laser-ablation target and of a growth substrate with metal nanocluster-catalyst particles. Alternatively, the catalyst material may be contained within the target, and may be prepared by sintering appropriate components. In general, laser wavelengths λ_e for LCG are chosen so that the optical absorption depth is shallow in order to maximize the effect of heating over a relatively small volume. Often, for growth of semiconductor NWs, an Nd:YAG ($\lambda_e = 1064$ nm, or frequency doubled or tripled) or an excimer laser (e.g., ArF, KrF, or XeF) is used. Ablation targets are typically placed at the upstream end, and in cases where heating of the target is needed, within the hot zone. Alternatively, if the required selected NW growth and furnace temperature is higher than the melting point of the target, targets are placed just outside the furnace.

The LCG method is a highly versatile route for obtaining a wide range of NW materials — many materials can be prepared via laser ablation. It is also attractive in that the laser ablation route eliminates the need for gaseous precursors, many of which are toxic and, in some cases, also pyrophoric. Many of the target materials are widely available or can be prepared, for example, by sintering of particle components. Moreover, the method can be used in conjunction with a vapor deposition process to yield NWs having solid solutions and periodic segments of different composition [15,16]. Finally, as suggested by Duan and Lieber [17], LCG methods can be pursued as a synthesis route even in the absence of detailed phase diagrams. Despite its versatility, a shortcoming of the LCG method for the growth of semiconducting NWs (when used alone) is that the targets themselves typically must contain all of the components required for the NW. This limits the ability to control the introduction of dopants or other dilute impurities during growth, or the ability to adjust the composition. Moreover, to date, precise control of diameter, diameter dispersion, and synthesis of ultranarrow (below ~10 nm) NWs has not been demonstrated with LCG.

7.2.3 Metal-Catalyzed Vapor–Liquid–Solid Growth

Chemical precursors may also be supplied in the gaseous phase for the growth of NWs via chemical vapor deposition (CVD), or VLS growth with metal nanocluster catalyst particles. Growth systems vary in their degree of complexity, but in their simplest forms consist a vacuum furnace or growth chamber at controlled and elevated temperature, into which gaseous precursors or bubbled-liquid-phase precursors and carrier gases are flowed at controlled mass flow rates.

Access to temperature–composition and temperature–pseudobinary phase diagrams can be extremely helpful (if not indispensable) in the design of a metal-catalyzed NW growth process. The conditions which favor NW growth over noncatalyzed deposition are determined from temperature–composition phase diagrams, and judicious tuning of the temperature and partial pressures through mass-flow rate(s), and background pressure is essential in achieving diameter selectivity and narrow diameter dispersion. In Figure 7.2, a representative schematic phase diagram for Ge–Au is shown; at the eutectoid temperature, the metal NC forms a miscible liquid alloy with the precursor. A basic schematic of the stages of the VLS mechanism is also shown in Figure 7.2. With sufficient partial pressure of the precursor, selection of temperatures just above the eutectoid enables precipitation from the alloy melt and thus nucleation of the NW. Wu and Yang [18] reported on *in situ* TEM studies of Ge and Si NW syntheses. The authors observed distinct nucleation and growth phases, and provided the direct evidence that for elemental NWs, the synthesis proceeds by the VLS mechanism. Though Wu and Yang observed secondary nucleation events during the earliest stages of NW growth for some of the alloy droplets, they report that sustained NW growth from a droplet generally proceeds along one direction.

The state of the art in small NW-diameter metal-catalyzed vapor-liquid-solid growth (MCG) growth is perhaps best represented in Si NWs approaching molecular scale as reported by Wu and coworkers [19] (Figure 7.3). The authors also report that large-diameter (20 to 30 nm) NWs grow principally along ⟨111⟩, whereas smallest diameter NWs (<10 nm) grow along ⟨110⟩, with an intermediate range of diameters (10 to 20 nm) exhibiting ⟨112⟩, reflecting important size-dependent differences in the surface energetics. The presence of these planes is explained by the following: during nucleation of NWs from eutectic droplets, the lowest energy growth plane is expected to be the ⟨111⟩ solid–liquid interface, leading to the initial formation of crystal planes. While ⟨111⟩ planes form initially, a ⟨110⟩ NW forms during axial elongation during unidirectional growth. When additional ⟨111⟩ nucleation sites form on the droplet, they drive growth towards a new plane with lower energy, the ⟨110⟩. The final direction ⟨112⟩ occurs as a so-called "transitional" plane between the ⟨111⟩ and ⟨110⟩ NWs, because the ⟨112⟩ plane is a step plane between ⟨111⟩ and ⟨110⟩.

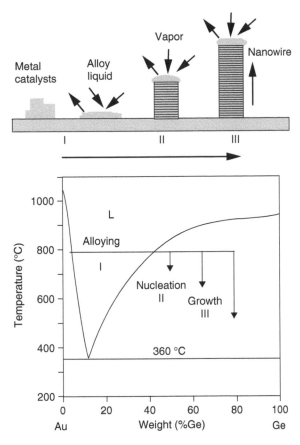

FIGURE 7.2 Upper panel: schematic of VLS growth; lower panel: binary Ge–Au composition phase diagram illustrating that control of temperature and partial pressure of Ge gaseous precursor can control different phases of NW growth. (Reproduced from Xia, Y. et al., *Adv. Mater.*, **15**, 353–389, 2003. With permission.)

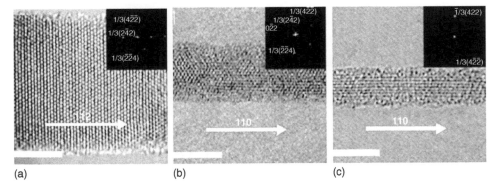

FIGURE 7.3 High-resolution transmission electron microscopy images of SiNWs with diameters of (a) 13.2, (b) 5.7, and (c) 3.5 nm. SiNWs were grown from 10-, 5-, and 2-nm gold nanoclusters catalysts, respectively. The scale bar is 5 nm in each panel. The NW growth axes are indicated by white arrows. (From Wu, Y. et al., *Nano Lett.*, **4**, 433–436, 2004. With permission.)

Some NW materials, such as those of Group III–V materials, can be prepared via MOVPE [12,20,21], in which deposition is carried out via the vapor phase, produced from bubbling of liquid-phase precursors. Like LCG, MOVPE enables access to a broader range of inorganic binary and ternary semiconductor materials and oxides than would be possible with gaseous sources alone.

Using MOVPE and chemical beam epitaxy (CBE), researchers from Lund University have demonstrated precise control of composition in the growth of III–V NWs, approaching monolayer sharpness in axial modulation [21–24]. A representative TEM image of an axially modulated InP–InA interface is shown in Figure 7.4. We also note that molecular beam epitaxy (MBE) has been used to synthesize high-quality semiconducting NWs [25]. Finally, NW synthesis by sublimation of solid sources has been used to produce Si NWs having diameters of ~15 nm [13], and, for example, nitrogen-doped GaN [26], AlN [27], and GaP [28] NWs. In what follows, we briefly review recent evidence for a different growth mechanism for the synthesis of metal nanocluster-catalyzed binary and other compound semiconductor NWs.

7.2.4   VAPOR–SOLID–SOLID GROWTH

As discussed above, the prevailing view of the mechanism of metal nanocluster-catalyzed NW growth for both elemental and binary semiconductor NWs has been the VLS route. Though this mechanism has been shown to be operative for metal-catalyzed elemental NW synthesis, such as Si and Ge, there have been a number of reports for the formation of NWs on metal nanocluster-catalyst particles at temperatures well below the eutectoid, and on nanoparticles having diameters much larger than those for which the melting temperature is expected to be significantly depressed [12,29,30]. Persson et al. [23] examined recently the question of the validity of the application of the VLS mechanism to describing binary compound NW synthesis. Using *in situ* TEM of GaAs NWs as a representative system Persson et al. demonstrated that a different mechanism is likely to be operative. The authors showed that NW nucleation and growth may proceed by a VSS route in which the NW elemental species are transported by solid-state diffusion. This work is particularly significant because an improved understanding of the growth mechanism should enable the synthesis of a broader range of combinations of NW material components. The Lund group has confirmed that this mechanism may also be operative in the InAs NW vapor-phase-assisted synthesis [31]. This mechanism may help explain the reported

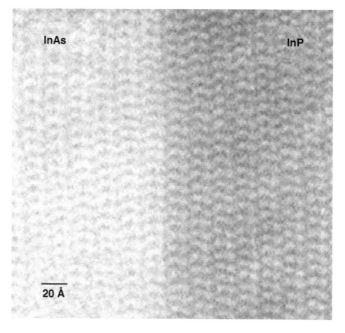

FIGURE 7.4   Scanning transmission electron microscopy (STEM) image of interface within atomically sharp axially modulated InAs-InP NW grown by CBE. The slow growth rate (~1 monolayer/sec) enables the formation of sharp interfaces. (From Samuelson, L. et al., *Physica E*, **25**, 313–318, 2004. With permission.)

growth of Si–Ge alloy NWs at moderate temperatures via MCG from the vapor phase, without laser ablation as reported by Lew et al. [32] and Redwing et al. [33].

7.2.5 CATALYST-FREE VAPOR-PHASE GROWTH

Metal-catalyzed NW growth process offers a significant degree of control in terms of diameter selectivity and diameter dispersion. However, many semiconductor and oxide NWs can also be grown from the vapor or solid phase in the absence of metal-nanocluster catalysts via a vapor–solid (VS) route. These methods include oxide-assisted (OA) growth, carbothermal growth, sublimation, PLD, and MOCVD. As the name implies, VS growth [9] involves a flowing vapor-phase precursor produced by thermal evaporation, laser ablation, MOCVD, sublimation, or other means. Nanowires of a number of binary-metal oxides have been grown epitaxially without metal nanocluster catalyst, based on the VS growth mechanism.

With respect to binary-semiconductor NWs, Stach and coworkers [34] observed directly that self-catalyzed NW growth can occur. The authors showed that by sublimation of a GaN film, liquid droplets of Ga form, providing nucleation sites for the condensation of the concurrently produced GaN vapor phase. Johnson et al. [35] have reported the synthesis of pure hexagonal wurtzite InN using indium and ammonia. In addition to syntheses of GaN [36,37] and CdS [38], Yang et al. [39] employed vapor–solid growth to produce 20 to 100-nm-diameter single-crystalline NWs of β-SiC on SiC substrates. Yang and Zhang [40] reported the synthesis of 20 to 100-nm-diameter ZnS NWs. Catalyst-free MOCVD has also been employed to produce aligned GaAs and InGaAs NWs on partially masked GaAs and InAs substrates [41]. Pan et al. [42] reported a generalizable synthesis for binary oxides, including ZnO, SnO_2, In_2O_3, and CdO. Among the other binary oxides produced by VS methods are CuO [43], WO_x [44], Ga_2O_3 [45], Al_2O_3 [46], and SiO_2 [47]. In addition, Lee et al. [48] demonstrated growth of ZnO NWs on Si(100). Zhao et al. [49] reported VS synthesis of SeO_2 NWs with diameters of 20 to 70 nm.

Using OA growth, Lee et al. [50] have shown that decomposition of an Si_xO ($x > 1$) vapor produced by either thermal evaporation or laser ablation of SiO_x targets results in a high yield of Si NWs. The authors reported that the Si nanoparticles, which precipitate as a result of this process, serve as nucleation sites for subsequent growth. A combination of the catalytic effect of the Si_xO ($x > 1$) on the NW tips, with the reduction of lateral growth by SiO_2 (formed by decomposition of SiO) on the sides of the Si NWs, produces sustained NW growth. Stacking faults along the growth direction of $\langle 112 \rangle$ and the availability of low-energy [51] surfaces promote the fast growth of NWs along $\langle 112 \rangle$, and chains of nanoparticles in the case of nuclei having nonpreferred orientations. Zhang et al. [52] have reported on a similar process developed for the synthesis of Ge NWs, producing Ge NWs sheathed by a germanium-oxide layer. The growth of GaAs [53] and the high-temperature superconductor yttrium–barium–copper oxide (YBCO) in nanorod form via laser ablation of a YBCO target [54] are also attributed to OA growth.

Carbothermal processes represent a method for producing a large variety of metal oxide, semiconducting metal nitride as well as metal carbide NWs. Carbothermal methods involve the heating of a metal oxide in the presence of activated or nanostructured carbon (e.g., carbon nanotubes [CNTs]), producing a metal suboxide and CO. The suboxide is subsequently reacted, for example, in O_2, N_2, or NH_3 to produce metal oxide or semiconducting-metal nitride NWs. This method has been utilized to produce a wide range of NW materials, including materials such as In_2O_3 [55], GeO_2 [56], and GaN [57] NWs.

7.2.6 CHEMICAL SOLUTION-BASED GROWTH

Chemical solution-based methods are attractive alternatives to vapor-phase methods: the preparations are often carried out at lower temperatures. Chemical-solution methods of promoting anisotropic crystal growth leading to 1-D nanostructures are distinguished by either kinetic control via supersaturation or by the use of coordinating ligands as capping agents. Precise control of size and shape can often be obtained. These strategies include solution–liquid–solid (SLS), solvother-mal, and hydrothermal methods.

The SLS method of NW growth is similar to VLS growth. Here, low melting-point metal nan-oclusters (e.g., Sn, In, Bi) are used, and metallorganic liquid-phase precursors are decomposed instead of vapor-phase precursors. Nanocluster-catalyst particles composed of metals with low melting temperatures are used, and the decomposition products form an alloy with the miscible NC-liquid droplet. Using this strategy, Buhro et al. [58–60] have synthesized single-crystalline III–V NWs (GaAs, InAs, and GaP) at low (~200°C) temperatures. A variation of this method using the so-called supercritical fluid-liquid solid (SFLS) method was employed by Holmes et al. [61] to produce high-quality, defect-free, single-crystalline Si NWs. The SFLS-produced Si NWs were seeded with 2.5-nm-diameter Au NCs and are nearly monodisperse (~4 to 5 nm in diameter).

Since the seminal work of Murray et al. [5] in synthesizing monodisperse, single-crystalline semiconducting II–VI NCs, significant progress has been made in extending chemical solution-based synthesis methods to produce other binary and ternary semiconducting and oxide-NC mate-rials, and to develop rational methods for producing these NCs as shape-anisotropic 1-D nanostructures in a wide range of material systems. Here we briefly outline the general progress in the area of chemical synthesis of NWs and nanorods. Briefly, NC synthesis from solution involves injection of organic precursors at elevated temperatures into a coordinating solvent. Nucleation occurs from precipitation of the desired stable compound, and adsorption and desorption of the sur-factant — typically one monolayer — enables deposition (and removal) of atoms on the nuclei. Kinetic control and focusing of size is regulated by monomer concentration: for monomer concen-tration above a solubility limit threshold, rapid growth of smaller NCs take place, and less rapid growth of larger NCs occurs. Once the monomer concentration falls below this threshold, Ostwald ripening takes place: NCs below a critical diameter become smaller and larger NCs grow. Capping by the coordinating ligands also prevents agglomeration. The capping also passivates the surface of NCs against oxidation, and improves the photoluminescence characteristics (intensity and mono-chromic emission) by reducing or eliminating unterminated-surface bonds which lead to nonradia-tive recombination of electron–hole pairs. Another key feature in attaining high-quality NCs — and highly relevant for nanorod and NW synthesis using this method — is homogeneous nucleation and its separation from the growth phase. In addition, maintaining a concentration of monomer above a given threshold ensures rapid growth of smaller NCs and less rapid growth of larger NCs, i.e., size focusing occurs.

Peng et al. [62] synthesized anisotropic NCs — quantum rods of CdSe — by taking advantage of the anisotropic growth of this wurtzite-structured material when driven by an extremely high monomer concentration. This anisotropy in growth is enhanced by the inherently anisotropic–wurtzite structure, with preferred growth along the c-axis. The authors reported the pur-poseful introduction of impurities to the coordinating ligand, trioctylphosphene oxide (TOPO). Numerous other syntheses of binary semiconductors and oxide materials have been reported, including ZnSe, Bi_2S_3, γ-Fe_2O_3, PZT, PbSe, PbS, ZnO, $BaCrO_4$, Fe_3O_4, BiS, Bi_2O_3, Bi_2Se_3, $MnFe_2O_4$, SnO_2, cerium phosphate ($CePO_4$), InAs, TiO_2, CdTe, CoTe, NiTe, $BaWO_4$, GaP, and europium oxide [7]. A comprehensive review of oxide NW and nanorod materials and their process routes is presented by Patzke et al. [63].

The synthesis of ternary oxide NWs, e.g., perovskite oxide NWs, has been particularly challeng-ing. Urban et al. [64–66] demonstrated the synthesis of single-crystalline perovskite ternary oxide ($BaTiO_3$ and $SrTiO_3$) NWs by decomposition of metal-alkoxide precursors in oleic acid, based on a synthetic strategy for producing monodisperse perovskite NCs developed by O'Brien et al. [67]. The $BaTiO_3$ NWs possess diameters as small as 5 nm and lengths exceeding 10 μm. High resolution TEM and converging-beam electron diffraction confirmed that the NWs are essentially defect-free, single-crystalline, and the c-axis is oriented along the long axis of the NW (Figure 7.5). Xu et al. [68] employed a hydrothermal synthesis in the presence of polyvinyl alcohol and polyacrylic acid, to

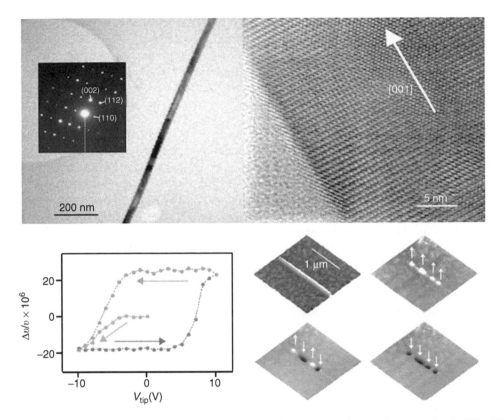

FIGURE 7.5 *Upper panel*: electron microscopy images of a representative single-crystalline BaTiO$_3$ NW. Results from high-resolution transmission electron microscopy, and converging beam electron diffraction (inset), collected at a number of locations along more than 50 NWs confirm that each NW is an essentially defect-free single-crystalline perovskite NW. Lower panel: FE hysteresis of individual FE domain oriented perpendicular to the axis of an 11-nm-diameter NW (left), and independent manipulation of FE polarization within an individual BaTiO$_3$ NW (right), measured by scanning probe microscopy. (From Urban, J.J. et al., *Adv. Mater.*, 423–426, 2002; Spanier, J.E. et al., in *Nanowire Materials*, Vol. 1, Wong, Z.L., Ed., Kluwer, Boston, MA, 2002; Yun, W.S. et al., *Nano Lett.*, **2**, 447–450, 2002. With permission.)

produce single-crystalline PbZr$_{0.52}$Ti$_{0.48}$O$_3$ NWs. Of particular significance is that the BaTiO$_3$ NWs produced by Urban et al. enabled more detailed investigations into ferroelectricity in nanostructured materials [65,66,69]. Electrostatic force microscopy of these NWs reveals that the NWs possess stable ferroelectric (FE) polarizations, which can be reproducibly reoriented perpendicular to the NW axes. As shown in Figure 7.5, these NWs exhibit FE hysteresis, and polarizations in these NWs can be independently induced and manipulated. The FE domains in these NWs represent the smallest volumes of stable FE reported to date. Subsequent studies have revealed distinct scaling behavior of the evolution of FE phase transition temperature, and further fundamental insights into the origins of FE stability in nanostructured and ultrathin-FE films. Our findings and the results of our investigations, combining density-functional theoretical simulations with experiments, will be reported in a forthcoming publication [70].

7.2.7 TEMPLATE-ASSISTED GROWTH

Another route for obtaining 1-D nanostructures involves the use of nanostructured templates, which act as a host material into or onto which NWs (or alternatively NTs) of a wide range of materials can be grown. For example, Ge NWs have been formed via deposition onto periodic corrugations

(V-grooves produced by anisotropic etching in Si[100]) [71]. Salamo and coworkers [72] have used MBE on high-index GaAs surfaces to promote the assembly of (In, Ga)As NWs and other nanostructures during deposition. Following the growth or deposition into templates, selective etchants can be used to isolate NWs. Among the advantages for this method is straightforward alignment of NWs using template material having columnar pores, and the wide flexibility in deposition methods, including metal-catalyzed growth, catalyst-free growth from the vapor or liquid phases, electrochemical methods, and solution-gelation (sol-gel). In addition, NW diameter is controlled by the template-pore diameter. A principal disadvantage, however, is that the NWs produced using template-assisted growth (TAG) are frequently not single crystalline. Moreover, the lower limit of diameter of NWs is governed by the template-pore diameter, and some of the hierarchal structures that can be produced via MCG may not be as easily prepared using TAG. In addition, aspect ratios of TAG may be limited by mass transport. Among templates, two types of nanoporous-membrane templates are frequently used: anodic aluminum oxide, (AAO) [73–76] and porous-polymer films [77]. Pathways for the production of inorganic NWs include CNTs [78] and zeolites [79]. To produce AAO templates, aluminum films are electrochemically oxidized, and local etching takes place producing pores that have local hexagonally close-packed ordering. Subsequent anodization of the film leads to columnar pores with relatively monodisperse diameters ranging from 5 nm to >100 nm, depending upon preparation conditions, e.g., electrolyte composition and concentration, and anodization voltage. Such template materials (e.g., Anodisc™) can be obtained commercially via Whatman™ or prepared. Shingubara [80] has presented a review of the use of AAO templates for the preparation of nanomaterials. Polymer films having track-etched channels are formed by irradiation of heavy ions from nuclear fissions [81]. In many cases, following deposition, the NWs can be isolated by selective etching of the template; often an appropriately selected acidic solution provides the necessary selective etching. A diverse number of materials, including metals, semiconductors, and binary oxides have been deposited in templates via vapor phase, electrochemical means, or sol-gel methods.

Lew et al. [82], Lew and Redwing [83], and Redwing et al. [33] demonstrated that high-quality, single-crystalline Si and SiGe NWs can be produced using AAO templates with a metal catalyst. In contrast to template-free MCG, in this case, the NW diameter is controlled by the template diameter. This process has several notworthy features: it provides one possible route for growth-aligned NWs; it enables investigation of growth mechanisms and kinetics in the absence of catalyst-particle size-driven considerations. The demonstration of SiGe NW growth suggests the possibility of producing more complex NW materials and compositions, such as Si_xGe_{1-x} NWs, whose composition is controlled. Template-based methods have not been limited to AAO and polymer-based templates: for example, Han et al. [78] used CNTs as templates to facilitate the syntheses of single-crystalline GaN and Si_3N_4 NWs.

Al-Mawlawi and coworkers [84] demonstrated synthesis of CdS NW arrays via electrodeposition using AAO. Electrochemical methods typically yield NWs, which are polycrystalline and not single crystalline; for many applications, however template-assisted electrochemical deposition offers a rather easy route for synthesizing 1-D structures. Often limitations due to mass transport — well understood from the electrodeposition through or in vias in top-down semiconductor device processing — can be ameliorated by use of pulsed or alternating current (AC) electrodeposition techniques. In selected cases, high-quality and even single-crystalline NWs have been produced using this technique. Another attractive (though not exclusive) feature of electrodeposition is that multiple species can be reduced and deposited at selected over potentials, enabling the formation of compositionally segmented NWs within a single electrochemical bath.

For several decades, sol-gel methods [85] have represented an attractive, low-cost method for producing both planar and nonplanar-thin films of a diverse number of materials, including binary, ternary, and quaternary oxides, and binary semiconductors. For more than a decade, sol-gel methods have been used with nanoporous templates to produce 1-D nanostructures, both NWs and NTs. In nearly all cases, these nanostructures are polycrystalline in nature with varying degrees of control of orientation or texture, though there have been some exceptions. For example, Miao et al. [86] reported the synthesis of single-crystal anataste TiO_2 NWs via an electrochemically enhanced sol-gel route. Frequently, the instability of polymeric precursors in air requires sol-gel methods to be carried out in an anhydrous environment. Here, the commercially available, air-stable polymeric precursors for sol-gel processing routes have been helpful in furthering the research in the synthesis and characterization of nanomaterials.

7.2.8 SELECTED OTHER METHODS

One particularly novel template-like method for producing a wide range of single-crystalline NW materials, developed by Belcher and coworkers [87], is through the use of a virus-based scaffold. Selected peptides that were shown to exhibit control of composition, size, and phase during nanoparticle nucleation were expressed on a highly ordered capsid of a bacteriophage, providing a template for directed synthesis of NWs. Variation of the synthetic route via substrate-specific peptides through the virus offers a new material tunability not previously available. Other routes are possible as well. For example, an additional route particularly suited for the preparation of metal oxide NWs is the thermal oxidation of metal NWs produced by template-assisted or chemical solution-based methods, and Tang et al. [88] have shown that dipole-dipole interactions can be used to facilitate the assembly of monodisperse CdTe NCs into highly crystalline NWs with uniform sizes.

7.3 HIERARCHAL COMPLEXITY IN 1-D NANOSTRUCTURES

Control of NW diameters, diameter dispersion, and crystallographic and physical orientation are important features for fundamental investigations of size-scaling effects and for enabling the use of NWs as building blocks for nanotechnology. Moreover, synthesis routes for preparing other quasi-1-D topologies may provide new insights into mechanisms of nanostructure growth and new material platforms for a range of technologies. In addition to the well-known effects of finite size on electronic-band structure, the effects of finite size on nonlinear-optical properties, electrical and thermal conductivity, ferroelectricity, ferromagnetism, superconductivity, phonons, and on the evolution of phase-transition temperatures and pressures, and susceptibility behavior, are all areas of intense investigation.

7.3.1 CONTROL OF DIAMETER AND DIAMETER DISPERSION

While several methods permit synthesis of NWs with lengths of many microns or even tens of microns, and with diameters of the order of 10 nm and higher, controlled and reproducible growth of NWs with smaller diameters (<5 nm) and with narrow diameter dispersion has been more challenging. Control of the relative growth rates in axial and radial directions is an important feature in enabling access to a range of NW architectures. As pointed out by Wagner [9] for the VLS mechanism, it is hypothesized that several factors are significant in determining the diameter of the NWs. The size(s) of the catalyst nanoclusters is important, since the eutectic-droplet size for nucleation is defined by these. Secondly, the temperature will profoundly affect the diameter (and in some cases also the shape), since the solubility of the precursor reactant in the nanocluster will alter the eutectic droplet size, which in turn will alter the NW diameter. Finally, a vapor pressure of the precursor reactant in excess of the critical value at which deposition of the precursor reactant onto the NW surface (as opposed to the eutectic droplet) takes place, will also have the effect of producing radial as well as axial NW growth.

Gudiksen and Lieber [89] and Gudiksen et al. [90] demonstrated that Si NWs with control of mean diameter and with relatively narrow size dispersions can be synthesized: 5 (4.9 ± 1.0), 10 (9.7 ± 1.5), 20 (19.8 ± 2.0), and 30 (30.0 ± 3.0) nm-diameter Au colloid was used to produce Si NWs with diameters of 6.4 ± 1.2, 12.3 ± 2.5, 20 ± 2.3, and 31.1 ± 2.7 nm, respectively. The authors point out that the dispersion in diameters for each size group is comparable to that of the catalyst particles, providing further evidence of the distinct role of nanocluster size in determining NW diameter.

Single-walled CNTs have attracted significant interest because they behave as 1-D metallic conductors or semiconductors with respect to electronic carriers. With regard to NWs, one challenge has been to produce "molecular wires" in which the diameter of the NWs is as small as possible — approaching a single chain of atoms. Such wires would find application, for instance, to the probing of size scaling and dimensionality effects in electronic transport, for extreme carrier or phonon confinement, or, for example, unique signatures in metal-insulating transitions associated with Peierls distortions in 1-D. In principle, the minimum NW radius is governed by equilibrium thermodynamics, namely that $r_{min} = 2\sigma_{LV}V_L/RT \ln \sigma$, where σ_{LV} is the liquid–vapor surface-free energy, V_L the molar volume of liquid, and σ the vapor-phase supersaturation. This suggests that nucleation and growth of ultranarrow-diameter NWs may be achieved using a higher partial pressure of precursor for the nucleation phase, followed by lower partial pressure for the growth phase. This has recently been applied by Wu et al. [19] in the VLS synthesis of molecular-scale Si NWs as shown in Figure 7.3.

In chemical solution-based NW synthesis, however, the control of diameter can be superior to those produced through VLS. For example, as of now, several investigators have applied size-selective precipitation methods originally developed by Murray et al. [5], to produce a growing number of nanostructured materials with size dispersion of the order of one monolayer. Peng et al. [62] have shown that the CdSe nanorods possess narrowsize dispersions, as demonstrated by the periodic ordering of these structures. Control of diameter and diameter dispersion has been shown in binary-oxide NW systems. For example, O'Brien and coworkers [91,92] reported on an aqueous solution-based chemical synthesis producing ZnO NWs with monodisperse-diameter NW yields, with diameters as narrow as 2 nm; the authors also reported on the measurement of strong excitonic confinement in these structures.

7.3.2 CONTROL OF SHAPE: NOVEL TOPOLOGIES

One of the significant features of inorganic-nanomaterial synthesis by surfactant-mediated chemical routes is the possibility of controlling the size and shape of nanostructures, thereby effectively manipulating their properties. Manna et al. [93] demonstrated the growth of a range of nanostructures by varying the relative concentrations of two types of surfactants–TOPO and hexylphosphonic acid (HPA) — and systematically varying the monomer concentration with time. The HPA serves to accentuate the difference between growth rates on different crystal faces. Different systematic variations yield long rods, arrow-shaped NCs, teardrop-shaped NCs, tetrapods, and dendritic tetrapods. Manna et al. have also reported [93] that differences between wurtzite and rocksalt free energies and surface energetics govern the evolution of some of these structures — particularly stacking faults in the tetrapods. They have explained the topology in terms of abrupt changes in the crystal structure, from wurtzite in the rod segments to rock salt in the sphere-like vertices. Very recent investigations by Manna and coworkers, however, have indicated that this multiphase description of the tetrapod, and the corresponding model of their formation may not be valid: rather, it seems that the tetrapod nanostructures may consist of single phase, and that it is likely that the formation of twins plays a significant role in the unique evolution of their topologies [94].

The control of the relative axial and radial growth rates is the basis for the synthesis of coaxial, or core–shell nanostructures as discussed above. In our laboratory at Drexel, we have demonstrated the growth of apex-angle-controlled crystalline Si nanocones (Si NCs) and Ge nanocones (Ge NCs) of diamond-hexagonal phase through simultaneous control of axial and radial growth rates using metal-catalyzed CVD; we have also synthesized single-crystalline Ge nanosprings and characterized its composition and crystallinity with electron microscopy and Raman scattering. As shown in Figure 7.6, the Si NCs are tapered polyhedra, possessing hexagonal cross sections. They are typically several

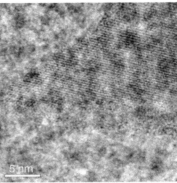

FIGURE 7.6 Silicon nanocones (Si NCs) formed by metal nanocluster-catalyzed CVD. The left panel shows an array of nanocones, which are tapered polyhedra, possessing hexagonal cross sections. These Si NCs are typically several microns in base diameter. As shown in the panel on the right, the radius of curvature at the tip is as small as 1 to 2 nm. Though not shown, control of conical angle and its dispersion has also been achieved. (From Cao, L. et al., *J. Am. Chem. Soc.* in press, 2005. With permission.)

microns in base diameter, and the radius of curvature at the tip is as small as 1 to 2 nm. Significantly, these nanocones are of the diamond-hexagonal polymorph, and the taper angles can be tuned during growth. These Si NCs may offer new opportunities as scanning probes and as central components in single-molecule sensing. Details on the synthesis and characterization of the unique structural, optical, and other properties will be presented in several forthcoming publications [95,96].

7.3.3 OTHER BINARY OXIDE 1-D NANOSTRUCTURES

An incredibly diverse range of 1-D nanostructure topologies composed of binary oxides has been developed by Z. L.Wong and coworkers at Georgia Tech, using solid-state thermal sublimation and other techniques. A representative group of these 1-D nanostructures is shown in Figure 7.7. Briefly, oxide powder is placed in the hot zone of a furnace, and ions from the vapor-phase condense onto a nearby substrate in a lower temperature zone of the furnace. The control of growth kinetics of the polar surface-dominated nanostructures as well as the use of templates and metal nanoclusters leads to the synthesis of a wide range of nanostructured topologies, including nanobelts, nanoribbons, nanorings and nanobows [97], nanosaws [98,99], and nanohelixes [100] — comprehensive reviews are presented in [101–104]. While this diverse class of nanostructures possess a common feature in that each is composed of a binary oxide with polar surfaces, we have recently identified CVD-based synthesis methods in our own laboratory to produce similar crystalline-elemental nanostructures, including Ge nanohelixes and tree-branched Ge nanostructures possessing sixfold branching symmetry [105]. Our findings of synthesis conditions producing crystalline nanosprings in a material without polar surfaces (e.g., Ge as shown in Figure 7.8) suggests that mechanisms other than polar surface-specific energies and piezoelectricity may also be operative in the formation of one or more of the binary-oxide nanostructures, this is an area of ongoing investigation by our group [105].

7.3.4 HIERARCHAL 1-D NANOSTRUCTURES

The integration of two or more semiconductor or other materials and the controlled incorporation of dopants within individual components of nanostructures are essential characteristics of the advancing development of NWs as building blocks for nanotechnology. Homo- and heterojunctions within individual NWs enable fabrication of bipolar devices within individual NWs, opening up possibilities for light-emitting diodes, photodetectors, and band gap-engineered devices within individual NWs, and even NW-based superlattices.

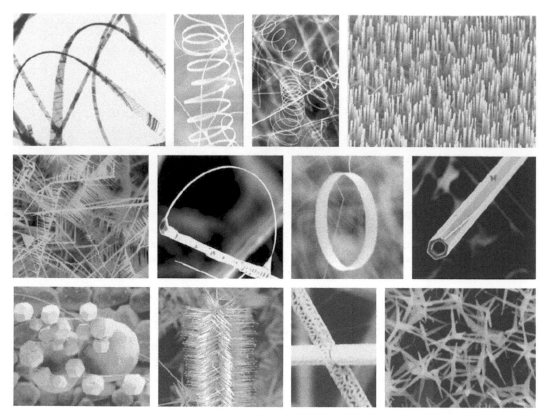

FIGURE 7.7 Representative polar surface-dominated, single-crystalline, zinc oxide 1-D nanostructures. (From Wang, Z.L., *Mater. Today*, 26–33, 2004. With permission.)

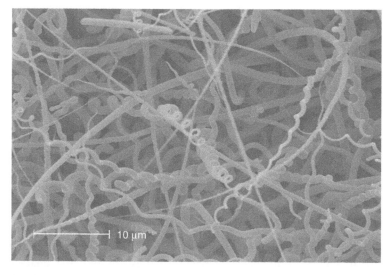

FIGURE 7.8 Scanning electron micrograph of crystalline germanium nanosprings. The formation of nanosprings having inversion symmetry may generate further debate on the formation mechanisms of these and similar structures, in both polar and nonpolar crystalline materials. (From Martin, R.R. et al., submitted. With permission.)

Some synthesis methods permit formation of hierarchal nanostructures, including NWs with segments composed of different materials. For most electronic or photonic device applications, synthesis of single-crystalline components with some degree of control of composition at interfaces is desired. To this end, synthesis of hierarchal nanostructures via metal-catalyzed and catalyst-free VLS and solid–liquid–solid methods has been demonstrated successfully in several semiconducting and functional-oxide material systems. Specifically, modulation of composition and doping in axial and radial directions in a number of material systems has been achieved.

7.3.5 AXIAL AND RADIAL MODULATION OF COMPOSITION AND DOPING

Groups at Harvard and Lund University independently reported the synthesis of semiconductor NWs with compositionally modulated segments. The Harvard group employed the LCG method in low vacuum to demonstrate the formation of GaAs/GaP heterojunctions as well as p–n junctions within Si and InP NWs [106]. The group at Lund University reported the synthesis of axially modulated III–V NWs via CBE and MOVPE [107,108]. The Lund University group also demonstrated synthesis of NW heterostructure interfaces with near-atomic sharpness.

Using a combination of MCG and LCG, Wu et al. [16] demonstrated the growth of segments of SiGe alloy within Si NWs. The SiGe segments were produced by periodic additional use of laser ablation of a Ge target at elevated furnace. As detailed by Dresselhaus et al. elsewhere in this monograph, the Seebeck coefficient, a measure of the figure of merit for thermoelectric materials, can be raised above bulk values in NWs through careful selection of diameters, compositions, carrier concentrations. Alternatively, engineered nanostructures, which combine some degree of carrier confinement, phonon scattering, and high electrical conductivity via such superlattices in nanostructures, may yet offer new routes for advanced high-performance thermoelectrics [109,110]. Advances in the science and technology of such systems would support further miniaturization of electronic devices through improved thermal management.

The formation of metal–semiconductor heterostructures within single-crystalline NWs can add to the array of device capabilities. Wu et al. [111] reported the synthesis of single-crystalline metal–semiconductor heterostructure NWs composed of segments of Si and nickel silicide. The synthesis relies on the VLS-growth method via metal-catalyzed CVD. Following the Si NW synthesis, NWs placed transverse to the Si NWs acted as a mask for evaporation of Ni. Following the segmented Ni deposition on the surface of the Si NWs, an annealing step was used to form the single-crystalline silicide segments.

Within 1 year of the first reports of axial modulation, Lauhon et al. [112] reported the synthesis of radially modulated semiconductor NWs. Such nanostructures are synthesized by adjustment of process conditions to strongly favor either selective decomposition on the metal alloy droplet or thermal decomposition and noncatalyzed deposition. In this experiment, following the synthesis of NW cores, the conditions were altered to favor noncatalyzed deposition. As reported in [112], this usually involves raising the process temperature and lowering the substrate temperature by adjusting its position in the furnace. In addition, deposited amorphous shell layers were also recrystallized with subsequent heating, producing high-quality epitaxial interfaces. Lauhon et al. [112] reported the synthesis of core–shell structures consisting of intrinsic (i) Si/p-type doped Si and of Si/Ge. They also extended the methods to produce core-multishell structures composed of i-Si/SiO$_x$/p-Si, of i-Ge/SiO$_x$/p-Ge and of Si/Ge/Si. As shown in Figures 7.9 and 7.10, the synthesis of radially modulated structures composed of p-type Si core with successive shell layers of i-Ge, SiO$_x$ and p-Ge was used to demonstrate coaxially gated NW transistors. Other researchers have demonstrated field-effect devices in other coaxial nanostructures, including the use of Ga$_2$O$_3$ as a gate oxide shell material on GaP NW cores [113].

Following the initial demonstration of coaxial semiconductor NW heterostructures, syntheses of a number of other radially modulated semiconductor and oxide heterostructures have been demonstrated, including preparations involving GaN, GaP [114,115], and CNT sheaths on InP cores

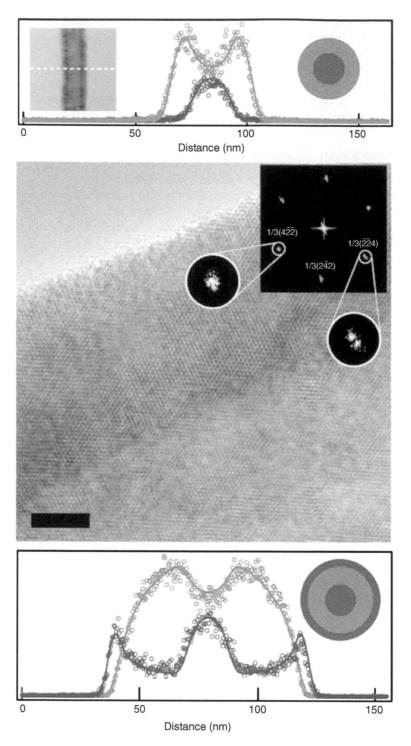

FIGURE 7.9 Radially modulated Si–Ge and Si–Ge–Si core–shell and core–multishell NWs. Upper panel: elemental mapping of the cross section, revealing a 21-nm-diameter Si core, a 10-nm-thick Ge shell and an interface of ~1 nm. Below this is shown a high-resolution TEM image of a representative NW from the same synthesis; the scale bar, 5 nm. The inset shows a 2-D Fourier transform of the real-space image showing the [111] zone axis. Bottom panel: cross-sectional elemental mapping of a double shell structure. (From Lauhon, L.J. et al., *Nature*, **420**, 57, 2002. With permission.)

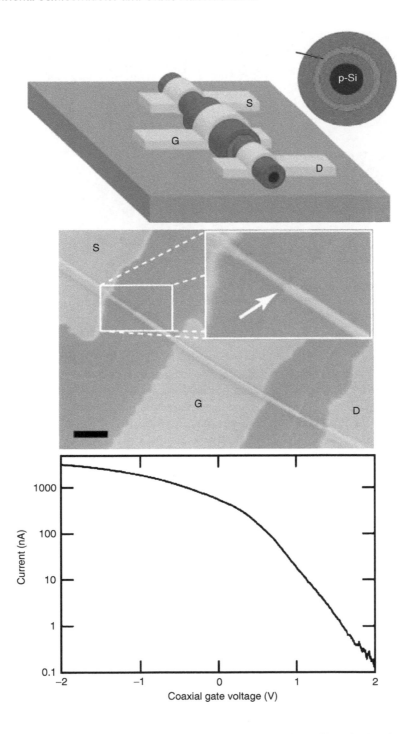

FIGURE 7.10 Coaxially gated NW transistors. At the top is the device schematic showing the transistor structure. The inset shows the cross section of the as-grown NW, starting with a p-doped Si core with subsequent layers of i-Ge (10 nm), SiO_x (4 nm), and p-Ge (5 nm). The source (S) and drain (D) electrodes are contacted to the inner i-Ge core, while the gate electrode (G) is in contact with the outer p-Ge shell, and electrically isolated from the core by the SiO_x layer. Scanning electron micrograph (SEM) of a coaxial transistor. Scale bar, 500 nm. Gate response of the coaxial transistor at $V_{SD} = 0.1$ V, showing a maximum transconductance of 1500 nA V21. Charge transfer from the p-Si core to the i-Ge shell produces a highly conductive and gateable channel. (From Lauhon, L.J. et al., *Nature*, **420**, 57, 2002. With permission.)

FIGURE 7.11 Branched and hyper-branched NWs composed of III–V NWs produced by sequential seeding of nanostructures with catalyst particles. Such nanostructures may find a number of applications, including the guiding of neuron growth. (From Dick, K.A. et al., *Nat. Mater.*, **3**, 380–384, 2004. With permission.)

[116]. One particularly important area of development has been efforts by the group of Chongwu Zhou at the University of Southern California in integrating functional oxides, including perovskites, and superconducting oxides, as epitaxial shells on binary-oxide (MgO and ZnO) NW cores. Han et al. [117] reported synthesis and characterization of such core–shell NWs with shells of FE piezoelectric single-crystalline lead zirconate titanate (PZT), and alternatively, superconducting YBCO, the colossal magnetoresistive material $La_{0.67}Ca_{0.33}MnO_3$ (LCMO) and of maghemite (Fe_3O_4). The authors also performed electronic transport characterizations of LCMO shells of various thicknesses, and demonstrated the persistence of metal–insulator transition and magnetoresistance, down to the nanoscale.

A number of other hierarchal 1-D nanostructures have been produced. For example, Zhang et al. [118] reported the synthesis of SiC/SiO$_x$ core–shell nano-helices [118], and Li et al. [79] reported on the synthesis of Mg_2Zn_{11}-MnO belt-like core–shell nano-cables via CVD. Li et al. [119] reported on Zn/ZnS nano-cabled coaxial structures. We also note that chemical solution methods have also been employed to produce core–shell nanorods. Manna et al. [120] and Mokari and Banin [121] reported the synthesis of core–shell CdSe/CdS/ZnS and CdSe/ZnS nanorods respectively, demonstrating increased quantum efficiencies in the photoluminescence associated with the passivation of the nanorod core surfaces.

Shape anisotropy and hierarchal complexity can enhance the array of interesting and technologically exciting optical properties of nanostructures. For example, the photoluminescence characteristics of GaP semiconducting NWs are observed to strong optical polarization parallel to the axis of the NWsnanowires as well as strong polarization sensitivity to the photoconductive and absorption response [122,123]. Significantly, NWs can be used as optical-resonance cavities. As reported by Yang and coworkers [124], ZnO and GaN NWs can be made to produce lasing emission despite the cross section of the NW cavity being well below the wavelength of light. The availability of core–shell architecture (e.g., GaN/Al GaN) with the shell material having a larger band gap than the core provides simultaneous confinement of the exciton and photon, providing the necessary wave-guiding function to the nanostructure. Nanowire-based lasers are key components in the development of highly adaptable and flexible nanostrucured solutions for the formation of optical networks and other device components [125].

Dick et al. [185] and others have used secondary seeding of Au nanoclusters on III–V NWs in order to produce branched and hyperbranched structures as shown in Figure 7.11. The authors

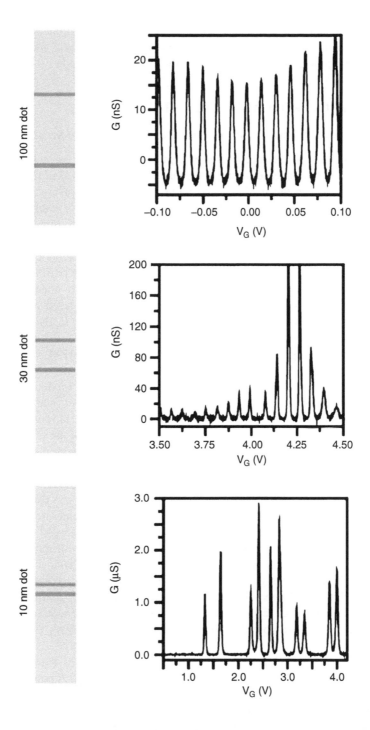

FIGURE 7.12 Effect if reduced quantum dot length. Top: gate characteristics of a single electron transistor with a 100 nm long dot. The oscillations are perfectly periodic and are visible up to 12 K. Middle: when the dot length is 30 nm, the energy level spacing at the Fermi energy is comparable to the charging energy and the Coulomb oscillations are no longer completely periodic. Bottom: a 10 nm dot results in a device depleted of electrons at zero gate voltage. By increasing the electrostatic potential electrons are added one by one. For some electron configurations the addition energy is larger, corresponding to filled electron shells. All data in this figure were recorded at 4.2 K. (From Bjork, M.T. et al., *Nano Lett.*, **4**, 1621–1625, 2002. With permission.)

demonstrated that the diameter, length, number, and chemical composition of each branch can be controlled. Such three-dimensional networked structures have the potential for a diverse number of applications, including the guiding of neuron growth.

7.4 SELECTED PROPERTIES AND APPLICATIONS

7.4.1 MECHANICAL AND THERMAL PROPERTIES AND PHONON TRANSPORT

It has long been known that the mechanical properties of materials are not scale-invariant. For example, for materials that undergo plastic deformation, the so-called Hall [127] and Petch [128] behavior — the increase in the hardness and yield strength of a polycrystalline material with reduction in grain size — involves the relative increase in the area of grain boundaries and the associated piling up of dislocations. The reduction in grain size leads to more impediments to dislocation motion and results in a scale-dependent toughness, though there exists a finite nanoscale size for which the toughness reaches a maximum. With respect to ceramic nanomaterials, relatively little has been reported on the scaling of mechanical properties (stiffness, modulus, toughness, hardness) in the semiconducting and oxide NW or nanobelt materials. Moreover, the synthesis of essentially defect-free 1-D *single-crystalline* nanostructured materials has created opportunities for investigating the theoretical strength of materials in the absence of defects. Lieber and coworkers [129] have measured the elasticity, strength, and toughness of individual SiC nanorods in fixed-free configurations via atomic force microscopy (AFM). The authors reported values in good agreement with theoretically predicted modulus for [110]-oriented SiC. Wang and coworkers [130,131] have developed a method of evaluating Young's modulus in individual NTs, NWs, nanobelts and other nanostructures. The authors relate electric-field-induced mechanical resonances in these nanostructures produced by electron beam in a TEM to values of elastic moduli. Despite this and other progress, a systematic experimental study of the scaling of modulus and other mechanical properties of single-crystalline semiconducting or oxide 1-D nanostructures has not been reported to date.

Thermal conduction via phonons is expected to be significantly reduced in narrow NWs. When the dimensions of a NW approach the mean free path of phonons, the thermal conductivity is reduced relative to bulk values by scattering of phonons by surfaces and interfaces. The miniaturization of electronic and photonic devices, e.g., the higher number density of transistors, is accompanied by new challenges in effective thermal management. A figure of merit for thermoelectric materials, however, involves the unusual combination of poor thermal conductivity with high electrical conductivity. Yang and his group at Berkeley have demonstrated that axial periodicity in Si/SiGe axial superlattice NWs may be highly effective in tuning NW thermal conductivities for thermoelectric applications; they reported the reduction in the thermoconductivity of Si/SiGe superlattice NWs by five times, compared with elemental undoped Si NWs of the same diameter [109,110].

Raman scattering is one of the most powerful experimental tools for characterizing lattice dynamics and corresponding thermodynamic properties in solids [132]. A number of investigations have been performed to characterize the effects of finite size and wire-like geometry on the confinement of phonons. Theoretical predictions indicate that for Si NWs of diameters $<\sim20$ nm, phonon dispersions become altered and group velocities are lowered relative to bulk values [133,134]. Reduction in NW diameter is accompanied by a downshifting and asymmetric broadening of the Raman line shape, and has been attributed to the confinement of optical phonon(s). In perfect bulk single crystals with no loss of translational symmetry, first-order Raman scattering is restricted to $q = 0$ phonons in accordance with momentum selection rules. The reduction of size, however, leads to the relaxation of these selection rules and necessitates an inclusion of a larger fraction of the dispersion curves away from the Γ point. A model for including and averaging $q \rightarrow 0$

phonons was developed and applied by Richter et al. [135] and by Herman and coworkers [136]. This Gaussian-correlation model has been used to describe the sampling of the phonon dispersion, and is widely used to model phonon confinement in nanostructures, including NWs. Enhancements of the model are included to incorporate known-size dispersion and the effects of strain via a Grüneisen parameter [137]. In addition, the effects of lattice strain at and near the NW surface defects, and coupling of free carriers to longitudinal optical phonons in degenerately doped semiconductors may also affect the Raman line shapes and zone-center phonon energies. In all cases reported to date, the measurements have been carried out on ensembles of NWs with varying degrees of diameter dispersion. One of the most systematic studies of the effects of finite size on Raman spectra in NWs was recently presented by Eklund and coworkers [138]. Though a number of papers have been published on Raman scattering from Si and other semiconducting and oxide NWs, a recent theoretical study of Si NWs [139] predicts the presence of Raman-active low-frequency breathing modes. This prediction and an experimental confirmation of these modes may be helpful in providing another means of characterizing the diameters of NWs. In future, as the use of hierarchal nanostructures becomes more prevalent in nanoelectronics and nanophotonics, Raman scattering will continue to play a unique role in the evaluation of 1-D nanostructure materials and devices, including, for example, crystal structure and quality, interfacial strain, thermal management, and strongly correlated electron behavior.

7.4.2 ELECTRONIC PROPERTIES OF NANOWIRES

Since the invention of modulation doping and the higher electron mobility transistors by Stormer et al. [140], precise control of the composition in semiconductors remains a critical component to 2-D electronic and photonic devices. In a seminal 1980 publication, Sasaki [141] pointed out that the restriction of electronic carriers to 1-D from 2-D or 3-D would result in significantly reduced carrier-scattering rates, owing to the reduction in the possible k-space points accessible to carriers. In general, electronic-carrier mobility in real systems can be affected by the scattering of carriers in a number of ways: scattering by other carriers, by surfaces, by interfacial roughness, by acoustic phonons, optical phonons, impurities, and by plasmons. Significant theoretical and experimental work involving electronic transport in CNTs has helped in distinguishing ballistic and diffusive modes of transport. However, for free-standing semiconductor NWs, experimental and theoretical consensus of carrier-scattering mechanisms that are most significant in single- and multicomponent coaxial semiconductor NWs is less clear. In some cases, carrier mobilities in Si NW-field effect transistors and transconductance values have been reported to exceed those associated with conventional Si planar technology devices [142].

Even if the mobility of carriers in semiconductor NWs in a given device is lower than that in the corrsponding bulk material, there are a number of other possible advantages to 1-D transistor devices over their 2-D counterparts. With respect to electronic properties, NWs represent an important link between bulk and molecular materials. The electronic properties of many bulk semiconductors, including oxide semiconductors, are well known, and the electronic-band structures have, in many cases, been modeled in great detail. Systematic control of NW diameter enables systematic investigation of the effects of dimensionality on electronic transport. In general, the diameter below which electronic transport is significantly altered is related to the degree of confinement of carriers and excitons, the Fermi wavelength and Coulomb interactions, although NWs with larger diameters still possess significant surface-to-volume ratios that can affect electronic transport via surface scattering. Semiconductor NWs, like CNTs, are currently being investigated for elements in nanoscale spintronic devices by introducing dilute concentrations of magnetic dopants. Several groups have reported successful doping of semiconductor NWs with manganese [143–146], using different methods. There are numerous reports on the field emission properties of transition metal oxide nanowires and semiconductor nanocones.

There have been a number of simulations reported of the electronic transport and properties of semiconductor NWs. The strategies can be classified into effective mass and $k \cdot p$ methods [147–150], tight-binding theory [151–153], and pseudopotential approaches [154–157]. In the past, the sizes of unit cells required to describe NWs, using first-principles density functional theoretical (DFT) methods, were too computationally expensive to pursue; this has changed of late. Musin et al. [158] have contributed a (DFT) study of the structural and electronic properties of [111]-oriented Si-Ge core–shell interface in very small-diameter (1 to 2 nm) NWs. Significantly, the authors found positive and negative deviations from Vegard's law for compressively strained Ge and tensile-strained Si cores respectively. They also found that the direct-to-indirect transition for the fundamental band gap is found in Ge-core/Si-shell NWs, while Si-core/Ge-shell NWs preserve a direct energy gap almost over the whole compositional range. We note an earlier study by Zhao et al. [159] in which the authors reported density-functional theoretical simulations of electronic-band structure and optical properties of Si NWs, predicting that NWs having a principal axis along ⟨110⟩ possess a direct gap at the Γ point for wires modeled up to 4.2 nm in diameter. These findings hold significant promise for the potential application of such Group IV single component and coaxial NW heterostructures for a broader range of photonic applications.

The formation of addressable 0-D semiconductor structures is important for fundamental studies of electronic transport phenomenon, and for future applications in single-electron transistors, quantum computing, and metrology. Axial modulation of the composition of NWs may be used to produce precisely located 0-D and 1-D segments within NWs (Figure 7.12) as shown by Samuelson and coworkers [108,160,161]. The authors reported the fabrication of single-electron transistors and resonant-tunneling diodes by producing heterostructural interfaces separated by as few as several monolayers or as many as large as hundreds of nanometers. In addition, the authors have recently used these so-called 1-D/0-D/1-D structures to fabricate nanowire-based, single-electron memory devices [162]. Superlattices in NWs based on these structures may enable extension of the quantum cascade laser as pioneered by Capasso [163]. Addressability of individual NWs for electronic and optical devices is an important consideration for practical applications: among the developments in this area, Zhong et al. [164] reported on the use of NW-crossbar arrays as address decoders.

Electronic transport in semiconductor-NW devices has been shown to be a highly sensitive means of selective binding and detection of molecules on NW surfaces functionalized with receptors. Nanowire-based detection of a number of analytes, including the prostate-specific antigen (pSA) [165], label-free [166], single virus detection [167] has been demonstrated. NW arrays with selectively functionalized NW surfaces represent one of the frontiers in high-sensitivity detection of a broad range of chemical and biological molecules and other species. An overview of this work has been presented in a recent review [168].

Among the most significant recent achievements pertaining to electronic transport are reports of coherent transport in molecular scale semiconductor NWs [169] and the report of formation of a 1-D hole gas in a Ge/Si NW heterostructure [170]. The authors reported room-temperature ballistic hole transport in semiconductor NWs with electrically transparent contacts, opening important further possibilities for new fundamental studies in low-dimensional, strongly correlated carrier gases and high-speed, high-performance nanoelectronic devices based on semiconductor NWs [171]. Recently, proximal superconductivity was reported in semiconductor NWs, in which the electronic transport characteristics of InAs NWs with closely spaced aluminum electrodes were measured at temperatures below the superconducting transition temperature of Al. The NW-electrode interfaces act as mesoscopic Josephson junctions with electrically tunable coupling [172].

7.4.3 Optical Properties of Nanowires

In NWs, like their 0-D nanostructure and 2-D ultrathin film counterparts, confinement of electronic carriers leads to altered electronic-band structures that are manifested as changes in band-to-band transition emission and absorption energies, and changes in other optical properties according to the

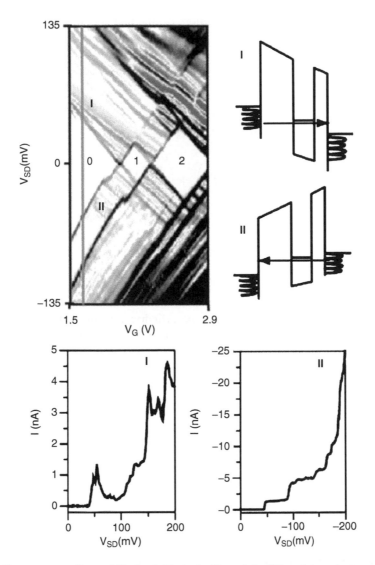

FIGURE 7.13 Resonant tunneling and Coulomb blockade. Upper left: differential conductance as a function of bias and gate voltage for zero, one, and two electrons on the dot for large bias voltages at 300 mK. The vertical line indicates the (italic: I–V) curves shown in (lower left panel) and (lower right panel). Lower left: for positive bias voltage (insert I) tunneling occurs to the dot via the thicker barrier and electrons then escape quickly to the collector. Hence, charging effects are weak. Resonances in the transport characteristics (dark diagonal lines) appear when emitter states are aligned with discrete states in the dot. Lower right: for negative bias voltages (insert II) electrons tunnel quickly to the dot through the thin barrier, whereas the tunneling rate out of the dot is lower, resulting in charging effects, i.e., a Coulomb staircase. (From Bjork, M.T. et al., *Nano Lett.*, **4**, 1621–1624, 2002. With permission.)

degree of confinement. In order to provide a basic description of the effect of the confinement of electronic carriers to within a NW, the electronic wave functions and bound-state energies are obtained by solving the Schrödinger equation under the effective mass model and with the assumption that the NW has a circular cross-section. The ground-state solution to the particle-in-a-cylinder [173] is given by the wave function

$$\Psi(r_{e,h}, z_{e,h}) = N_R N_L \cdot J_0\left(\alpha_{01} \frac{r_{e,h}}{R}\right)\sin\left(\frac{\pi z_{e,h}}{L}\right) \qquad (7.1)$$

where $J_0(\alpha_{01} \cdot r/R)$ is the zero-order Bessel function, α_{01} the first zero of the zero-order Bessel function, L the length of the cylinder, and

$$N_R = \left(2\pi \int_0^R r J_0^2\left(\alpha \frac{r_{e,h}}{R} \right) dr \right)^{-1/2} \tag{7.2}$$

and $N_L = (2/L)^{1/2}$ are normalization factors for the radial and axial portions of the wave function. The confinement energy increase in the band gap is then given by

$$\Delta E_{QC} = \frac{h^2}{2m^*}\left[\left(\frac{\alpha_{01}}{R} \right)^2 + \left(\frac{\pi}{L} \right)^2 \right] \tag{7.3}$$

where m^* is the reduced effective excitonic mass, $m^* = m_e m_h/(m_e + m_h)$, and Q is the Planck's constant.

In cases involving electronic excitations (e.g., those involving optical transitions), an additional contribution to the energy from the excitonic Coulomb interaction must also be considered. This is typically a small correction relative to the confinement energy, except for small (~10 nm) diameters. The contribution for QDs is exactly solvable, and results in the additional term ~$1.8e^2/\varepsilon R$ [174,175]. However, an analytical solution for the cylindrical geometry is not possible. Gudiksen [176] has compared the results of a numerical evaluation of the required matrix element to experimental data of size-dependent excitonic photoluminescence from III–V NWs. Wang and Gudiksen [122] also reported that the shape anisotropy of NWs produces strong effects in the optical properties, namely that the luminescence emission is strongly polarized, and that the photoconductivity of NW-based photodetectors is highly polarization sensitive. Though these features are explained on the basis of classical electromagnetic theory for an anisotripic dielectric material, these findings were the first to suggest possibilities for single-photon polarized emission and detection based on NW arrays.

One of the most exciting developments involving the optical properties of NWs is the demonstration of room-temperature lasing emission in the ultraviolet from individual NWs. As reported by Yang and coworkers, well-faceted ZnO and GaN [124] NWs produced using VLS growth methods and having diameters from 100 to 500 nm support predominantly axial Fabry–Perot waveguide modes, which are separated by $\Delta\lambda = \lambda^2/[2Ln(\lambda)]$, where L is the cavity length and $n(\lambda)$ the group index of refraction. Structures with smaller diameters are prevented from lasing due to diffraction, and photoluminescence emission is lost in the form of surrounding radiation. Lieber and coworkers [177] have reported on lasing in CdS NW optical cavities. The flexibility of the hierarchal structure of NWs has also been applied to NW lasers. Using the core–shell GaN/Al$_{0.75}$Ga$_{0.25}$N configuration, Yang and coworkers [178] were able to provide simultaneous exciton and photon confinement in NW cores having diameters as small as 5 nm — much smaller than the normal minimum to avoid the diffraction effects that prevent lasing. The Berkeley group has extended these concepts to a family of NW photonic components, and taken advantage of the ability of these NWs to sustain significant deformation without fracture in order to develop NW-based optical networks and devices [125]. In fact, the large nonlinear optical response of ZnO NWs, as reported by the Yang group [179], may enable these to be used in optical frequency conversion for nanoscale optical circuitry.

Semiconductor nanostructures with wide band gaps (e.g., TiO$_x$) have emerged as important components in a new class of photovoltaic devices. Highly conductive wide band gap transition metal oxide nanostructures have been used in dye-sensitized solar cells to provide a conductive path for the collection of photoexcited carriers. As shown recently by Law et al. [180], and Baxter and Aydil [181], dye-sensitized solar cells based on aligned ZnO NWs offer features of a large specific surface area, connectivity for carriers, and the desired optical response.

7.5 CONCLUDING REMARKS

Presently, significant new results involving the development of novel 1-D semiconductor and oxide nanostructure materials and devices, experimental characterizations, and theoretical predictions and simulations are being reported every few weeks. The wider dissemination and use of methods for controllably producing materials of selected composition and hierarchal topologies in 1-D nano-structured form as introduced here undoubtedly will enable the advancement of material technologies, and facilitate wider adoption of 1-D-based semiconductor and oxide NW technologies in commercial, consumer, and defense sectors. Significant challenges involving the controlled growth, processing, and assembly of 1-D nanostructures remain to be addressed before these materials displace the prevalence of top-down fabricated devices. However, this situation may change rather soon: for example, the recent demonstration of monolithic integration of III–V NWs with standard Si technology [182,183] represents a significant achievement in advancing the next generation of optoelectronics devices and technologies. Nevertheless, issues regarding biocompatibility of these nanomaterials, unknown environmental and health hazards pertaining to their production, handling, use and disposal, and intellectual property considerations have not been addressed in a systematic way. These and other challenges and opportunities for applying controlled growth, the flexibility of working with nontraditional substrates and platforms, and advantages in the electronic, thermal, and optical response and sensing capability of many 1-D nanostructured materials over their bulk counterparts will continue to attract researchers and technologists across many disciplines, as well as investors from the public and private sectors worldwide. It seems increasingly likely that the prospects for NWs and related nanostructures to impact the quality of our everyday lives will continue its steady rise.

ACKNOWLEDGMENTS

I am grateful to the students in my group for their dedication and hard work, and to my colleagues Bahram Nabet and Yury Gogotsi of Drexel, and Jon Lai and Brian Lim of Atomate Corporation for supportive and stimulating discussions. I would also like to acknowledge the U.S. Army Research Office under Young Investigator Award (W911NF-04-100308), the NSF under DMR-0216343, and Drexel University for support during the preparation of this manuscript.

REFERENCES

1. A.N. Goldstein, C.M. Echer, and A.P. Alivisatos, Melting in semiconductor nanocrystals, *Science*, **256**, 1425–1427 (1992).
2. S.H. Tolbert, A.B. Herhold, L.E. Brus, and A.P. Alivisatos, Pressure-induced structural transformations in Si nanocrystals: surface and shape effects, *Phys. Rev. Lett.*, **76**, 4384–4387 (1996).
3. C.-C. Chen, A.B. Herhold, C.S. Johnson, and A.P. Alivisatos, Size dependence of structural metastability in semiconductor nanocrystals, *Science*, **276**, 398–401 (1997).
4. P.L. McEuen, Nanotechnology — carbon-based electronics, *Nature*, **393**, 15 (1998).
5. C.B. Murray, D.J. Norris, and M.G. Bawendi, Synthesis and characterization of nearly monodisperse CdE (E=S, Se, Te) semiconductor nanocrystallites, *J. Am. Chem. Soc.*, **115**, 8706–8715 (1993).
6. Ivan Amato, 10 Emerging Technologies that Will Change Your World, *MIT Technology Review*, **107**, 38–40 (2004).
7. C.N.R. Rao, F.L. Deepak, G. Gundiah, and A. Govindaraj, Inorganic nanowires, *Prog. Solid State Chem.*, **31**, 5–147 (2003).
8. Y. Xia, P. Yang, Y. Sun, Y. Wu, B. Mayers, B. Gates, Y. Yin, F. Kim, and H. Yan, One-dimensional nanostructures: synthesis, characterization, and applications, *Adv. Mater.*, **15**, 353–389 (2003).
9. R.S. Wagner and W.C. Ellis, Vapor liquid solid mechanism of single-crystal growth, *App. Phys. Lett.*, **4**, 89 (1964).

10. K. Hiruma, T. Katsuyama, K. Ogawa, M. Koguchi, H. Kakibayashi, and G.P. Morgan, Quantum size microcrystals grown using organometallic vapor phase epitaxy, *Appl. Phys. Lett.*, **59**, 431 (1991).

11. M. Yazawa, M. Koguchi, A. Muto, M. Ozawa, and K. Hiruma, Effect of one monolayer of surface gold atoms on the epitaxial growth of InAs nanowhiskers, *Appl. Phys. Lett.*, **61**, 2051–2053 (1992).

12. K. Hiruma, M. Yazawa, T. Katsuyama, K. Ogawa, K. Haraguchi, M. Koguchi, and H. Kakibayashi, Growth and optical properties of nanometer-scale GaAs and InAs whiskers, *J. Appl. Phys.*, **77**, 447–462 (1995).

13. D.P. Yu, Z.G. Bai, Y. Ding, Q.L. Hang, H.Z. Zhang, J.J. Wang, Y.H. Zou, W. Qian, G.C. Xiong, H.T. Zhou, and S.Q. Feng, Nanoscale silicon wires synthesized using simple physical evaporation, *Appl. Phys. Lett.*, **72**, 3458–3460 (1998).

14. A.M. Morales and C.M. Lieber, A laser ablation method for the synthesis of crystalline semiconductor nanowires, *Science*, **279**, 208–211 (1998).

15. M.S. Gudiksen, L.J. Lauhon, J. Wang, D. Smith, and C.M. Lieber, Growth of nanowire superlattice structures for nanoscale photonics and electronics, *Nature*, **415**, 617–620 (2002).

16. Y. Wu, R. Fan, and P. Yang, Block-by-block growth of single-crystalline Si/SiGe superlattice nanowires, *Nano Lett.*, **2**, 83–86 (2002).

17. X. Duan and C.M. Lieber, *Adv. Mater.*, **12**, 298 (2001).

18. Y. Wu and P. Yang, Direct observation of vapor-liquid-solid nanowire growth, *J. Am. Chem. Soc.*, **123**, 3165–3166 (2001).

19. Y. Wu, Y. Cui, L. Huynh, C.J. Barrelet, D.C. Bell, and C.M. Lieber, Controlled growth and structures of molecular-scale silicon nanowires, *Nano Lett.*, **4**, 433–436 (2004).

20. K. Hiruma, M. Yazawa, K. Haraguchi, K. Ogawa, T. Katsuyama, M. Koguchi, and H. Kakibayashi, *J. Appl. Phys.*, **74**, 3162 (1993).

21. M. Borgstrom, K. Deppert, L. Samuelson, and W. Seifert, *J. Cryst. Growth*, **260**, 18 (2004).

22. B.J. Ohlsson, M.T. Bjork, M.H. Magnusson, K. Deppert, L. Samuelson, and L.R. Wallenberg, *Appl. Phys. Lett.*, **79**, 3335 (2001).

23. A.I. Persson, M.W. Larsson, S. Stenstrom, B.J. Ohlsson, L. Samuelson, and L.R. Wallenberg, Solid-phase diffusion mechanism for GaAs nanowire growth, *Nature Mater.*, **3**, 677–681 (2004).

24. P.J. Poole, J. Lefebvre, and J. Fraser, Spatially controlled, nanoparticle-free growth of InP nanowires, *Appl. Phys. Lett.*, **83**, 2055–2057 (2003).

25. Z.H. Wu, M. Sun, X.Y. Mei, and H.E. Ruda, Growth and photoluminescence characteristics of AlGaAs nanowires, *Appl. Phys. Lett.*, **85**, 657–659 (2004).

26. J.Y. Li, Z.Y. Qiao, X.L. Chen, Y.G. Cao, Y.C. Lan, and C.Y. Wang, Morphologies of GaN one-dimensional materials, *Appl. Phys. A (Mater. Sci. Proc.)* **A71**, 587–588 (2000).

27. H. Chen, X.K. Lu, S.Q. Zhou, X.H. Hao, and Z.X. Wang, Fabrication and characteristics of AlN nanowires, *Modern Phys. Lett. B*, **15**, 1455–1458 (2001).

28. H.W. Seo, S.Y. Bae, J. Park, M.-I. Kang, and S. Kim, Nitrogen-doped gallium phosphide nanowires, *Chem. Phys. Lett.*, **378**, 420–424 (2003).

29. T.I. Kamins, R.S. Williams, D.P. Basile, T. Hesjedal, and J.S. Harris, Ti-catalyzed Si nanowires by chemical vapor deposition: microscopy and growth mechanisms, *J. Appl. Phys.*, **89**, 1008–1016 (2001).

30. B.J. Ohlsson, M.T. Bjork, A.I. Persson, C. Thelander, R.L. Wallenberg, M.H. Magnusson, K. Deppert, and L. Samuelson, Growth and characterization of GaAs and InAs nano-whiskers and InAs/GaAs heterostructures, *Physica E*, **13**, 1126–1130 (2002).

31. K.A. Dick, K. Deppert, T. Mårtensson, B. Mandl, L. Samuelson, and W. Seifert, Failure of the vapor-liquid-solid mechanism in Au-assisted MOVPE growth of InAs nanowires, *Nano Lett.*, **5**, 761–674 (2005).

32. K.-K. Lew, L. Pan, E. Dickey, and J.M. Redwing, Vapor-liquid-solid growth of silicon-germanium nanowires, *Adv. Mater.*, **15**, 2073–2076 (2003).

33. J.M. Redwing, K.-K. Lew, T.E. Bogard, L. Pan, E.C. Dickey, A.H. Carim, Y. Wang, M.A. Cabassi, and T.S. Mayer, Synthesis and properties of Si and SiGe/Si nanowires, *Proc. SPIE — Int. Soc. Opt. Eng.*, **5361**, 52–59 (2004).

34. E.A. Stach, P.J. Pauzauskie, T. Kuykendall, J. Goldenberger, R. He, and P. Yang, Watching GaN nanowires grow, *Nano Lett.*, **3**, 867–869 (2003).

35. M.C. Johnson, C.J. Lee, E.D. Bourret-Courchesne, S.L. Konsek, S. Aloni, W.Q. Han, and A. Zettl, Growth and morphology of 0.80 eV photoemitting indium nitride nanowires, *Appl. Phys. Lett.*, **85**, 5670–5672 (2004).

36. S.M. Zhou, Y.S. Feng, and L.D. Zhang, A physical evaporation synthetic route to large-scale GaN nanowires and their dielectric properties, *Chem. Phys. Lett.*, **369**, 610–614 (2003).

37. H.Y. Peng, N. Wang, X.T. Zhou, Y.F. Zheng, C.S. Lee, and S.T. Lee, Control of growth orientation of GaN nanowires, *Chem. Phys. Lett.*, **359**, 241–245 (2002).

38. C. Ye, G. Meng, Y. Wang, Z. Jiang, and L. Zhang, On the growth of CdS nanowires by the evaporation of CdS nanopowders, *J. Phys. Chem. B*, **106**, 10338–10341 (2002).

39. W. Yang, H. Araki, Q. Hu, N. Ishikawa, H. Suzuki, and T. Noda, In situ growth of SiC nanowires on RS-SiC substrate(s), *J. Cryst. Growth*, **264**, 278–283 (2004).

40. Y. Yang and W. Zhang, Preparation and photoluminescence of zinc sulfide nanowires, *Mater. Lett.*, **58**, 3836–3838 (2004).

41. J. Motohisa, J. Noborisaka, J. Takeda, M. Inari, and T. Fukui, Catalyst-free selective-area MOVPE of semiconductor nanowires on (111) B oriented substrates, *J. Crystl. Growth*, **272**, 180–185 (2004).

42. Z.W. Pan, Z.R. Dai, and Z.L. Wang, *Science*, **291**, 1947 (2001).

43. L.S. Huang, S.G. Yang, T. Li, B.X. Gu, Y.W. Du, Y.N. Lu, and S.Z. Shi, Preparation of large-scale cupric oxide nanowires by thermal evaporation method, *J. Cryst. Growth*, **260**, 130–135 (2004).

44. G. Gu, B. Zheng, W.Q. Han, S. Roth, and J. Liu, Tungsten oxide nanowires on tungsten substrates, *Nano Lett.*, **2**, 849–851 (2002).

45. H.Z. Zhang, Y.C. Kong, Y.Z. Wang, X. Du, Z.G. Bai, J.J. Wang, D.P. Yu, Y. Ding, Q.L. Hang, and S.Q. Feng, Ga_2O_3 nanowires prepared by physical evaporation, *Solid State Commn.*, **109**, 677–682 (1999).

46. X.S. Peng, L.D. Zhang, G.W. Meng, X.F. Wang, Y.W. Wang, C.Z. Wang, and G.S. Wu, Photoluminescence and infrared properties of α-Al_2O_3 nanowires and nanobelts, *J. Phys. Chem. B*, **106**, 11163–11167 (2002).

47. Y. Zhang, N. Wang, S. Gao, T. He, S. Miao, J. Liu, J. Zhu, and X. Zhang, A simple method to synthesize Si_3N_4 and SiO_2 nanowires from Si or Si/SiO_2 mixture, *J. Cryst. Growth*, **233**, 803–808 (2001).

48. J.-S. Lee, M.-I. Kang, S. Kim, M.-S. Lee, and Y.-K. Lee, Growth of zinc oxide nanowires by thermal evaporation on vicinal Si(100) substrate, *J. Cryst. Growth*, **249**, 201–207 (2003).

49. Q. Zhao, H.Z. Zhang, B. Xiang, X.H. Luo, X.C. Sun, and D.P. Yu, Fabrication and microstructure analysis of SeO_2 nanowires, *Appl. Phys. A (Mater. Sci. Proc.)*, **A79**, 2033–2036 (2004).

50. S.T. Lee, Y.F. Zhang, N. Wang, Y.H. Tang, I. Bello, C.S. Lee, and Y.W. Chung, Semiconductor nanowires from oxides, *J. Mater. Res.*, **14**, 4503–4507 (1999).

51. H. Alouach and G.J. Mankey, Epitaxial growth of copper nanowire arrays grown on H-terminated Si(110) using glancing-angle deposition, *J. Mater. Res.*, **19**, 3620–3625 (2004).

52. Y.F. Zhang, Y.H. Tang, N. Want, C.S. Lee, I. Bello, and S.T. Lee, Germanium nanowires sheathed with an oxide layer, *Phys. Rev. B*, **61**, 4518–4521 (2000).

53. W.S. Shi, Y.F. Zheng, N. Wang, C.S. Lee, and S.T. Lee, Oxide-assisted growth and optical characterization of gallium-arsenide nanowires, *Appl. Phys. Lett.*, **78**, 3304–3306 (2001).

54. Y.F. Zhang, Y.H. Tang, X.F. Duan, Y. Zhang, C.S. Lee, N. Wang, I. Bello, and S.T. Lee, Yttrium-barium-copper-oxygen nanorods synthesized by laser ablation, *Chem. Phys. Lett.* **323**, 180–184 (2000).

55. X.C. Wu, J.M. Hong, Z.J. Han, and Y.R. Tao, Fabrication and photoluminescence characteristics of single crystalline In_2O_3 nanowires, *Chem. Phys. Lett.*, **373**, 28–32 (2003).

56. X.C. Wu, W.H. Song, B. Zhao, Y.P. Sun, and J.J. Du, Preparation and photoluminescence properties of crystalline GeO_2 nanowires, *Chem. Phys. Lett.*, **349**, 210–214 (2001).

57. L.D. Zhang, G.W. Meng, and F. Phillipp, Synthesis and characterization of nanowires and nanocables, *Mater. Sci. Eng. A (Struct. Mater.: Prop., Microstruct. Proc.)*, **A286**, 34–38 (2000).

58. T.J. Trentler, K.M. Hickman, S.C. Goel, A.M. Viano, P.C. Gibbons, and W.E. Buhro, Solution-liquid-solid growth of crystalline III–V semiconductors: an analogy to vapor-liquid-solid growth, *Science*, **270**, 1791–1794 (1995).

59. W.E. Buhro, K.M. Hickman, and T.J. Trentler, Turning down the heat on semiconductor growth: solution-chemical synthesis and the solution-liquid-solid mechanism, *Adv. Mater.*, **8**, 685–688 (1996).

60. T.J. Trentler, S.C. Goel, K.M. Hickman, A.M. Viano, M.Y. Chiang, A.M. Beatty, P.C. Gibbons, and W.E. Buhro, Solution-liquid-solid growth of indium phosphide fibers from organometallic precursors: elucidation of molecular and nonmolecular components of the pathway, *J. Am. Chem. Soc.* **119**, 2172–2181 (1997).

61. J.D. Holmes, K.P. Johnston, R.C. Doty, and B.A. Korgel, Control of thickness and orientation of solution-grown silicon nanowires, *Science*, **287**, 1471–1473 (2000).

62. X. Peng, L. Manna, W. Yang, J. Wickham, E. Scher, A. Kadavanich, and A.P. Alivisatos, Shape control of CdSe nanocrystals, *Nature*, **404**, 59–61 (2000).

63. G.R. Patzke, F. Krumeich, and R. Nesper, Oxide nanotubes and nanorods — anisotropic modules for a future nanotechnology, *Angew. Chem. Ind. Ed.*, **41**, 2446–2461 (2002).

64. J.J. Urban, W.S. Yun, Q. Gu, and H. Park, Synthesis of single-crystalline perovskite nanorods composed of barium titanate and strontium titanate, *J. Am. Chem. Soc.*, **124**, 1186–1187 (2002).

65. J.J. Urban, J.E. Spanier, L. Ouyang, W.S. Yun, and H. Park, Single-crystalline barium titanate nanowires, *Adv. Mater.*, 423–426 (2002).

66. J.E. Spanier, J.J. Urban, L. Ouyang, W.S. Yun, and H. Park, in *Nanowire Materials*, Vol. 1, Z.L. Wong, Ed., Kluwer, Boston, MA, 2002.

67. S. O'Brien, L. Brus, and C.B. Murray, Synthesis of monodisperse nanoparticles of barium titanate: toward a generalized strategy of oxide nanoparticle synthesis, *J. Am. Chem. Soc.*, **123**, 12085–12086 (2001).

68. G. Xu, Z. Ren, P. Du, W. Weng, G. Shen, and G. Han, Polymer-assisted hydrothermal synthesis of single-crystalline tetragonal perovskite $PbZr_{0.52}Ti_{0.48}O_3$ nanowires, *Adv. Mater.*, **17**, 907–910 (2005).

69. W.S. Yun, J.J. Urban, Q. Gu, and H. Park, Ferroelectric properties of individual barium titanate nanowires investigated by scanned probe microscopy, *Nano Lett.*, **2**, 447–450 (2002).

70. J.E. Spanier, A. Kolpak, I. Grinberg, J. Urban, L. Ouyang, W.S. Yun, A.M. Rappe, and H. Park, Adsorbate-induced stabilization of ferroelectricity in perovskite nanowires, submitted (2005).

71. T. Muller, K.-H. Heinig, and B. Schmidt, Formation of Ge nanowires in oxidized silicon V-grooves by ion beam synthesis, *Nucl. Instr. Methods Phys. Res., B*, **175–177**, 468–473 (2001).

72. S. Seydmohamadi, Z.M. Wang, and G.J. Salamo, Self assembled (In,Ga) as quantum structures on GaAs (411) A, *J. Cryst. Growth*, **269**, 257–261 (2004).

73. C.R. Martin, Nanomaterials: a membrane-based synthetic approach, *Science*, **266**, 1961–1966 (1994).

74. D. Routkevitch, A.A. Tager, J. Haruyama, D. Almawlawi, M. Moskovits, and J.M. Xu, Nonlithographic nano-wire arrays: fabrication, physics, and device applications, *IEEE Trans. Electron Devices*, **43**, 1646–1658 (1996).

75. J.P. O'Sullivan and G.C. Wood, The morphology and mechanism of porous anodic films on aluminum, *Proc. Roy. Soc. Lond. A.*, **317**, 511–543 (1970).

76. H. Masuda and K. Fukuda, Ordered metal nanohole arrays made by a two-step replication of honeycomb structures of anodic alumina, *Science*, **268**, 1466–1468 (1995).

77. V.M. Cepak, J.C. Hulteen, G. Che, K.B. Jirage, B.B. Lakshmi, E.R. Fisher, and C.R. Martin, Fabrication and characterization of concentric-tubular composite micro- and nanostructures using the template-synthesis method, *J. Mater. Res.*, **13**, 3070–3080 (1998).

78. W. Han, S. Fan, Q. Li, and Y. Hu, Synthesis of gallium nitride nanorods through a carbon nanotube-confined reaction, *Science*, **277**, 1287–1289 (1997).

79. C.P. Lee, X.H. Sun, N.B. Wong, C.S. Lee, S.T. Lee, and B.K. Teo, Ultrafine and uniform silicon nanowires grown with zeolites, *Chem. Phys. Lett.*, **365**, 22–26 (2002).

80. S. Shingubara, Fabrication of nanomaterials using porous alumina templates, *J. Nanopart. Res.*, **5**, 17–30 (2003).

81. R.L. Fleicher, P.B. Price, and R.M. Walker, *Nuclear Tracks in Solids*, University of California Press, Berkeley, CA (1975).

82. K.-K. Lew, C. Reuther, A.H. Carim, J.M. Redwing, and B.R. Martin, Template-directed vapor-liquid-solid growth of silicon nanowires, *J. Vac. Sci. Technol. B*, **20**, 389–392 (2002).

83. K.-K. Lew and J.M. Redwing, Growth characteristics of silicon nanowires synthesised by vapor-liquid-solid growth in nanoporous alumina templates, *J. Cryst. Growth*, **254**, 14–22 (2003).

84. D. Al-Mawlawi, C.Z. Liu, and M. Moskovits, Nanowires formed in anodic oxide nanotemplates, *J. Mater. Res.*, **9**, 1014–1018 (1994).

85. J.D. MacKenzie, Sol-gel research-achievements since 1981 and prospects for the future, *J. Sol-Gel Sci. Technol.*, **26**, 23–27 (2003).

86. Z. Miao, D. Xu, J. Ouyang, G. Guo, X. Zhao, and Y. Tang, Electrochemically induced sol-gel preparation of single-crystalline TiO_2 nanowires, *Nano Lett.*, **2**, 717–720 (2002).

87. C. Mao, D.J. Solis, B.D. Reiss, S.T. Kottmann, R.Y. Sweeney, A. Hayhurst, B. Gerogiou, B. Iverson, and A.M. Belcher, Virus-based toolkit for the directed synthesis of magnetic and semiconducting nanowires, *Science*, **303**, 213–217 (2004).

88. Z. Tang, N.A. Kotov, and M. Giersig, Spontaneous organization of single CdTe nanoparticles into luminescent nanowires, *Science*, **279**, 237–240 (2002).

89. M.S. Gudiksen and C.M. Lieber, Diameter-selective synthesis of semiconductor nanowires, *J. Am. Chem. Soc.*, **122**, 8801–8802 (2000).
90. M.S. Gudiksen, J. Wang, and C.M. Lieber, Synthetic control of the diameter and length of single crystal semiconductor nanowires, *J. Phys. Chem. B*, **105**, 4062–4064 (2001).
91. M. Yin, Y. Gu, I.L. Kuskovsky, T. Andelman, Y. Zhu, G.F. Neumark, and S. O'Brien, Zinc oxide quantum rods, *J. Am. Chem. Soc.*, **126**, 6206–6207 (2004).
92. Y. Gu, I.L. Kuskovsky, M. Yin, S. O'Brien, and G.F. Neumark, Quantum confinement in ZnO nanorods, *Appl. Phys. Lett.*, **85**, 3833–3835 (2004).
93. L. Manna, E.C. Sher, and A.P. Alivisatos, Synthesis of soluble and processable rod-, arrow-, teardrop- and tetrapod-shaped CdSe nanocrystals, *J. Am. Chem. Soc.*, **122**, 12700–12706 (2000).
94. L. Manna, private communication (2005).
95. L. Cao, L. Laim, C. Ni, B. Nabet, and J.E. Spanier, Diamond-hexagonal semiconductor nanocones with controllable apex angle, *J. Am. Chem. Soc.*, in press (2005).
96. L. Cao and J.E. Spanier, submitted (2005).
97. W.L. Hughes and Z.L. Wang, Formation of piezoelectric single-crystal nanorings and nanobows, *J. Am. Ceram. Soc.*, **126**, 6703–6709 (2004).
98. C. Ma, Y. Ding, D. Moore, X.D. Wang, and Z.L. Wang, Single-crystal CdSe nanosaws, *J. Am. Ceram. Soc.*, **126**, 708–709 (2004).
99. Y. Ding, C. Ma, and Z.L. Wong, Self-catalysis and phase transformation in the formation of CdSe nanosaws, *Adv. Mater.*, **16**, 1740–1743 (2004).
100. R. Yang, Y. Ding, and Z.L. Wang, Deformation free single crystal nanohelixes of polar nanowires, *Nano Lett.*, **4**, 1309–1312 (2004).
101. Z.L. Wang, X.Y. Kong, Y. Ding, P. Gao, W.L. Hughes, R. Yang, and Y. Zhang, Semiconducting and piezoelectric oxide nanostructures induced by polar surfaces, *Adv. Funct. Mater.*, **14**, 943–956 (2004).
102. Z.L. Wang, *Mater. Today*, **7**, 26–33 (2004).
103. Z.L. Wang, Functional oxide nanobelts — materials, properties and potential applications in nanosystems and biotechnology, *Ann. Rev. Phys. Chem.*, **55**, 159–196 (2004).
104. G.-C. Yi, C. Wang, and W.I. Park, ZnO nanorods: synthesis, characterization and applications, *Semicond. Sci. Technol.*, **20**, 22–34 (2005).
105. R.R. Martin, L. Cao, S. Nonnenmann, and J.E. Spanier, submitted (2005).
106. M.S. Gudiksen, L.J. Lauhon, J. Wang, D.C. Smith, and C.M. Lieber, Growth of nanowire superlattice structures for nanoscale photonics and electronics, *Nature*, **415**, 617–620 (2002).
107. M.T. Bjork, B.J. Ohlsson, T. Sass, A.I. Persson, C. Thelander, M.H. Magnusson, K. Deppert, L. Wallenberg, and L. Samuelson, One-dimensional heterostructures in semiconductor nanowhiskers, *Appl. Phys. Lett.*, **80**, 1058–1060 (2002).
108. M.T. Bjork, B.J. Ohlsson, T. Sass, A.I. Persson, C. Thelander, M.H. Magnusson, K. Deppert, L.R. Wallenberg, and L. Samuelson, One-dimensional steeplechase for electrons realized, *Nano Lett.*, **2**, 87–89 (2002).
109. D. Li, Y. Wu, P. Kim, L. Shi, P. Yang, and A. Majumdar, Thermal conductivity of individual silicon nanowires, *Appl. Phys. Lett.*, **83**, 2934–2936 (2003).
110. D. Li, Y. Wu, R. Fan, P. Yang, and A. Majumdar, Thermal conductivity of Si/SiGe superlattice nanowires, *Appl. Phys. Lett.*, **83**, 3186–3188 (2003).
111. Y. Wu, J. Uang, C. Yang, W. Lu, and M.L. Charles, Single-crystal metallic nanowires and metal/semiconductor nanowire heterostructures, *Nature*, **430**, 61–65 (2004).
112. L.J. Lauhon, M.S. Gudiksen, D. Wang, and C.M. Lieber, Epitaxial core-shell and core-multishell nanowire heterostructures, *Nature*, **420**, 57 (2002).
113. B.-K. Jim, J.-J. Kim, J.-O. Lee, K.-J. Kong, H.J. Seo, and C.J. Lee, Top-gated field-effect transistor and rectifying diode operation of core-shell structured GaP nanowire devices, *Phys. Rev. B Condens. Matter Mater. Phys.*, **71**, 153313–1 (2005).
114. F. Qian, Y. Li, S. Gradecak, D. Wang, C.J. Barrelet, and C.M. Lieber, Gallium nitride-based nanowire radial heterostructures for nanophotonics, *Nano Lett.*, **4**, 1975–1979 (2004).
115. H.-M. Lin, Y.-L. Chen, J. Yang, Y.-C. Liu, K.-M. Yin, J.-J. Kai, F.-R. Chen, L.-C. Chen, Y.-F. Chen, C.-C. Chen, Synthesis and characterization of core-shell GaP@GaN and GaN@GaP nanowires, *Nano Lett.*, **3**, 537–541 (2003).
116. L.W. Yin, Y. Bando, Y.C. Zhu, and M.S. Li, Controlled carbon nanotube sheathing on ultrafine InP nanowires, *Appl. Phys. Lett.*, **84**, 5314–5316 (2004).

117. S. Han, C. Li, Z. Liu, D. Zhang, W. Jin, X. Liu, T. Tang, and C. Zhou, Transition metal oxide core-shell nanowires: generic synthesis and transport studies, *Nano Lett.*, **4**, 1241–1246 (2004).

118. H.-F. Zhang, C.-M. Wang, and L.-S. Wang, Helical crystalline SiC/SiO$_2$ core-shell nanowires, *Nano Lett.*, **2**, 941–944 (2002).

119. Y.B. Li, Y. Bando, and D. Golberg, Mg$_2$Zn$_{11}$-MgO belt-like nanocables, *Chem. Phys. Lett.*, **375**, 102–105 (2003).

120. L. Manna, E.C. Scher, L.-S. Li, and A.P. Alivisatos, Epitaxial growth and photochemical annealing of graded CdS/ZnS shells on colloidal CdSe nanorods, *J. Am. Chem. Soc.*, **124**, 7136–7145 (2002).

121. T. Mokari and U. Banin, Synthesis and properties of CdSe/ZnS core/shell nanorods, *Chem. Mater.*, **15**, 3955–3960 (2003).

122. J. Wang, M.S. Gudiksen, X. Duan, Y. Cui, and C.M. Lieber, Highly polarised photoluminescence and photodetection from single indium phosphide nanowires, *Science*, **293**, 1455–1457 (2001).

123. J. Wang, M. Gudiksen, X.F. Duan, Y. Cui, and C.M. Lieber, Highly polarized photoluminescence and polarization-sensitive photodetectors from single indium phosphide nanowires, *Science*, **293**, 1455 (2001).

124. J.C. Johnson, H.-J. Choi, K.P. Knutsen, R.D. Schaller, P. Yang, and R.J. Saykally, Single gallium nitride nanowire lasers, *Nat. Mater.*, **1**, 106–110 (2002).

125. M. Law, D.J. Sirbuly, J.C. Johnson, J. Goldberger, R.J. Saykally, and P. Yang, Nanoribbon waveguides for subwavelength photonics integration, *Science*, **305**, 1269 (2004).

126. J. Wang, E. Polizzi, and M. Lundstrom, paper presented at the *IEEE International Electron Devices Meeting*, Washington DC, 2003.

127. E.O. Hall, *Proc. Phys. Soc. London B*, **64**, 747 (1951).

128. N.J. Petch, *J. Iron Steel Inst.*, **174**, 25 (1953).

129. E.W. Wong, P.E. Sheehan, and C.M. Lieber, Nanobeam mechanics: elasticity, strength and toughness of nanorods and nanotubes, *Science*, **277**, 1971–1975 (1997).

130. Z.L. Wang, Z.R. Dai, R.P. Gao, Z.G. Bai, and J.L. Gole, Side-by-side silicon carbide-silica biaxial nanowires: synthesis, structure, and mechanical properties, *Appl. Phys. Lett.*, **77**, 3349–3351 (2000).

131. Z.L. Wang, R.P. Gao, Z.W. Pan, and Z.R. Dai, Nanoscale mechanics of nanotubes, nanowires, and nanobelts, *Adv. Eng. Mater.*, **3**, 657–661 (2001).

132. I. Herman, *Optical Diagnostics for thin Film Processing*, Elsevier/Academic Press, San Diego, 1996.

133. A. Buldum, S. Ciraci, and C.Y. Fong, Quantum heat transfer through an atomic wire, *J. Phys. Condens. Mater.*, **12**, 3349 (2000).

134. K. Schwab, E.A. Henriksen, J.M. Worlock, and M.L. Roukes, Measurement of the quantum of thermal conductance, *Nature*, **404**, 974 (2000).

135. H. Richter, Z.P. Wang, and L. Ley, The one phonon Raman scattering in microcrystalline silicon, *Solid State Commun.* **39**, 625–629 (1981).

136. Z. Sui, P.P. Leong, I.P. Herman, G.S. Higashi, and H. Tempkin, Raman analysis of light-emitting porous silicon, *Appl. Phys. Lett.*, **60**, 2086 (1992).

137. J.E. Spanier, R.D. Robinson, F. Zheng, S.W. Chan, and I.P. Herman, Size-dependent properties of CeO2-y nanoparticles as studied by Raman scattering, *Phys. Rev. B*, **64**, 245407/1–8 (2001).

138. K.W. Adu, H.R. Gutierrez, U.J. Kim, G.U. Sumanasekera, and P.C. Eklund, Confined phonons in Si nanowires, *Nano Lett.*, **5**, 409–414 (2005).

139. T. Thonhauser and G.D. Mahan, Predicted Raman spectra of Si[111] nanowires, *Phys. Rev. B: Condens. Matter*, **71**, 81307 (2005).

140. H.L. Stormer, K. Baldwin, A.C. Gossard, and W. Wiegmann, Modulation-doped field-effect transistor based on a two dimensional hole gas, *Appl. Phys. Lett.*, **44**, 1062 (1984).

141. H. Sasaki, Scattering suppression and high-mobility effect of size-quantized electrons in ultrafine semiconductor wire structures, *Jpn. J. Appl. Phys.*, **19**, L735 (1980).

142. Y. Cui, Z. Zhong, D. Wang, W.U. Wang, and C.M. Lieber, High performance silicon nanowire field effect transistors, *Nano Lett.*, **3**, 149–152 (2003).

143. F.L. Deepak, P.V. Vanitha, A. Govindaraj, and C.N.R. Rao, Photoluminescene spectra and ferromagnetic properties of GaMnN nanowires, *Chem. Phys. Lett.*, **374**, 314 (2003).

144. C. Ronning, P.X. Gao, Y. Ding, and Z.L. Wang, Manganese-doped ZnO nanobelts for spintronics, *Appl. Phys. Lett.*, **84**, 783 (2004).

145. D.S. Han, J. Park, K.W. Rhie, S. Kim, and J. Chang, Ferromagnetic Mn-doped GaN nanowires, *Appl. Phys. Lett.*, **86** (2005).

146. P.V. Radovanociv, C.J. Barrelet, S. Gradecak, F. Qian, and C.M. Lieber, General synthesis of manganese-doped II-VI and III-V semiconductor nanowires, *Nano Lett.*, **5**, 1407–1411 (2005).

147. G. Bastard, J.A. Brum, R. Ferreria, Eds., *Solid State Physics*, **44**, 1991.

148. C.R. Proetto, Self-consistent electronic structure of a cylindrical quantum wire, *Phys. Rev. B: Condens. Matter*, **45**, 11911 (1992).

149. A. Gold and A. Ghazali, Analytical results for semiconductor quantum-well wire: plasmons, shallow impurity states, and mobility, *Phys. Rev. B: Condens. Matter*, **41**, 7626–7640 (1990).

150. T. Ogawa and T. Takagahara, Optical absorption and Sommerfeld factors of one-dimensional semiconductors: an exact treatment of excitonic effects, *Phys. Rev. B (Condens. Matter)*, **44**, 8138 (1991).

151. Y. Arakawa, T. Yamauchi, and J.N. Schulman, Tight binding analysis of energy band structures in quantum wires, *Phys. Rev. B (Condens. Matter)*, **43**, 4732 (1991).

152. Y.M. Niquet, C. Delerue, G. Allan, and M. Lanoo, Method for tight-binding parametrization: application to silicon nanostructures, *Phys. Rev. B (Condens. Matter)*, **62**, 5109 (2000).

153. M.P. Persson and H.Q. Xu, Electronic structure of nanometer-scale GaAs whiskers, *Appl. Phys. Lett.*, **81**, 1309 (2002).

154. A. Franceschetti and A. Zunger, Comparison between direct pseudopotential and single-band truncated-crystal calculations, *J. Chem. Phys.*, **104**, 5572 (1996).

155. A. Franceschetti and A. Zunger, Free-standing versus AlAs-embedded GaAs quantum dots, wires and films: the emergence of a zero-confinement state, *Appl. Phys. Lett.*, **68**, 3455 (1996).

156. D.W. Wood and A. Zunger, Successes and failures of the k–p method: a direct assessment for GaAs/AlAs quantum structures, *Phys. Rev. B (Condens. Matter)*, **53**, 7949 (1996).

157. J.-B. Xia and K.W. Cheah, Quantum confinement effect in thin quantum wires, *Phys. Rev. B (Condens. Matter)*, **55**, 15688 (1997).

158. R.N. Musin and X.-Q. Wang, Structural and electronic properties of epitaxial core-shell nanowire heterostructures, *Phys. Rev. B (Condens. Matter Mater. Phys.)*, **71**, 155318/1–4 (2005).

159. X. Zhao, C.M. Wei, L. Yang, and M.Y. Chou, Quantum confinement and electronic properties of silicon nanowires, *Phys. Rev. Lett.*, **92**, 236805 (2004).

160. C. Thelander, B.J. Ohlsson, M.T. Bjork, T. Martensson, A.I. Persson, K. Deppert, M.W. Larsson, L.R. Wallenberg, and L. Samuelson, Paper presented at the *IEEE International Symposium on Compound Semiconductors*, San Diego, CA, 25–27 August 2003.

161. C. Thelander T. Martensson, M.T. Bjork, B.J. Ohlsson, M.W. Larsson, L.R. Wallenberg, and L. Samuelson, Single-electron transistors in heterostructure nanowires, *Appl. Phys. Lett.*, **83**, 2052–2054 (2003).

162. C. Thelander, H.A. Nilsson, L.E. Jensen, and L. Samuelson, Nanowire single-electron memory, *Nano. Lett.*, **5**, 635–638 (2005).

163. F. Capasso, C. Gmachl, D.L. Silvo, and A.Y. Cho, *Phys. Today*, **55**, 34 (2002).

164. Z. Zhong, D. Wang, Y. Cui, M.W. Bockrath, and C.M. Lieber, Nanowire crossbar arrays as address decoders for integrated nanosystems, *Science*, **302**, 1377–1379 (2003).

165. Y. Cui, Q. Wei, H. Park, and C.M. Lieber, Nanowire nanosensors for highly sensitive, selective and integrated detection of biological and chemical species, *Science*, **293**, 1289 (2001).

166. W.U. Wang, C. Chen, K. Lin, Y. Fang, and C.M. Lieber, Label-free detection of small molecule-protein interactions by using nanowire nanosensors, *Proc. Natl. Acad. Sci.*, **102**, 3208–3212 (2005).

167. F. Patolsky, G. Zheng, O. Hayden, M. Lakadamyali, X. Zhuang, and C.M. Lieber, Electrical detection of single viruses, *Proc. Natl. Acad. Sci.*, **101**, 14017–14022 (2004).

168. F. Patolsky and C.M. Lieber, *Mater. Today*, **8**, 20–28 (2005).

169. Z. Zhong, Y. Fang, W. Lu, and C.M. Lieber, Coherent single charge transport in molecular scale silicon nanowires, *Nano Lett.*, **5**, 1143–1146 (2005).

170. W. Lu, J. Xiang, B.P. Timko, Y. Wu, and C.M. Lieber, One-dimensional hole gas in germanium silicon nanowire heterostructures, *PNAS*, **102**, 10046–10051 (2005).

171. R.S. Friedman, M.C. McAlpine, D.S. Ricketts, D. Ham, and C.M. Lieber, High speed integrated nanowire circuits, *Nature*, **434**, 1085 (2005).

172. Y.-J. Doh, J.A.V. Dam, A.L. Roest, E.P.A.M. Bakkers, L.P. Kouwenhoven, and S. De Franceschi, Tunable supercurrent through semiconductor nanowires, *Science*, **309**, 272–275 (2005).

173. G.B. Arfken and H.J. Weber, *Mathematical Methods for Physicists*, Academic Press, San Diego, 1995.

174. L.E. Brus, Electron–electron and electron–hole interactions in small semiconductor crystallites: the size dependence of the lowest excited electronic state, *J. Chem. Phys.*, **80**, 4403 (1984).

175. M.L. Steigerwald and L.E. Brus, Semiconductor crystallites — a class of large molecules, *Acc. Chem. Res.*, **23**, 183–188 (1990).

176. M.S. Gudiksen, Semiconductor nanowires and nanowire heterostructures: development of complex building blocks for nanotechnology, Ph.D. thesis, Harvard University, Harvard, 2003.

177. R. Agarwal, C.J. Barrelet, and C.M. Lieber, Lasing in single cadmium sulfide optical cavities, *Nano Lett.*, **5**, 917–920 (2005).

178. H.-J. Choi, J.C. Johnson, R. He, S.-K. Lee, F. Kim, P. Pauzauskie, J. Goldberger, R.J. Saykally, and P. Yang, *J. Phys. Chem. B*, **107**, 8721 (2003).

179. J.C. Johnson, H. Yan, R.D. Schaller, P.B. Petersen, P. Yang, and R.J. Saykally, Near-field imaging of nonlinear optical mixing in single zinc oxide nanowires, *Nano Lett.*, **2**, 279–283 (2002).

180. M. Law, L.E. Greene, J.C. Johnson, R. Saykally, and P.D. Yang, Nanowire dye-sensitized solar cells, *Nat. Mater.*, **4**, 455–459 (2005).

181. J.B. Baxter and E.S. Aydil, Nanowire-based dye-sensitized solar cells, *Appl. Phys. Lett.*, **86** (2005).

182. E.P.A.M. Bakkers, *Nat. Mater.*, **3**, 769 (2004).

183. T. Martensson, *Nano Lett.* **4**, 1987 (2004).

184. L. Samuelson, C. Thelander, M.T. Bjork, M. Borgstrom, K. Deppert, K.A. Dick, A.E. Hansen, T. Martensson, N. Panev, A.I. Persson, W. Siefert, N. Skold, M.W. Larsson, and L.R. Wallenberg, Semiconductor nanowires for 0-D and 1-D physics and applications, *Physica E*, **25**, 313–318 (2004).

185. K.A. Dick, K. Deppert, M.W. Larsson, T. Martensson, W. Siefert, L.R. Wallenberg, and L. Samuelson, Synthesis of branched "nanotrees" by controlled seeding of multiple branching events, *Nat. Mater.*, **3**, 380–384 (2004).

186. M.T. Bjork, C. Thelander, A.E. Hansen, L.E. Jensen, M.W. Larsson, L.R. Wallenberg, and L. Samuelson, Few-electron quantum dots in nanowires, *Nano Lett.*, **4**, 1621–1625 (2002).

8 Nanofiber Technology

Frank K. Ko
Department of Materials Science and Engineering,
Drexel University, Philadelphia, Pennsylvania

CONTENTS

8.1   Introduction 233
8.2   The Electrospinning Process 234
8.3   Key Processing Parameters 235
8.4   Nanofiber Yarns and Fabrics Formation 238
8.5   Potential Applications of Electrospun Fibers 238
       8.5.1   Nanofibers for Tissue Engineering Scaffolds 239
       8.5.2   Nanofibers for Chemical/Bio Protective Membranes 239
       8.5.3   Nanocomposite Fibers for Structural Applications 242
8.6   Summary and Conclusions 243
References 244

8.1   INTRODUCTION

Nanofiber technology is a branch of nanotechnology whose primary objective is to create materials in the form of nanoscale fibers in order to achieve superior functions. The unique combination of high specific surface area, flexibility, and superior directional strength makes such fibers a preferred material form for many applications ranging from clothing to reinforcements for aerospace structures. Although the effect of fiber diameter on the performance and processibility of fibrous structures has long been recognized, the practical generation of fibers at the nanometer scale was not realized until the rediscovery and popularization of the electrospinning technology by Professor Darrell Reneker almost a decade ago [1]. The ability to create nanoscale fibers from a broad range of polymeric materials in a relatively simple manner using the electrospinning process, coupled with the rapid growth of nanotechnology in recent years have greatly accelerated the growth of nanofiber technology. Although there are several alternative methods for generating fibers in a nanometer scale, none matches the popularity of the electrospinning technology due largely to the simplicity of the electrospinning process. Electrospinning can be carried out from polymer melt or solution. A majority of the published work on electrospinning has been focused on solution-based electrospinning rather than on melt electrospinning due to higher capital investment requirements and the difficulty in producing submicron fibers by melt electrospinning. We will concentrate on solution-based electrospinning in this chapter. Specifically, after an introduction to the processing principles the relative importance of the various processing parameters in solution electrospinning is discussed. The structure and properties of the fibers produced by the electrospinning process are then examined. Recognizing the enormous increase in specific fiber surface, bioactivity, electroactivity, and enhancement of mechanical properties, numerous applications have been identified, including filtration, biomedical, energy storage, electronics, and multifunctional structural composites.

8.2 THE ELECTROSPINNING PROCESS

Electrostatic generation of ultrafine fibers or "electrospinning" has been known since the 1930s [2]. The rediscovery of this technology has stimulated numerous applications including high-performance filters [3] and as scaffolds in tissue engineering [4], which utilize the unique characteristics of surface area as high as 10^3 m²/g provided by nanofibers. In this nonmechanical, electrostatic technique, a high electric field is generated between a polymer fluid contained in a glass syringe with a capillary tip and a metallic collection target. When the voltage reaches a critical value, the electric field strength overcomes the surface tension of the deformed droplet of the suspended polymer solution formed on the tip of the syringe, and a jet is produced. The electrically charged jet undergoes a series of electrically induced bending instabilities during its passage to the fiber collection screen or drum that results in the hyperstretching of the jet. This stretching process is accompanied by the rapid evaporation of the solvent molecules that reduces the diameter of the jet, in a cone-shaped volume called the "envelope cone." The dry fibers are accumulated on the surface of the collection screen resulting in a nonwoven random fiber mesh of nano- to micron diameter fibers. The process can be adjusted to control the fiber diameter by varying the electric field strength and polymer solution concentration. By proper control of the electrodes, aligned fibers can also be produced. A schematic drawing of the electrospinning process and the random and aligned nanofibers are shown in Figure 8.1.

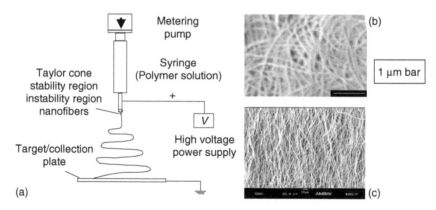

FIGURE 8.1 (a) Schematic drawing of the electrospinning process. (b) (top) random; (c) (bottom) aligned.

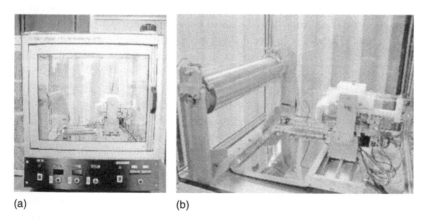

FIGURE 8.2 (a) The first commercially available nanofiber electrospinning unit (NEU) produced by the Kato-Tech Co. (b) Close-up of the NEU system including a metering pump, a syringe containing the spinning dope, the syringe needle is connected to a power supply (not shown), and a fiber collection drum connecting to the electrical ground.

Figure 8.2 shows a commercial nanofiber electrospinning unit known as NEU produced by the Kato-Tech Co. The NEU system consists of a metering pump, which controls the volume flow rate of the spinning dope in a syringe. An electrical potential is applied by a power supply between the steel syringe needle and the fiber collection ground in the form of a metallic screen or a drum. The temperature and humidity of the spinning chamber can be controlled.

8.3   KEY PROCESSING PARAMETERS

A key objective in electrospinning is to generate fibers of nanometer diameter consistently and reproducibly. Considerable effort has been devoted to understand the parameters affecting the spinnability and more specifically the diameter of the fibers resulting from the electrospinning process. Many processing parameters have been identified, which influence the spinnability and the physical properties of nanofibers. These parameters include electric field strength, polymer concentration, spinning distance, and polymer viscosity.

According to Rutledge et al. [5], the diameter of electrospun fibers is governed by the following equation:

$$d = \left[\gamma \varepsilon \frac{Q^2}{I^2} \frac{2}{\pi (2\ln\chi - 3)} \right]^{1/3} \tag{8.1}$$

where d is the fiber diameter, γ the surface tension, ε the dielectric constant, Q the flow rate, I the current carried by the fiber, and χ the ratio of the initial jet length to the nozzle diameter.

One can control the fiber diameter by adjusting the flow rate, the conductivity of the spinning line, and the spinneret diameter. Fiber diameter can be minimized by increasing the current carrying capability of the fiber through the introduction of a conductive filler such as carbon black, carbon nanotube, metallic atoms, or mixing with an inherently conductive polymer. For example, an increase of current carrying capability, I, of the fiber by 32 times will bring about 10-fold decrease in fiber diameter. Alternatively, if the current I is kept constant, one can bring about a 10-fold decrease in the fiber diameter by reducing the flow rate by 32 times. Reducing the spinneret nozzle diameter can also bring about a reduction in fiber diameter. When the value of χ is increased from 10 to 1000, the diameter of the fiber will decrease by approximately 2 times.

Experimental evidence has shown that the diameter of the electrospun fibers is influenced by molecular conformation that is related to the molecular weight and the concentration of the polymer in the spinning dope [6]. It was found that the diameter of fibers spun from dilute polymer solutions can be expressed in terms of the Berry number B, a dimensionless parameter and a product of intrinsic viscosity η, and polymer concentration, C. This relationship has been observed in a large number of polymers. An example for this relationship is illustrated using polylactic acid (PLA). The relationship between polymer concentration and fiber diameter of a PLA of different molecular weight in chloroform are shown in Figure 8.3. It can be seen that fiber diameter increases as polymer concentration increases. The rate of increase in fiber diameter is greater at higher molecular weight. Accordingly, one can tailor the fiber diameter by proper selection of polymer molecular weight and polymer concentrations.

Expressing fiber diameter as a function of B, as shown in Figure 8.4, a pattern emerges illustrating four regions of B vs. diameter relationships. At region (I), $B<1$, characterizing a very dilute polymer solution with molecular chains barely touching each other. It is almost impossible to form fibers by electrospinning of such solution, since there is not enough chain entanglement to form a continuous fiber, and the effect of surface tension will make the extended conformation of a single molecule unstable. As a result, only polymer droplets are formed. In region (II), $1<B<3$, fiber diameter increases slowly as B increases within the range ~100 to ~500 nm. In this region, the degree of molecular entanglement is just sufficient for fiber formation. The coiled macromolecules of the dissolved polymer are transformed by the elongational flow of the polymer jet into orientated

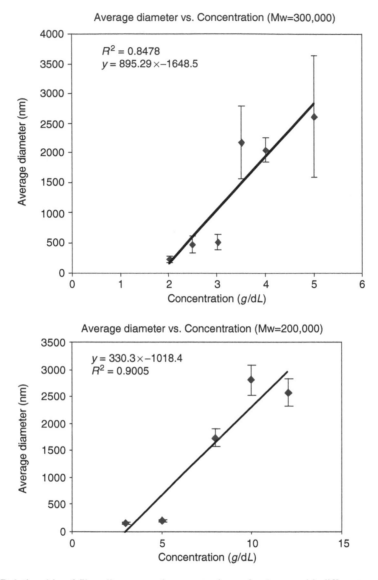

FIGURE 8.3 Relationship of fiber diameter and concentrations of polymer with different molecular weights.

molecular assemblies with some level of inter- and intramolecular entanglement. These entangled networks persist as the fiber solidifies. In this region, some bead formations are observed as a result of polymer relaxation and surface tension effect. In region (III), $3<B<4$, fiber diameter increases rapidly with B in the range from ~ 1700 to ~ 2800 nm. In this region, the molecular chain entanglement becomes more intensive, contributing to an increase in polymer viscosity. Because of the intense level of molecular entanglement, it requires a higher level of electric field strength for fiber formation by electrospinning. In region (IV), $B>4$, the fiber diameter is less dependent on B. With a high degree of inter- and intramolecular chain entanglement, other processing parameters such as electric field strength and spinning distance become dominant factors affecting fiber diameter. A schematic illustration of the four Berry regions is shown in Table 8.1, showing the corresponding relationship between fiber morphology and Berry numbers.

More recently, we further demonstrated, as shown in Figure 8.5, that the diameter, d, of electrospun fibers is related to the Berry number B for three amorphous polymers with different

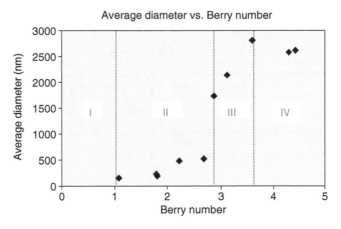

FIGURE 8.4 The relationship between Berry number and fiber diameter.

TABLE 8.1
Schematic of Polymer Chain Conformation and Fiber Morphology Corresponding to Four Regions of Berry Number

	Region I	Region II	Region III	Region IV
Berry Number	B<1	1<B<2.7	2.7B<3.6	B>3.6
Polymer chain conformation in solution				
Fiber morphology				
Average fiber diameter	(Only droplets formed)	~100–500 nm	1700–2800 nm	~2500–3000 nm

molecular weights. It was shown that the relationship of fiber diameter to Berry number can be expressed by $d = a\,B^c$, where a and c are experimental coefficients. The molecular weight of the three polymers polystyrene, polybutadiene, and SBS are 48,000, 60,000, and 100,000 dalton, respectively. It is of interest to note that the slope of the diameter vs. Berry number curves are approximately the same, with a value of 5, reflecting the degree of crystallinity of the polymer. The coefficient q is thought to be related to the molecular weight, radius of gyration of molecular chains, and entanglement of the molecular chains. From this relationship, the polymer concentration necessary to produce nanofiber can be predicted.

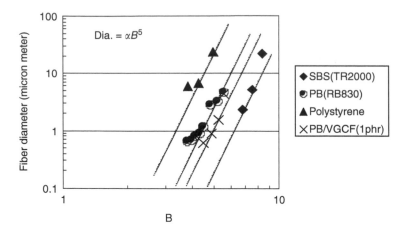

FIGURE 8.5   The relationship between fiber diameter and the Berry number.

8.4   NANOFIBER YARNS AND FABRICS FORMATION

In order to translate nanoeffect to the macroscopic structural level, the nanofibers must be organized into linear, planar, and 3-D assemblies. The formation of nanofiber assemblies provides a means to connect nanofiber effect to macrostructure performance. Electrospun fiber structures can be assembled by direct fiber-to-fabric formation to create a nonwoven assembly and by the creation of a linear assembly or a yarn from which a fabric can be woven, knitted, or braided. Linear fiber assemblies can be aligned mechanically or by electrostatic field control. Alternatively, a self-assembled continuous yarn can be formed during electrospinning by proper design of the ground electrode. Self-assembled yarn can be produced in continuous length with appropriate control of electrospinning parameters and conditions. The fibers are allowed to accumulate until a tree-like structure is formed. Once a sufficient length of yarn is formed, the accumulated fibers attach themselves to the branches and continue to build up. A device such as a rotating drum can be used to spool up the self-assembled yarn in continuous length as shown in Figure 8.6. This method produces partially aligned nanofibers yarn bundles of continuous length.

8.5   POTENTIAL APPLICATIONS OF ELECTROSPUN FIBERS

The Donaldson Company was the first to realize the commercial value of electrospinning taking advantage of the enormous availability of the specific surface of electrospun fibers and the ultrafine nature of the fibers. They introduced the Ultra-Web® cartridge filter for industrial dust collection in 1981 and more recently the Hollingsworth & Vose Company introduced the Nanoweb® for automotive and truck filter applications. To date, industrial filter remains the primary commercial use of electrospun fibers. The rediscovery and popularization of the electrospinning process in the 1990s have changed the dynamics of the field of nanofiber technology. Electrospinning has been transformed from an obscure technology to a word common in academia and one gradually being recognized by industry and government organizations. This was most evident in the Polymer Nanofiber Symposium held in New York at the American Chemical Society (ACS) in September 2003, wherein 75 papers and posters were presented [7]. On the basis of the papers presented in the ACS symposium and the rapidly growing number of publications, one can categorize the potential applications of electrospun fibers into (1) biomedical applications — tissue engineering scaffolds, wound care, superabsorbent media, drug delivery carrier, etc. [8,9]; (2) electronic applications — electronic packaging, sensors, wearable electronics, actuators, fuel cells, etc. [10,11]; (3) industrial applications — filtration, structural toughening/reinforcement, chemical/bio-protection, etc.

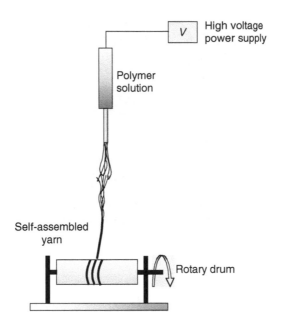

FIGURE 8.6 Schematic illustration of the formation of self-assembled yarns by the electrospinning process.

[12,13]. It is of interest to note that these applications invariably made use of the large surface area created by the fineness of the fibers. A few examples will be used herein to illustrate the potential applications of nanofibers.

8.5.1 NANOFIBERS FOR TISSUE ENGINEERING SCAFFOLDS

It is well recognized, as articulated by Vacanti and Mikos [14], that the key challenges in tissue engineering are the synthesis of new cell adhesion-specific materials and the development of fabrication methods to produce reproducible three-dimensional synthetic or natural biodegradable polymer scaffolds with tailored properties. These properties include porosity, pore size distribution and connectivity, mechanical properties for load-bearing applications, and rate of degradation. Of particular interest in tissue engineering is the surface adhesion of the cells to the scaffold. Considering the importance of surfaces for cell adhesion and migration, experiments were carried out in our laboratory using osteoblasts isolated from neonatal rat calvarias and grown to confluence in Ham's F-12 medium (GIBCO), supplemented with 12% sigma fetal bovine on PLAGA-sintered spheres, 3-D-braided 20 μm filament bundles, and nanofibrils. Four matrices were fabricated for the cell culture experiments. These matrices include (1) 150–300 μm PLAGA-sintered spheres, (2) unidirectional bundles of 20 μm filaments, (3) 3-D-braided structure consisting of 20 bundles of 20 μm filaments, and (4) nonwoven consisting of nanofibrils. The cell proliferation, as shown in Figure 8.7, is expressed in terms of the amount of [³H]-thymidine uptake as a function of time. It can be seen that there is a consistent increase in cell population with time for all the matrices. However, the nanofibrous structures demonstrated the most proliferate cell growth, whereas the cell growth in the structures consisting large diameter fibers and spheres were the least proliferate. Taking advantage of this cell-friendly characteristic of nanofiber scaffolds, nanofibers are being evaluated for the regeneration of skin, vascular grafts, ligament, cartilage, bone, and many other tissues and organs.

8.5.2 NANOFIBERS FOR CHEMICAL/BIO PROTECTIVE MEMBRANES

Nanofiber assemblies are excellent structure barrier systems for dust filtration [12] and for the prevention of the penetration of chemical and biological agents [13]. Considering the various barrier concepts shown in Figure 8.8, it is concluded that neither the impermeable nor the permeable

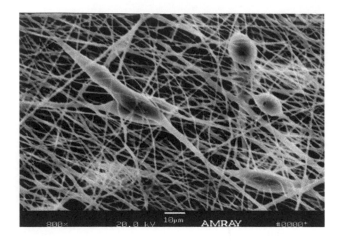

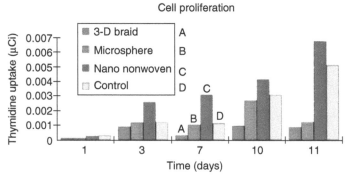

FIGURE 8.7 Directed cell growth (top) and cell proliferation (bottom) behavior in PLA nanofiber scaffold.

barrier systems are suitable because of the unacceptable level of heat generation in the impermeable barriers and the porous nature of the permeable systems. The semi-permeable barrier provides
incremental improvement but the addition of the sorptive layer may serve as a reservoir for bacteria growth as well as adding bulk to the garment. Accordingly, the current design favors a system
that is carbon-free with selective permeable capability and which has the multiple functions of
repelling CB agents while allowing the body moisture vapor to escape. As an example, we illustrate
the feasibility of using the electrospinning process to construct a selective permeable membrane
using the polymers formulated and provided by Utility Development Cooperation (UDC).

An example of an solvent-based electrospun polymer is shown in Figure 8.9 showing a three-
dimensional interconnected fiber network. The morphology of the polymer was found to be controllable by varying polymer concentration.

The electrospun membranes were tested at UDC for moisture vapor transmission by using the
Thwing Albert permeation cups. The testing was performed at 110°F to simulate hot weather conditions. The moisture vapor transmission results of dip-coated nylon fabric with the same polymer as the
electrospun polymer formulation were compared to uncoated nylon fabric. These dip-coated fabrics
were about 6 mil thick compared to electrospun membrane at 10 mil thick. As shown in Table 8.2,
moisture vapor transmission through electrospun membrane was equivalent to uncoated nylon fabric
and there was no detectable moisture vapor transmission through the dip-coated nylon fabric. The test
duration was 5 h. Considering that the water vapor permeability of the state-of-the-art protective
membranes are of the order of 0.1 to 0.25 g/cm^2, these results indicated that the electrospun membranes meet the requirements for wearing comfort over an extended period. This favorable comparison with the performance of the state-of-the-art membranes appears to be consistent with the findings

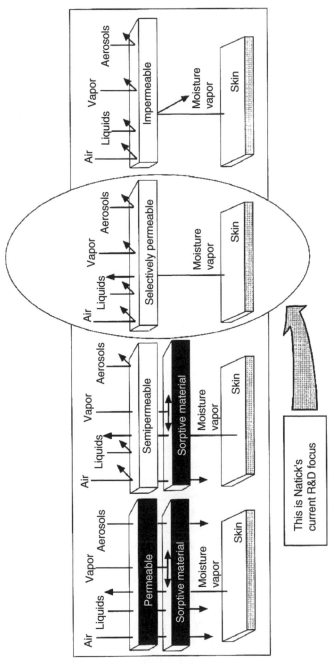

Selectively semi-permeable barrier construction. *Source:* Wilusz, 1998.

FIGURE 8.8 Concepts of barrier systems for chemical/biological agent protection.

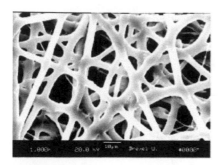

FIGURE 8.9 Electrospun UDC843A polymer showing an interconnected fiber network.

TABLE 8.2
Water Vapor Loss of Various Membranes (g/cm²)

Uncoated Nylon	Electrospun Membrane on Nylon Cloth	Dip Coated Nylon Cloth
0.402	0.348	0.055
0.323	0.105	0.012
0.478	0.235	0.035

of Gibson et al. [15], who showed that the water-transport properties of electrospun nonwovens are superior to that of the state-of-the-art commercial membranes.

Chemical agent testing was performed at Geomet Technology using HD chemical agents for these tests in accordance with MIL-STD-282 static diffusion test. As indicated in the report from Geomet, the UDC-formulated product has more than 24 h of HD chemical agent resistance exceeding the performance requirement for state-of-the-art chemical/bio protective membranes.

8.5.3 NANOCOMPOSITE FIBERS FOR STRUCTURAL APPLICATIONS

Carbon nanotubes (CNTs) [16] are seamless graphene tubule structures with nanometer-size diameters and high aspect ratios. This new class of one-dimensional material is shown to have exceptional mechanical, thermal, and novel electronic properties. The elastic moduli of the CNTs are in the range of 1 to 5 TPa [17–19] and fracture strains of 6 to 30%—both parameters about an order of magnitude better than those of commercial carbon fibers, which typically have 0.1 to 0.5 TPa elastic moduli and 0.1 to 2% fracture strains [20]. The factor of 10 enhancement in strength implies that, for the same performance, replacing or augmenting commercial carbon fibers with CNTs will lead to significant reduction in the volume and weight of the structural composites currently used in space applications.

Studies carried out in our laboratory have demonstrated that composite nanofibers containing 1 to 10 wt% single-wall nanotube (SWNT) can be produced by electrospinning [21,22]. Fiber diameters in the range of 40 to 300 nm have been obtained by varying the solution concentration and electrospinning parameters. The alignment of SWNT in the nanofibers was confirmed by Raman spectroscopy and TEM. Bulk mechanical properties of the green (as-spun) composite nanofibers were characterized in the random mat form. With proper dispersion of SWNT and preparation of the spinning dope, a twofold increase in strength and modulus is observed for the polyacrylonitrile (PAN)/SWNT fiber assemblies containing 1 wt% SWNT. In addition, a continuous process for the conversion of nanofibers into continuous yarn has also been successfully demonstrated. These continuous yarns provide an effective means to translate the properties of the SWNT to higher order textile structures suitable for advanced composites and numerous other applications.

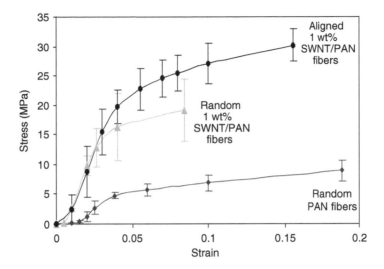

FIGURE 8.10 Tensile properties of electrospun pristine PAN and 1 wt% SWNT/PAN fibers in random and aligned.

Figure 8.10 shows the stress–strain properties of electrospun pristine PAN and 1 wt% SWNT/PAN nanofibers. As can be seen in the random fiber mats, a more than twofold increase in tensile strength and elastic modulus was obtained with the addition of 1 wt% SWNT. For the aligned 1 wt% SWNT/PAN, a 33% increase in tensile strength and almost double strain were obtained as compared with the random composite nanofibers. However, the increase in elastic modulus is not significant. A large increase in the area under the stress–strain curve of SWNT/PAN indicates the effectiveness of SWNT in toughening and strengthening the fibers. Fibril as well as SWNT alignments play an important role in maximizing properties translation from the nanoscopic level to higher order structure. Even without any posttreatment, remarkable improvement in properties was achieved in the electrospun composite nanofibrils. It is anticipated that posttreatment such as mechanical drawing further aligns the SWNT in the fiber axis. Drawing of the fibers also induces molecular orientation resulting in much higher properties. Furthermore, heat treatment enhances interaction between the reinforcement and matrix leading to better load transferring across reinforcement–matrix interface.

The stress–strain behavior of the SWNT-loaded PAN suggested toughening and strengthening of the fiber with the incorporation of 1 wt% SWNT.

8.6 SUMMARY AND CONCLUSIONS

The combination of high-specific surface area, flexibility, and superior directional strength makes fiber a preferred material form for many applications ranging from clothing to reinforcements for aerospace structures. The availability of fibers in nanoscale will greatly expand the performance limit of materials and thus create new marketing opportunities leading to the birth of a new nanofiber-based industry. Of the processing methods capable of fabricating nanofibers, electrospinning has received the most intensive attention due to its simplicity and the potential for generating continuous nanofiber assemblies. With the enormous increase in available surface area per unit mass, nanofibers provide a remarkable capacity for the attachment or release of functional groups, absorbed molecules, ions, catalytic moieties, and nanometer scale particles of many kinds. As a result, these nanofiber assemblies provide favorable conditions for cell adhesion and cell migration for bioactive nanofibers thus creating new opportunities for nanofiber-based scaffolds for tissue engineering and for ultra-sensitive biosensors. Other areas expected to be impacted by nanofiber-based technology include drug delivery systems, wires, capacitors, transistors, and diodes for information technology, systems for energy transport, conversion and storage, such as batteries and fuel cells, and structural composites for aerospace structures. In order to capitalize on these new opportunities new processing technology is needed, precise characterization tools for single nanofibers are required, and nanoscale manipulation machines must be developed. These technologies must be supported by well-verified engineering design tools and analytical methods. It is expected that nanofiber technology will provide a critical link between nanoscale effect and macroscopic phenomena and become the enabling pathway for the conversion of nanomaterials into practical large-scale applications.

REFERENCES

1. Reneker, D.H. and Chun, I., *Nanotechnology*, 7, 216, 1966.
2. Formhals, A., US Patent 1,975,504, October 2, 1934.
3. Donaldson Co., US Patent 6,716,274; 6,673,136.
4. Ko, F.K., Laurencin, C.T., Borden, M.D., and Reneker, D., *Proceedings, Annual Meeting*, Biomaterials Research Society, San Diego, April 1998.
5. Rutledge, G., Fridrikh, S. et al., Texcomp 6, Philadelphia, PA, September 2002.
6. Ko, F.K., Nanofiber technology: bridging the gap between nano and macro world, in *NATO ASI on Nanoengineeered Nanofibrous Materials*, Guceri, S., Gogotsi, Y. and Kuznetsov, V., NATO Series II, Vol. 169, 2004.
7. Reneker, D.H. and Fong, H., Eds., Polymer nanofibers, *Polymer Preprints*, 44, September 2003, American Chemical Society.
8. Li, W.J., Laurencin, C.T., Caterson, E.J., Tuan, R.S., and Ko, F.K., Electrospun nanofibrous structure: a novel scaffold for bioengineering, *J. Biomed. Mater. Res.*, 58, 613–621, 2002.
9. Boland, E.D., Simpson, D.G., Wnek, G. E., and Bowlin, G.L., Electrospinning of biopolymers for tissue engineering scaffolds, *Polymer Pre-preprints*, 44, 92, 2003.
10. Kwoun, J., Lec, R.M., Han, B., and F. Ko. K., A novel polymer nanofiber interface for chemical sensor applications, *Proceedings of the IEEE International Frequency Control Symposium*, May 7–9, 2000, Kansas City, MS, pp. 52–57.
11. MacDiarmid, A.G., Jones, W.E., Jr., Norris, I.D., Gao, J., Johnson, A.T., Jr., Pinto, N.J., Hone, J., Han, B., Ko, F.K., Okusaki, H., and Llanguno, M., Electrostatically-generated nanofibers of electronic polymers, *Synth. Met.*, 119, 27–30, 2001.
12. Lifshutz, N. and Binzer, J.C., Nanofiber coated filter materials for automotive and other applications, *Proceedings of First International Congress on Nanofiber Science and Technology*, June 28, 2004, The Society of Fiber Science and Technology, Tokyo, Japan, pp. 54–62.
13. Ko, F. K., Yang, H., Argawal, R., and Katz, H., Electrospinning of improved CB protective fibrous materials, *Proceedings*, March 30–31, 2004, Techtextil, Atlanta.
14. Vacanti, C.A. and Mikos, A.G., Letter from the editors, *Tiss. Eng.*, 1, 1, 1995.
15. Gibson, P.W., Schreuder-Gibson, H.L., and Rivin, D., Electrospun fiber mats:transport properties, *AIChE J.*, 45, 190–195, 1999.
16. Ijimi, S., Helical microtubules of graphitic carbon, *Nature*, 354, 53–58, 1991.
17. Treacy, M.M., Ebbesen, T.W., and Gibson, J.M., *Nature*, 381,67–80, 1996.
18. Wang, E.W., Sheehan, P.E., and Lieber, C.M., Nanobeam mechanics: elasticity, strength and toughness of nanorods and nanotubes, *Science*, 277, 1971–1975, 1997.
19. Lu, J.P., Elastic properties of single and multilayered nanotubes. *Phys. Rev. Lett.*, 179, 1297, 1997.
20. Peebles, L.H., *Carbon Fibers: Formation, Structure, and Properties*. CRC Press, Boca Raton, FL, 1995.
21. Ko, F., Gogotsi, Y., Ali, A., Naguib, N., Ye, H., Yang, G., Li, C., and Willis, P., Electrospinning of continuous carbon nanotube-filled nanofiber yarns, *Adv. Mater.*, 15, 1161–1165, 2003.
22. Ko, F.K., Lam, H., Titchenal, N., Ye. H., and Gogotsi, Y., Co-electrospinning of carbon nanotube reinforced nanocomposite fibrils, in *Science and Technology of Electrospinning*, Reneker, D. and Fong, H., Eds., Wiley, New York, chapter X, in press.

INDEX

A

Acid–base titration method, 43
Acid treatment, 42
Alewives, 56
Armchair NT, 160
Amperometric sensors, 70
Anti-ferromagnetic resonance (AFMR), 28, 29
Arc discharge and melting method, BNNTs, 164–166
Artificial muscle, 13
Atomic force microscopy (AFM), 220

B

Ball milling and annealing, 168–169
    BNNTs, 168–169
Basal planes, 158
$BaTiO_3$, preparation, 208, 209
Berry number and fiber diameter relationship, 235, 237
Binary oxide 1-D nanostructures, 213
Biological redox processes, 71
Biosensors, 70–72
    bacterial binding, 71
    organophosphorus detection, 71
BNNTs, electron diffraction, 161
BNNTs, structure, 158–164
    chirality, TEM studies, 161–164
    hexagonal, 158
BNNTs, synthesis method, 164–173
    arc discharge and melting, 164–166
    ball milling and annealing, 168–169
    chemical vapor deposition, 172–173
    CNT substitution, 169–172
    laser-assisted method, 166–168
Boron carbon nitride nanotubes (BCN NTs), 171
Boron nitride nanotubes, *see* BNNTs
Bottom-up synthesis methods, 200
Brownian rods, 56

C

$C_{48}N_{12}$ aza-fullerene, schematic representation, 142
Carbon nanotube, 179, 180
    1-D electronic structure, 179
    decoration, 73–75
    electronic properties, 18–30
    gas sensors, 69–70
    inner crystals structure, 65–66
    inner reactions, 64–65
    laser ablation, 141
    morphology and structure, 39–40
    magnetic properties, 30–31
    opening of, 41–43
    superconducting properties, 30–31

types, 179
    unique feature, 2
Carbon nanotube, chemistry
    as templates, 72–75
    biomolecules attachment, 70–72
    functionalization, 42–59
    gases adsorption and storage, 66–70
    "guest" moieties intercalation, 75–77
    inner cavity filling, 59–66
    morphology and structure, 39–40
    opening of, 41–42
    synthesis, 40–41
Carbon nanotube, functionalization, 42–59
    amidation, 48–50
    carboxylic groups reactions, 43–46
    covalent bonding, 50–53
    dispersions, in oleum, 56
    fiber formation, 56
    films, 56
    fluorination, 47–48
    noncovalent bonding, 53–56
    oxidic groups attachment, 43
    self-assembly, 56–59
Carbon nanotube, inner cavity filling, 59–66
    inner crystals structure, 65–66
    inner reactions, 64–65
    in situ filling, 60–61
    post-processing filling, 61–64
Carbon nanotube, mechanical properties, 12–13
    Yakobson's simulation, 12
Carbon nanotube, physical properties
    electronic, 18–30
    magnetic and superconducting, 30–31
    mechanical, 12–13
    thermal, 13–17
Carbon nanotube, post-processing filling
    gas phase, 63–64
    liquid media, 61–63
Carbon nanotube, structure, 3–11
    bundles, 3
    crystalline ropes, 3
    fibers, 7
    filled tubes, 7–10
    macroscopic nanotube materials, 5–7
    multiwall tubes, 5
    nanotube suspension, 10–11
    single-wall tubes, 3
Carbon nanotube, thermal properties, 13–17
    phonon–phonon (Umklapp) collisions, 14
    pulsed laser vaporization (PLV), 17
Carbon nitride nanotube (CNNT), 171

Carbon NT, *see* Carbon nanotube
Catalyst-free vapor-phase growth mechanism, 207
Charge–discharge cycling method, 67
Chemical/bio protective membranes, nanofibers, 239–242
Chemical reduction, NT decoration, with nickel, 73
Chemical solution-based growth mechanism, 207–209
Chemical stability, of GPCs, 130–131
Chemical vapor deposition (CVD), 5, 157, 173
"Chemie douce" process, 137
    hydrothermal synthesis, 140
CNT, *see* Carbon nanotube
Conduction electron spin resonance (CESR), 28–29
Cones, 113–116
Covalent bonds, 180
Crystalline germanium nanosprings, scanning electron
        micrograph, 214

D
Debye temperature, 17
Decoration, 73
Density functional theoretical (DFT) methods, 222
Diameter dispersion, nanowires
    control, 211–212
Double-resonance Raman scattering, 123
Double-wall nanotubes, 9
Drude plasmon, 24
Drude scattering, 25

E
Electrical conductivity, 180, 183, 190–192
    percolation theory, 190
Electrochemical deposition process, NT decoration, with
        nickel, 73
Electroless plating method, 73
Electron beam evaporation, 73
Electron diffraction, 161
    from individual BNNTs, 163
    MWNTs, 161
Electronic band structure, of GPCs, 128–129
Electrospinning process, 234
    key processing parameters, 235–237
    nanofiber electrospinning unit (NEU), 234, 235
    schematic drawing, 234
Electrospun fibers, potential applications, 238
    chemical/bio protective membranes, 239–242
    industrial filter, 238
    nanocomposite fibers, structural applications, 242–243
    tissue engineering scaffolds, 239
Envelope cone, 234
Epoxy, 49, 53, 182, 187, 193
Euler's theorem, 118

F
Fermi energy, 18, 28
Feynman crystals, 65
Fiber diameter, 233
    and Berry number, relationship, 235, 237
    and polymer concentrations, relationship, 235
"Flash CVD" process, 111
Fluorinated SWNTs, 47–48
Fullerene cones, 117

G
Gaseous thermal oxidation, 42
Gases adsorption and storage
    CNT gas sensors, 69–70
    hydrogen problem, 67–69
Gas phase filling, 63–64
Gaussian "bugle horn," 7
Gaussian-correlation model, 221
Gibbs–Thomson effect, 203
Graphene, 1
Graphite anode doping, 60
Graphite layers, zipping, *see* "Lip–lip" interactions
Graphite polyhedral crystals (GPCs), 109, 123–131
    properties and application, 128–131
    structure, 125–128
    synthesis, 123–125
Graphite whiskers and cones, 110–123
    natural occurrence, 116–117
    properties and applications, 121–123
    structure, geometrical considerations, 117–120
    synthetic whiskers and cones, 111–116
"Guest" moieties intercalation, 75–77

H
Hall and Petch behavior, 220
"Herring-bone" structure, 131
Hexagonal boron nitride (h-BN), 141, 157
    atom configuration, 160
    ball-stick model, 158, 159
High-energy ball milling (HEBM), grinding energy of,
        168
$H_2Ti_3O_7$, schematic representation, 143
Hydrogen problem, 67–69

I
Inorganic-nanomaterial synthesis, 212
Inorganic nanotubes, 135
    first-principle calculations, 145
    and fullerene-like structure, 144–148
    properties and applications, 148–150
    synthetic approaches, 141
    synthesis, 138–144
Inorganic nanotubes and fullerene-like structures
    computational methods, 144–148
    potential applications, 148

K
Kramers–Kronig transform, 23
Kratschmer–Huffman arc-discharge method, 123–124

L
Langmuir–Blodgett (LB) films, 38, 58
Laser ablation, 167–168
Laser-assisted metal-catalyzed nanowire growth (LCG),
        203–204
Laser-assisted method, 166–168
    BNNTs, 167–168
Laser heating, 167
Lennard–Jones potential, 17
"Lip–lip" interactions, 128, 164
Liquid media filling, 61–63

M

Macroscopic nanotube materials, 5–7
Mechanical stability, of GPCs, 130–131
Metal-catalyzed VLS growth mechanism, 204–206
Metal dichalcogenide compound, 135, 146
  fullerene-like materials, 144–148
  inorganic nanotubes, 144–148
  related layers, 135
  weak van der Waals forces, 135
Metal nanoclusters, 1-D nanostructure growth, 202–203
Metallorganic vapor-phase epitaxy (MOVPE), 202
Microwave-plasma-assisted CVD method, 115
Mineral kaolinite, schematic representation, 138
Minimal-lithography, 58
Molecular dynamic simulations, 62, 66
Molecular hydrogen sensor, 70
MS_2 nanotubes, electronic structure, 144
Multifunctional polymer nanocomposites, nanotubes, 179
  electrical conductivity, 190–192
  mechanical properties, 185–187
  nanocomposite fabrication, nanotube alignment,
    181–185
  thermal and rheological properties, 187–189
  thermal conductivity, flammability, 192–193
Multi-walled nanotube, *see* MWNT
MWNT, 1, 5, 39, 75, 179, 183
  electron diffraction, 162

N

Nanobeam diffraction (NBD) technique, nanotubes,
    chirality, 163
Nanocomposite fabrication
  drop casting, 181
  melt mixing, 181
  solvent casting, 181
  spin casting, 182
  surfactants, 183
  x-ray scattering, 184
Nanodiffraction, Kratschmer–Huffman arc-discharge
    method, 123–124
Nanofiber electrospinning unit (NEU), 234, 235
Nanofiber technology, 233
  electrospinning process, 234–235
  electrospun fibers, potential applications, 238–243
  key processing parameters, 235–237
  nanofiber yarns and fabrics formation, 238
Nanofiber yarns and fabrics formation
  fiber-to-fabric formation, 238
  self-assembled yarns, 239
Nanoscroll, 142
Nanotube alignment, 181–185
  melt spinning, 185
  Raman spectroscopy, 5184
Nanotube-based nanocomposite, 181–185, 192–193
  coagulation method, 182
  heat transport, 192
  in situ polymerization method, 182
Nanotube functionalization, 183
Nanotubes, 179
  electrical conductivity, 190–192
  thermal conductivity, 192–193

Nanotubes, chirality, NBD technique, 163
Nanotubes, as templates
  carbon atoms substitution, 72–73
  CNTs decoration, 73–75
Nanotube suspensions, 10–11
Nanowires, electronic properties
  carrier mobility, in real systems, 221
  density functional theoretical method, 222
  Vegard's law, deviations, 222
Nanowires, optical properties, 222–224
  Berkeley group, 224
  Schrödinger equation, 223
Noninteracting rods, 56

O

1-D nanostructure
  annual growth, in numbers, 201
  properties and applications, 220–224
  semiconductor and oxide, 201, 203
1-D nanostructure, hierarchal complexity, 211–220
  binary oxide, 213
  composition and doping, 215–220
  control of shape, novel topologies, 212–213
  diameter control, 211–212
  diameter dispersion, 211–212
1-D nanostructure synthesis, strategies, 201–211
  catalyst-free vapor-phase growth, 207
  chemical solution-based growth, 207–209
  LCG, 203–204
  metal catalyzed VLS growth, 204–206
  metal nanoclusters, 202–203
  nucleation, 201
  selected methods, 211
  template-assisted growth, 209–211
  VSS growth, 206–207
1-D single crystalline nanostructured materials, 220
Oxide nanotubes, sol–gel synthesis, 141

P

π-stacking, 42
Pauli paramagnetism, 28
Percolation theory, 190
Phonons, 13, 14, 179, 193, 201, 211
Phonon transport, mechanical and thermal properties,
    220–221
Plasma-jet evaporation, 166
Plasma polymerization technique, 75
Polygonal multiwall tubes, *see* Graphite polyhedral
    crystals
Polyhedral nanoparticle formation, 144
Polylactic acid (PLA), 235
Pulse laser deposition (PLD), 203

R

Raman scattering, 7, 122–123, 184
  degree of alignment, 6, 25
Raman spectra, 184, 221
  graphite polyhedral crystals, 129–130
  graphite whiskers and cones, 122–123
Raman spectroscopy, 122, 129, 184, 185, 187, 242
Reactive magnetron sputtering method, 141

Rice HiPco process, 5
Roll-up vector, 2
"Russian doll" structures, 5, 6, 117

S
Scanning electron microscopy (SEM), 182
Seebeck coefficient, 215
Self-assembly, carbon NT, 56
Single-walled carbon nanotube, *see* SWNT
Spin coating, 182
Supercritical fluid-liquid solid (SFLS), 208
SWNT, 1
    dispersions in oleum, 56
    Langmuir–Blodgett method, 58
    molecular hydrogen sensor, 70
    packing behavior, 66
Synthetic whiskers and cones, electronic properties, 121–122

T
Template-assisted growth (TAG) mechanism, 209–211
TEM, 142, 158
    BNNT chirality, 161–164
    electron diffraction (ED), MWNTs, 161
    and $H_2Ti_3O_7$, schematic representation, 143
    of WS_2 nanotube, 137

Thermal stability, of GPCs, 130–131
Tissue engineering, scaffolds, nanofibers, 234
Titania nanotubes, 74, 141–142
Transitional plane, 204
Tubular graphite cones (TGCs), 115, 116
    MWCVD method, 115

V
van der Waals attraction, 4, 180
van der Waals forces, 135, 136, 159
van der Waals gaps, 39
van der Waals interaction, 65, 142
Vapor–liquid–solid (VLS) growth, 141, 202, 204
Vapor–solid–solid (VSS) growth mechanism, 206–207

W
Whiskers, 111–113

Y
Yakobson's simulation, 12
Yttrium–barium–copper oxide (YBCO), and GaA growth, 207

Z
Zeeman effect, 28

Other Related Titles of Interest by Taylor & Francis

Carbon Nanotubes: Properties and Applications
Michael J. O'Connell
ISBN: 0849327482

Microfabrication and Nanomanufacturing
Mark J. Jackson
ISBN: 0824724313

Nano- and Micro-Electromechanical Systems: Fundamentals of Nano- and Microengineering, Second Edition
Sergey Edward Lyshevski
ISBN: 0849328381

Nanostructure Control of Materials
H.J. Hannink and A.J. Hill
ISBN: 0849334497

And Coming Soon

Nanomanufacturing Handbook
Ahmed Busnaina
ISBN: 0849333261